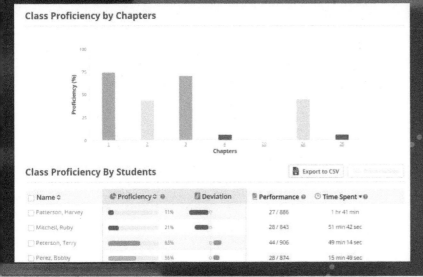

Psychology

Catherine Sanderson
Amherst College

Karen Huffman
Palomar College

A WileyPLUS Learning Space Course

VICE PRESIDENT, EDUCATION	Tim Stookesberry
VICE PRESIDENT & EDITORIAL DIRECTOR	George Hoffman
EDITORIAL DIRECTOR	Veronica Visentin
ASSISTANT DEVELOPMENT EDITOR	Emma Townsend-Merino
EDITORIAL ASSISTANT	Ashley Patterson
MARKETING MANAGER	Glenn Wilson
DESIGN DIRECTOR	Harry Nolan
SENIOR CONTENT MANAGER	Dorothy Sinclair
SENIOR PRODUCTION EDITOR	Sandra Rigby
PRODUCT DESIGN ASSOCIATE	Wendy Ashenberg
SENIOR DESIGNER	Tom Nery
PRODUCTION SERVICE	Jeanine Furino

This book was typeset in 9.5/11.5 Source Sans Pro at Aptara, Inc. Printed and bound by Quad/Graphics.

This book is printed on acid free paper. ∞

Founded in 1807, John Wiley & Sons, Inc. has been a valued source of knowledge and understanding for more than 200 years, helping people around the world meet their needs and fulfill their aspirations. Our company is built on a foundation of principles that include responsibility to the communities we serve and where we live and work. In 2008, we launched a Corporate Citizenship Initiative, a global effort to address the environmental, social, economic, and ethical challenges we face in our business. Among the issues we are addressing are carbon impact, paper specifications and procurement, ethical conduct within our business and among our vendors, and community and charitable support. For more information, please visit our website: www.wiley.com/go/citizenship.

Evaluation copies are provided to qualified academics and professionals for review purposes only, for use in their courses during the next academic year. These copies are licensed and may not be sold or transferred to a third party. Upon completion of the review period, please return the evaluation copy to Wiley. Return instructions and a free of charge return shipping label are available at www.wiley.com/go/returnlabel. If you have chosen to adopt this textbook for use in your course, please accept this book as your complimentary desk copy. Outside of the United States, please contact your local representative.

ISBN 978-1-118-97805-4

The inside back cover will contain printing identification and country of origin if omitted from this page. In addition, if the ISBN on the back cover differs from the ISBN on this page, the one on the back cover is correct.

Printed in the United States of America.

V10007578_011719

TO THE STUDENT

Your *WileyPLUS Learning Space* course includes video lessons that bring the chapter concepts to life through high-interest stories and vivid, real-world examples. The individual video lectures, expert interviews, and interactive media are enhanced by additional engaging visual elements—such as graphics and definitions—to help activate your curiosity and deepen your understanding of the material.

The video lessons and the chapter reading content are coupled together to provide you with a more meaningful learning experience. In general, video lessons present an overview of the major concepts as well as real-world examples and applications. The accompanying etext provides greater detail and more in-depth coverage of the chapter concepts, and may include media like videos and animations. For the best course experience and mastery of the concepts you will want to utilize the video and media materials alongside the etext and this printed course companion.

The video player allows you to experience the material at your own speed, stopping the video delivery to study graphics and media more closely, and then resuming the lesson when you're ready. The player also includes an editing tool that allows your instructor to customize your course, by adding additional questions, comments, and even more video.

How to Use this Print Companion

This Print Companion includes all of the text passages in the online course and will direct you on where to find more information in *WileyPLUS Learning Space*. This study tool, a secondary source for the reading, will reinforce your conceptual understanding and will help you make a deeper connection to the content. For ease of use, you'll find the following elements throughout:

- **Boldface type** is used to indicate figures, tables, and other elements.
- Marginal notes and small "thumbnail" images help you know where you can find art, figures, additional media, and Concept Check questions.
- Icons help direct you to all the resources in *WileyPLUS Learning Space*.

 View the large version of the figure.

 See the table or boxed feature.

 Play the video or animation.

 Answer the Concept Check questions.

TO THE INSTRUCTOR

Psychology, a *WileyPLUS Learning Space* course, couples comprehensive core content with a wealth of engaging digital assets to provide a dynamic teaching and learning environment for you and your students. The carefully crafted original video segments for each chapter's learning objectives use the power of personal storytelling via current examples to bring the chapter concepts to life for students. This model gives professors an excellent framework for flipping the class and providing a more interactive in-class experience. By assigning these videos, instructors will make the content personally relevant to their students and increase their motivation in the course.

Video segments feature the authors, who coach students through various topics using memorable examples and study tips. Each chapter also features academic discussions with real psychology students, demonstrating how the material relates to their everyday lives and future careers. In addition, the embedded video player includes an editing tool that allows you to customize your course, by adding your own questions, comments, and even more video.

Designed to engage today's student, *WileyPLUS Learning Space* will transform any course into a vibrant, collaborative, learning community.

WileyPLUS Learning Space is class tested and ready-to-go for instructors. It offers a flexible platform for quickly organizing learning activities, managing student collaboration, and customizing courses—including choice of content as well as the amount of interactivity between students. An instructor using *WileyPLUS Learning Space* is able to easily:

- Assign activities and add special materials
- Guide students through what's important by easily assigning specific content
- Set up and monitor group learning
- Assess student engagement
- Gain immediate insights to help inform teaching

WileyPLUS Learning Space now includes ORION, a personal, adaptive learning experience that assists students in building their proficiency on learning objectives and using their study time more effectively, especially before quizzes and exams. By tracking students' work, ORION provides instructors with insights into students' work, without having to ask. Efficacy research shows that *WileyPLUS Learning Space* improves student outcomes by as much as one letter grade.

Sanderson & Huffman *Psychology* also includes:

- Key Term Flashcards and Crossword Puzzles
- Media-Enriched PowerPoint Presentations
- Web Resources
- Tutorial Videos
- Virtual Field Trips
- Visual Drag-and-Drop Activities

CONTENTS

INTRODUCTION AND RESEARCH METHODS

1

WP LS Go to your WileyPLUS Learning Space course for video episodes, examples, art, tables, Concept Checks, practice, and other pedagogical resources that will help you succeed in this course.

1.1 INTRODUCING PSYCHOLOGY

The term **psychology** is most commonly defined as the *scientific study of behavior and mental processes. Scientific* is a key feature of the definition because psychologists follow strict scientific procedures, when collecting and analyzing our data. *Behavior* is everything we do (such as crying, hitting, and sleeping) that can be directly observed. *Mental processes* are private, internal experiences that cannot be directly observed (like feelings, thoughts, and memories).

For many psychologists, the most important part of the definition of psychology is the word *scientific*. Psychology places high value on *empirical evidence* that can be objectively tested and evaluated. Psychologists also emphasize **critical thinking**, *the process of objectively evaluating, comparing, analyzing, and synthesizing information* (Halpern, 2014; Hughes & Lavery, 2015).

Be careful not to confuse psychology with *pseudopsychologies*, which are based on common beliefs, folk wisdom, or superstitions. (*Pseudo* means "false.") They sometimes give the appearance of science, but do not follow the basics of the scientific method. Examples include supposed psychic powers, horoscopes, mediums, palmists, and "pop psych" statements, such as "I'm mostly right-brained" or "We use only 10% of our brain." For some, horoscopes or palmists are simple entertainment. Unfortunately, some true believers seek guidance and waste large sums of money on charlatans purporting to know the future (e.g., Wilson, 2015). Broken-hearted families also have lost valuable time and emotional energy on psychics claiming they could locate their missing children. As you can see, distinguishing scientific psychology from pseudopsychology is vitally important (Lilienfeld et al., 2010, 2015; Loftus, 2010).

Psychology—Past and Present

Mankind has always been interested in human nature. Most of the great historical scholars, from Socrates and Aristotle to Bacon and Descartes, asked questions that we would today call psychological. What motivates people? How do we think and problem solve? Where do our emotions and reason reside? Do our emotions control us, or are they something we can control? Interest in such topics remained largely among philosophers, theologians, and writers for several thousand years. However, in the late nineteenth century, psychology began to emerge as a separate scientific discipline.

In the next section, we will briefly describe psychology's past, and then explore how the modern field of psychology, and its seven major perspectives, developed. As a student, you may find these multiple (and sometimes contradictory) approaches frustrating and confusing. However, diversity and debate have always been the life blood of science and intellectual progress.

Psychology's Past

Wilhelm Wundt (1832–1920), a German philosopher-physician, conducted the first experiments and measurements of the workings of the human brain (Benjamin, 2014). Thanks to his experiments, and the establishment of a lab utilizing scientific methods, Wundt moved the study of behavior and mental processes from a mixture of philosophy and biology to a separate and distinct field of its own. Today, Wundt is most commonly identified as the "father" of psychology.

See **FIGURE 1.1, William James (1842–1910)**

Bettmann/Corbis Images

See **TABLE 1.1, Modern Psychology's Seven Major Perspectives**

See **FIGURE 1.2, Why do we need multiple perspectives?**

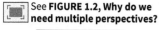

Bettmann/Corbis Images

See **FIGURE 1.3, B.F. Skinner (1904–1990)**

Nina Leen/Time Life Pictures/Getty Images

Wundt and his collaborators sought to identify the basic building blocks, or "structures," of the mind, and founded the aptly named perspective, called *structuralism*. One of their chief research methods was *introspection*, which literally means "looking within," to try to describe our memories, motivations, and other cognitive processes. Because introspection could not be used to study nonhuman animals, children, or complex mental disorders, structuralism failed as a working psychological approach. Although short-lived, it did establish a model for studying mental processes scientifically.

In contrast to structuralism's focus on structures of the mind, *functionalism*, studied the way the mind *functions* to enable humans and nonhuman animals to adapt to their environment. William James was the leading force in the functionalist school (■ **Figure 1.1**). Although functionalism also eventually declined, it expanded the scope of psychology to include research on emotions and observable behaviors, initiated the psychological testing movement, and influenced modern education and industry.

During the late 1800s and early 1900s, while functionalism was still prominent in the United States, the **psychoanalytic perspective** was forming in Europe. Its founder, Austrian physician Sigmund Freud, believed that a vital part of the human mind, the *unconscious*, contains thoughts, memories, and desires that lie outside personal awareness, yet still exert great influence. For example, according to Freud, a man who is attracted to another woman might slip up and say, "I wish you were her," when he consciously planned to say, "I wish you were here." Such seemingly meaningless, so-called "Freudian slips" supposedly reveal a person's true unconscious desires and motives.

Freud also believed many psychological problems are caused by unconscious sexual or aggressive motives, and conflicts between "acceptable" and "unacceptable" behaviors (Chapter 11). His psychoanalytic perspective led to a system of therapy known as *psychoanalysis* (Chapter 13).

Modern Psychology

As summarized in ■ **Table 1.1**, contemporary psychology reflects seven major perspectives: *psychodynamic, behavioral, humanistic, cognitive, biological, evolutionary*, and *sociocultural*. Although there are numerous differences among these seven perspectives, most psychologists recognize the value of each orientation, and agree that no one view has all the answers (see ■ **Figure 1.2**).

Why aren't structuralism, functionalism, and the psychoanalytic approach included as part of the seven modern perspectives? The first two approaches basically lost their following and died out. Although Freud's nonscientific approach and emphasis on sexual and aggressive impulses have long been controversial, his views remain as the foundation for the modern **psychodynamic perspective**. Like Freud, the general goal of psychodynamic psychologists is to explore unconscious *dynamics*, internal motives, conflicts, and past experiences. However, they focus more on social and cultural factors—and less on sexual drives.

In the early 1900s, another major perspective appeared that dramatically shaped the course of modern psychology. Unlike earlier approaches, the **behavioral perspective** emphasizes objective, observable environmental influences on overt behavior. Behaviorism's founder, John B. Watson, rejected the practice of introspection and the influence of unconscious forces. Instead, he adopted Russian physiologist Ivan Pavlov's concept of *conditioning* (Chapter 6) to explain behavior as a result of observable stimuli and responses (behavioral actions). To behaviorists, the mind is like a black box that's sealed, and you can't see inside. What's important, and scientifically accessible, is the relationship between stimulus and response (S-R).

One of the best-known behaviorists, B. F. Skinner, was convinced that all learning is based on the consequences (reinforcement or punishment) that follow certain behaviors. Given the difficulty of completely controlling a human's environment, Skinner often used nonhuman animals in his research (■ **Figure 1.3**). As you'll discover in Chapters 6 and 13, therapeutic techniques rooted in the behavioristic perspective have been most successful in treating observable behavioral problems in humans, such as those related to phobias and alcoholism (Donovan et al., 2015; Kiefer & Dinter, 2013; May et al., 2013).

Although the psychoanalytic and behavioral perspectives dominated U.S. psychology for some time, in the 1950s a new approach emerged—the **humanistic perspective**, which stresses *free will* (voluntarily chosen behavior) and *self-actualization* (an inborn drive to develop all of one's talents and capabilities). According to Carl Rogers and Abraham Maslow, two key figures within this perspective, human nature is naturally positive and growth seeking. Therefore, all individuals naturally strive to develop, and move toward self-actualization. Like psychoanalysis, humanistic psychology developed an influential theory of personality, and its own form of psychotherapy (Chapters 11 and 13). The humanistic approach also led the way to a contemporary research specialty known as **positive psychology**—the study of optimal human functioning (Diener, 2013; Lopez et al., 2015; Seligman, 2003, 2015).

One of the most influential modern approaches, the **cognitive perspective**, recalls psychology's earliest days in that it emphasizes the mental processes we use in thinking, knowing, remembering, and communicating (Galotti, 2014; Goldstein, 2015). These processes include perception, memory, imagery, concept formation, problem solving, reasoning, decision making, and language. Many cognitive psychologists also use an *information-processing approach*. They compare the mind to a computer that sequentially takes in information, processes it, and then produces a response. Like the psychodynamic, behavioral, and humanistic perspectives, the cognitive approach also has developed its own personality theories and methods of psychotherapy (Chapters 11 and 13).

During the past few decades, scientists have explored the role of biological factors in almost every area of psychology. Using sophisticated tools and technologies, scientists who adopt this **biological perspective** examine behavior through the lens of genetics and biological processes in the brain, and other parts of the nervous system. For example, research shows that genes influence many aspects of our behavior, including whether we finish high school and college, how kind we are to other people, and even whom we vote for in elections (Beaver et al., 2011; Chang et al., 2014; Garrett, 2015).

The **evolutionary perspective** stresses natural selection, adaptation, and the evolution of behavior and mental processes (Buss, 2011, 2015; Goldfinch, 2015; Workman & Reader, 2014). Its proponents argue that natural selection favors behaviors that enhance an organism's reproductive success. According to the evolutionary perspective, there's even an evolutionary explanation for the longevity of humans over other primates—it's grandmothers! Without them, a mother who has a two-year-old, and then gives birth to another child, would have to devote her limited time and resources to the newborn at the expense of the older child. Grandmothers help the family's genes survive due to their role as supplementary caregivers.

Finally, the **sociocultural perspective** emphasizes social interactions, and cultural determinants of behavior and mental processes. Although we are often unaware of their influence, factors such as ethnicity, religion, occupation, and socioeconomic class, all have an enormous psychological impact (Berry et al., 2011; Cohen, 2014; Leong et al., 2014). For example, in countries with low levels of gender equality, women are more likely to be attracted by their partner's resources, and men by physical attractiveness (Zentner & Mitura, 2012).

Gender and People of Color

One of the first women to be recognized in the field of psychology was Mary Calkins. Her achievements are particularly noteworthy, considering the significant discrimination she overcame. During the late 1800s and early 1900s, most colleges and universities provided little opportunity for women and people of color, either as students or as faculty members. For example, married women could not be teachers or professors in co-educational settings. In Mary Calkins' case, even after she completed all the requirements for a Ph.D. at Harvard University in 1895, and was described by William James as his brightest student, the university refused to grant the degree to a woman. Nevertheless, Calkins went on to perform valuable research on memory, and in 1905 served as the first female president of the American Psychological Association (APA). The first woman to receive her Ph.D. in psychology was Margaret Floy Washburn from Cornell University in 1894. She served as a professor at Vassar for 36 years, and also as the second female president of the APA.

See **FIGURE 1.4, Kenneth Clark and Mamie Phipps Clark**

Library of Congress Prints and Photographs Division

Francis Cecil Sumner became the first African American to earn a Ph.D. in psychology from Clark University in 1920. Sumner later chaired one of the country's leading psychology departments at Howard University. In 1971, one of Sumner's students, Kenneth B. Clark, became the first person of color to be elected APA president. Clark's research with his wife, Mamie Clark, documented the harmful effects of prejudice, and directly influenced the Supreme Court's landmark 1954 ruling against racial segregation in schools, Brown v. Board of Education (■ **Figure 1.4**).

Calkins, Washburn, Sumner, and Clark, along with other important people of color and women, made significant and lasting contributions to psychology's development. Today, women earning doctoral degrees in psychology greatly outnumber men, but, unfortunately, people of color are still underrepresented (American Psychological Association, 2014; Willyard, 2011).

Biopsychosocial Model

As mentioned before, each of these seven perspectives offers its own unique and valuable insights. However, most contemporary psychologists do not adhere to one single intellectual perspective. Instead, a more integrative, unifying theme—the **biopsychosocial model**—has gained wide acceptance. This model views biological processes (genetics, neurotransmitters, evolution), psychological factors (learning, personality, motivation), and social forces (culture, gender, ethnicity) as interactive and interrelated. It also sees all three factors as influences inseparable from the seven major perspectives (■ **Figure 1.5**).

See multipart **FIGURE 1.5, The biopsychosocial model**

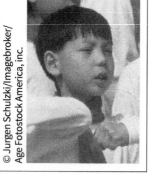

© Bonnie Jacobs/iStockphoto

© Jurgen Schulzki/Imagebroker/ Age Fotostock America, inc.

Why is the biopsychosocial model so important? As the old saying goes, "A fish doesn't know it's in water." Similarly, as individuals living alone inside our own heads, we're often unaware of the numerous, interacting factors that affect us—particularly cultural forces. For example, most people in the United States, Canada, and Western Europe are raised to be very individualistic. And we're often surprised to learn that over 70% of the world's population lives in collectivistic cultures. As you can see in ■ **Table 1.2**, in *individualistic cultures*, the needs and goals of the individual are emphasized over those of the group. When asked to complete the statement "I am . . .," people from individualistic cultures tend to respond with personality traits ("I am shy"; "I am outgoing"), or their occupation ("I am a teacher"; "I am a student").

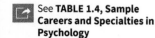

See **TABLE 1.2, A Comparison Between Individualistic and Collectivistic Cultures**

In *collectivistic cultures*, however, the person is defined and understood primarily by his or her place in the social unit (Conway et al., 2014; Greenfield & Quiroz, 2013; Saucier et al., 2015). Relatedness, connectedness, and interdependence are valued, as opposed to separateness, independence, and individualism. When asked to complete the statement "I am . . .," people from collectivistic cultures tend to mention their families or nationality ("I am a daughter"; "I am Chinese"). Keep in mind, however, that these sample countries, and their related values, exist on a continuum. Within each country there is a wide range of individual differences.

Careers and Specialties in Psychology

Many people think of psychologists only as therapists, and it's true that the fields of clinical and counseling psychology do make up the largest specialty areas. However, many psychologists have no connection with therapy. Instead, we work as researchers, teachers, or consultants in education, business, industry, and government, or in a combination of settings. What can you do with a major in psychology? As you can see in ■ **Table 1.3**, there are several career paths, and valuable life skills, associated with a bachelor's degree in psychology. Of course, your options are even greater if you go beyond the bachelor's degree, and earn your master's degree, Ph.D., or Psy.D.—see ■ **Table 1.4**. For more information about what psychologists do—and how to pursue a career in psychology—check out the websites of the American Psychological Association (APA), which is www.apa.org, or the Association for Psychological Science (APS), which can be found at www.psychological science.org.

See **TABLE 1.3, Top Careers with a Bachelor's Degree In Psychology**

See **TABLE 1.4, Sample Careers and Specialties in Psychology**

Q Answer the **Concept Check** questions.

1.2 THE SCIENCE OF PSYCHOLOGY

In science, research strategies are generally categorized as either *basic* or *applied*. **Basic research** is most often conducted to advance core scientific knowledge, whereas **applied research** is generally designed to solve practical ("real world") problems (■ **Figure 1.6**). As you'll see in Chapter 6, classical and operant conditioning principles evolved from numerous *basic research* studies designed to advance the general understanding of how human and nonhuman animals learn. In Chapter 13, you'll also discover how *applied research* based on these principles has been used to successfully treat psychological disorders, such as phobias. Similarly, in Chapter 7, you'll see how basic research on how we create, store, and retrieve our memories has led to practical applications in the legal field, such as a greater appreciation for the fallibility of eyewitness testimony.

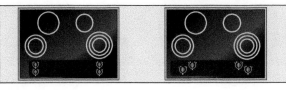

See multipart **FIGURE 1.6, Applied research in psychology**

Keep in mind that basic and applied research approaches are not polar opposites. Instead, they frequently share similar goals, and their outcomes interact, with one building on the other.

The Scientific Method

Like scientists in any other field, psychologists follow strict, standardized procedures so that others can understand, interpret, and repeat or test their findings. Most scientific investigations consist of six basic steps, collectively based on the **scientific method** (**Process Diagram 1.1**).

To help you understand this scientific method, and how it might apply to your student life, consider how it could be used to answer questions like: "Are the Learning Objectives, Test Yourself, Key Terms, and other forms of practice testing provided in each chapter of this text worth my time?" "Will I get a better grade on my exams if I do these exercises?" Starting with Step 1, *Question and Literature Review,* you could go to professional journals, and read up on the research about practice testing that already exists. To complete Step 2, *Testable Hypothesis,* you would first form an educated guess based on your literature review in Step 1. You would then turn this guess into a statement, called a **hypothesis**, which provides a tentative and testable explanation about the relationship between two or more variables. You would also need to explicitly state how each of the variables in your hypothesis will be **operationally defined** (observed, manipulated, and measured). For example, a better grade on exams might be operationally defined as earning one letter grade higher than the letter grade on a previous exam.

> This Process Diagram contains essential information NOT found elsewhere in the text, which is likely to appear on quizzes and exams. Be sure to study it CAREFULLY!

Process Diagram 1.1

The Scientific Method

Scientific knowledge is constantly evolving and self-correcting through application of the scientific method. As soon as one research study is published, the cycle almost always begins again. The ongoing, circular nature of theory building often frustrates students. In most chapters, you will encounter numerous, and sometimes conflicting, scientific theories and research findings. You'll be tempted to ask: Which one is right?" But, like most aspects of behavior, the "correct" answer is almost always a combination and interaction of multiple theories and competing findings. The good news is that such complex answers lead to a fuller and more productive understanding of behavior and mental processes.

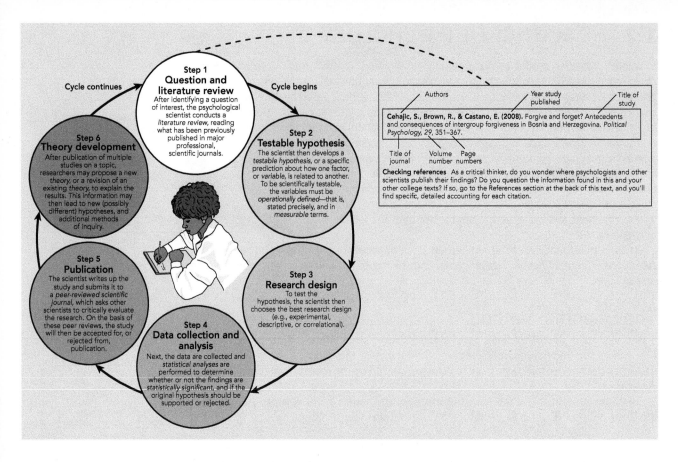

Step 1
Question and literature review
After identifying a question of interest, the psychological scientist conducts a *literature review*, reading what has been previously published in major professional, scientific journals.

Cycle begins

Step 2
Testable hypothesis
The scientist then develops a *testable hypothesis*, or a specific prediction about how one factor, or *variable*, is related to another. To be scientifically testable, the variables must be *operationally defined*—that is, stated precisely, and in *measurable* terms.

Step 3
Research design
To test the hypothesis, the scientist then chooses the best research design (e.g., experimental, descriptive, or correlational).

Step 4
Data collection and analysis
Next, the data are collected and *statistical analyses* are performed to determine whether or not the findings are *statistically significant*, and if the original hypothesis should be supported or rejected.

Step 5
Publication
The scientist writes up the study and submits it to a *peer-reviewed scientific journal*, which asks other scientists to critically evaluate the research. On the basis of these peer reviews, the study will then be accepted for, or rejected from, publication.

Step 6
Theory development
After publication of multiple studies on a topic, researchers may propose a new *theory*, or a revision of an existing *theory*, to explain the results. This information may then lead to new (possibly different) hypotheses, and additional methods of inquiry.

Cycle continues

Authors — Year study published — Title of study

Cehajic, S., Brown, R., & Castano, E. (2008). Forgive and forget? Antecedents and consequences of intergroup forgiveness in Bosnia and Herzegovina. *Political Psychology, 29*, 351–367.

Title of journal — Volume number — Page numbers

Checking references   As a critical thinker, do you wonder where psychologists and other scientists publish their findings? Do you question the information found in this and your other college texts? If so, go to the References section at the back of this text, and you'll find specific, detailed accounting for each citation.

Using the initial question about the value of the practice testing exercises, your testable hypothesis, and operational definition, might be: "Students who spend two hours studying Chapter 1 in this text, and one hour completing all the practice testing options, will earn higher scores on a standard academic exam than students who spend three hours using free-choice study techniques, without completing the practice testing options."

For Step 3, *Research Design*, you could then choose an experimental research design, and solicit 100 volunteers from various classes. Fifty of these volunteers should be randomly assigned to Group 1 (practice testing), and the other 50 to Group 2 (no practice testing). After both groups study (and/or practice test) for three hours, you would present and score a 20-point quiz, followed by a statistical analysis (Step 4—*Data Collection and Analysis*) to determine whether the difference in test scores between the two groups is **statistically significant**. To be statistically significant, the difference between the groups must be large enough that the result is probably not due to chance. In Step 5, *Publication*, you could publish your research.

During the final, Step 6, *Theory Development*, you could investigate additional study techniques, and perhaps come up with a large, overarching theory. Note that In common usage, the term **theory** is often assumed to mean something is only a hunch or someone's personal opinion. In reality, scientific theories are well-substantiated explanations for phenomena, or a group of facts, that have been repeatedly confirmed by previous research. [You'll be interested to know that research does exist on the superiority of practice testing (Test Yourself) on retention of material and improved exam scores (Bourne & Healy, 2014; Dunlosky et al., 2013; Karpicke & Smith, 2012), which is why we've emphasized it throughout this text.]

Before going on, note also in **Process Diagram 1.1** that the scientific method is cyclical and cumulative. Scientific progress comes from repeatedly challenging and revising existing theories, and building new ones. If numerous scientists, using different procedures or participants in varied settings, can repeat, or *replicate*, a study's findings, there is increased scientific confidence in the findings. If the findings cannot be replicated, researchers look for other explanations, and conduct further studies. When different studies report contradictory findings, researchers may average or combine the results of all such studies, and reach conclusions about the overall weight of the evidence, using a popular statistical technique called **meta-analysis**. For example, one recent meta-analysis found that a healthy diet, high-quality preschool, and interactive reading with parents can all lead to measurable increases in children's intelligence (Protzko et al., 2013).

Psychology's Four Main Goals

In contrast to *pseudopsychologies*, which we discussed earlier, and which rely on unsubstantiated beliefs and opinions, psychology is based on rigorous scientific methods. When conducting their research, psychologists have four major goals—to *describe, explain, predict,* and *change* behavior and mental processes:

1. Description Description tells *what* occurred. In some studies, psychologists *describe*, or name and classify, particular behaviors by making careful scientific observations. Description is usually the first step in understanding behavior. For example, if someone says, "You're being too aggressive," what does that mean? Psychologists would be more precise, and likely describe aggression as, "behavior designed to harm another living being."

2. Explanation An explanation tells *why* a behavior or mental process occurred, which requires us to discover and understand the causes. One of the most enduring debates in science is the **nature–nurture controversy** (Chang & Ota Wang, 2014; Gruber, 2013). Are we controlled by biological and genetic factors (the nature side), or by the environment and learning (the nurture side)? As you will see throughout the text, psychology (like allother sciences) generally avoids "either or" positions and focuses instead on *interactions*. For example, research suggests numerous interacting causes or explanations for aggression, including culture, learning, genes, brain damage, and testosterone (Caprara et al., 2014; Longino, 2013; Pournaghash-Tehrani, 2011).

3. Prediction Psychologists generally begin with description and explanation (answering the "whats" and "whys"). Then they move on to the higher-level goal of *prediction*, identifying "when," and under what conditions, a future behavior or mental process is likely to occur. For instance, knowing that alcohol leads to increased aggression (Parker & McCaffree, 2013; Shorey et al., 2014; Zinzow & Thompson, 2015), we can predict that more fights will erupt in places where alcohol is consumed than in places where it isn't.

4. Change For some people, change as a goal of psychology brings to mind evil politicians, or cult leaders brainwashing unknowing victims. However, to psychologists, *change* means applying psychological knowledge to prevent unwanted outcomes, or bring about desired goals. In almost all cases, change as a goal of psychology is positive. Psychologists help people improve their work environment, stop addictive behaviors, become less depressed, improve their family relationships, and so on. Furthermore, as you may know from personal experience, it is very difficult (if not impossible) to change someone's attitude or behavior against her or his will. (*Here is an old joke*: Do you know how many psychologists it takes to change a light bulb? *Answer*: None. The light bulb has to want to change.)

In sum, the goal of psychology is to answer four basic questions about behavior and mental processes:

1. *What* is their nature? **(Description)**

2. *Why* do they occur? **(Explanation)**

3. *When* will they occur? **(Prediction)**

4. How can we *change* them? **(Change)**

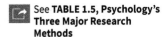

Answer the **Concept Check** questions.

1.3 RESEARCH METHODS

Having studied the scientific method, and psychology's four main goals, we can now examine how psychologists conduct their research. Psychologists generally draw on three major research methods—*descriptive, correlational*, and *experimental* (◧ **Table 1.5**). Each of these approaches has advantages and disadvantages, and psychologists often use variations of all three methods to study a single problem. In fact, when multiple approaches lead to similar conclusions, scientists have an especially strong foundation for concluding that one variable does affect another in a particular way (Cohen, 2014; Morling, 2015).

See **TABLE 1.5, Psychology's Three Major Research Methods**

Descriptive Research

See **TUTORIAL VIDEO,** Applying Research Methods

Almost everyone observes and describes others in an attempt to understand them. However, when conducting **descriptive research**, psychologists do so systematically and scientifically. This type of research observes and records behavior and mental processes in order to answer *who, what, when* and *where* questions. However, by itself, it cannot answer questions about causation. For that, we need experiments—the topic of our upcoming third method research.

The key types of descriptive research are *naturalistic observation, survey, case study*, and *archival research*.

Naturalistic Observation

See **VIRTUAL FIELD TRIP:** Yerkes Primate Research Center

When conducting **naturalistic observation**, researchers systematically observe and record participants' behavior in their natural setting, without interfering. Many settings lend themselves to naturalistic observation, from supermarkets to airports to outdoor settings. For example, Jane Goodall's classic naturalistic observations of wild chimpanzees provided invaluable insights into their everyday lives, such as their use of tools, acts of aggression, demonstrations of affection, and, sadly, even the killing of other chimps' babies (infantacide). In Chapter 5, you'll read about a study in which researchers used naturalistic observation

See **FIGURE 1.7,** Laboratory observation

to examine whether distracted driving (e.g., eating, texting, reaching for a phone) leads to a higher risk of car crashes among teen drivers versus older, more experienced drivers (Klauer et al., 2014). Can you guess what they found?

The chief advantage of naturalistic observation is that researchers can obtain data about natural behavior, rather than about behavior that is a reaction to an artificial experimental situation. But naturalistic observation can be difficult and time-consuming, and the lack of control by the researcher makes it difficult to conduct observations for behavior that occurs infrequently.

For a researcher who wants to observe behavior in a more controlled setting, *laboratory observation* has many of the advantages of naturalistic observation, but with greater control over the variables (**Figure 1.7**).

Jeffrey Greenberg/Photo Researchers

Survey

Psychologists frequently use **surveys** to ask people a series of questions about their behavior and mental processes (Cohen, 2014; Evans & Rooney, 2014). Surveys are often conducted with anonymous questionnaires, on the Internet, on the telephone, and in person in the form of interviews. In Chapter 6 you'll read about a study in which researchers interviewed 3- to 6-year-old girls about their preference for using game pieces depicting thinner or heavier women when playing children's board games to measure preference for the thin ideal (Harriger et al., 2010). In Chapter 10, you'll read about the use of survey research to study how a person's first sexual experience may impact subsequent sexual relationships (Smith & Shaffer, 2013).

Keep in mind that *psychological testing* is often considered another form of the survey method. The SAT or ACT tests commonly taken in high school, and the intelligence and personality tests described in Chapters 8 and 11, are all psychological tests designed to measure your specific behaviors or individual characteristics.

A key advantage of the general survey method is that researchers can gather a large amount of data from many more people than is generally possible with other research designs. Unfortunately, most surveys rely on self-reported data, and not all participants are honest. As you might imagine, people are especially motivated to give less-than-truthful answers when asked about highly sensitive topics, such as infidelity, drug use, and pornography.

Case Study

What if a researcher wants to investigate photophobia (fear of light)? In such a case, it would be difficult to find enough participants to conduct an experiment, or to use surveys or naturalistic observation. For rare disorders or phenomena, researchers try to find someone who has the problem. They then study him or her intensively. This type of in-depth study of a single research participant, or a small group of individuals, is called a **case study** (Yin, 2014). In Chapter 9, we'll share a fascinating case study that examined the impact of severe neglect during childhood on language acquisition. Can you see why this study could not be conducted using surveys or naturalistic observation due to the rarity of such severe deprivation?

Archival Research

The fourth type of descriptive research is **archival research**, in which researchers study previously recorded data. For example, archival data from 30,625 Himalayan mountain climbers from 56 countries found that expeditions from countries with *hierarchical* cultures, which believe that power should be concentrated at the top and followers should obey leaders without question, had more climbers reach the summit than did climbers from more *egalitarian* cultures (Anicich et al., 2015). Sadly, they also had more climbers die along the way. The researchers concluded that hierarchical values impaired performance by preventing low-ranking team members from sharing their valuable insights and perspectives. (If you're wondering about how America ranked, we're a little below midpoint in hierarchical values.)

Interestingly, the new "digital democracy," based on spontaneous comments on Twitter or Facebook, may turn out to be an even better method of research than the traditional random sampling of adults. Researchers, who used a massive archive of billions of stored data from Twitter, found "tweet share" predicted the winner in 404 out of 435 competitive races in the U.S. House of Representatives elections in 2010 (DiGrazia et al., 2013). Apparently, just the total amount of discussion—good or bad—is a very good predictor of ultimate voting behavior.

Correlational Research

As mentioned before, data collected from descriptive research provides invaluable information on behavior and mental processes because they describe the dimensions of a phenomenon or behavior, in terms of who, what, when, and where it occurred. However, if we want to know whether and how two or more variables change together, we need **correlational research**. As the name implies, the purpose of this approach is to determine whether any two variables are *co-related*, meaning a change in one is accompanied by a change in the other. If one variable increases, does the other variable also increase or decrease?

Data collected from descriptive research can indicate not only the direction of a relationship (increasing or decreasing), but also about the strength of that relationship. For example, a careful analysis of over 30 years of descriptive data, using archival research, discovered a link between happiness and age in a sample of more than 5,000 people across the United States (Sutin et al., 2013). This study revealed that, contrary to common stereotypes, increases in age are accompanied by increases in happiness!

Correlational research is also very popular because it allows us to make predictions about one variable based on the knowledge of another. For instance, suppose scientists noted a relationship between the hours of television viewing and performance on exams. The researchers could then predict exam grades based on the amount of TV viewing. The researchers also could determine the direction and strength of the relationship using a statistical formula that gives a **correlation coefficient**, which is a number from −1.00 to +1.00 that indicates the direction and strength of a relationship between two variables.

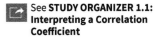 See **STUDY ORGANIZER 1.1: Interpreting a Correlation Coefficient**

Limits of Correlations

As you've just seen, correlations are sometimes confusing, or not particularly useful. In addition, sometimes a mathematical correlation can be found between two events or variables that have no direct connection—yet people may wrongly infer that they do. Therefore, it's very important to note two major cautions concerning correlations.

1. Correlation does NOT prove causation! Cities with a higher number of churches have a higher crime rate. Does this mean that an increase in churches leads to more crime? Of course not! Instead, a *third variable* (increased population) is the real source of the link

See **FIGURE 1.8, Correlation versus causation**

between more churches and more crime. Similarly, sales of ice cream are higher when the rate of drownings in swimming pools is highest. Once again, a third variable (hot weather) leads to more ice cream sales, and to more drownings. This mistake of confusing correlation with causation is referred to as the **third-variable problem**, which refers a situation in which a variable that has not been measured accounts for a relationship between two or more other variables. Would you like a less obvious and more commonly confused example? See **Figure 1.8**.

2. Correlations are sometimes illusory—meaning they don't exist! In this second problem, there is NO factual connection between two variables—the relationship is the result of random coincidence and/or misperception. Popular beliefs, such as infertile couples often conceiving after an adoption, or that certain slot machines are more likely to pay off than others, are called **illusory** (false) **correlations** (Bleske-Rechek et al., 2015; Lilienfeld et al., 2015).

If you're confused about these two problems with correlations, note that with the *third-variable problem*, an actual correlation does exist between two or more variables, but a third factor might be responsible for their connection. In contrast, with an *illusory correlation* there is NO factual connection between two variables—the illusory connection is totally FALSE.

Interestingly, superstitions, such as breaking a mirror supposedly leading to 7 years of bad luck, or sports fans who always wears their lucky team sports jackets because they believe it will bring the team good luck, are additional examples of illusory correlations. We mistakenly perceive an association that factually does not exist. Unfortunately, these and other well-known superstitions (**Table 1.6**) persist despite logical reasoning and scientific evidence to the contrary.

See **TABLE 1.6, Superstitions as Illusory Correlations**

Why are beliefs in illusory correlations so common? As you'll discover in upcoming chapters, we tend to focus on the most noticeable (salient) factors when explaining the causes of behavior. Paying undue attention to the dramatic (but very rare) instances when infertile couples conceive after adoption, or when a gambler wins a large payout on one specific slot machine, are both examples of the *saliency bias* (see Chapter 14). In addition to this saliency bias, we also more often note and remember events that confirm our expectations and ignore the "misses." This is known as the *confirmation bias*.

The important thing to remember while reading research reports in this or any textbook, or reports in the popular media, is that observed correlations may be illusory and that correlational research can NEVER provide a clear cause and effect relationship between variables. Always consider that a third factor might be a better explanation for a perceived correlation. To find causation, we need the experimental method.

The Value of Correlations

After discussing the limits of correlational research, it's important to point out that it's still an incredibly valuable research method. A correlation can tell us if a relationship exists between variables, as well as it's strength and direction. It we know the value of one variable, we can predict the value of the other.

In addition to providing more accurate predictions, correlational studies often point to *possible* causation, which can then be followed up with later experiments. For example, smoking cigarettes and drinking alcohol while pregnant are highly correlated with birth defects (Doulatram et al., 2015; Mason & Zhou, 2015; Popova et al., 2014). Conducting experiments on pregnant women would be obviously impossible and illegal. However, evidence from this strong correlation, and other research, has helped convince women to avoid these drugs while pregnant—thus preventing many birth defects.

Experimental Research

As you've just seen, both descriptive and correlational studies are important because they reveal important data, insights, and practical applications. However, to determine *causation* (what causes what), we need **experimental research**. Experiments are considered the "gold standard" for scientific research because only through an experiment can researchers manipulate variables to determine cause and effect (Cohen, 2014; Goodwin & Goodwin, 2013; Morling, 2015).

To understand all the important key terms, and the general set up for an experiment, it helps to imagine yourself as a psychologist interested in determining how texting while driving an automobile might affect the number of traffic accidents. You begin by reviewing the **Process Diagram 1.1** for the scientific method, which we discussed earlier in this chapter. After *reviewing the literature* and developing your *testable hypothesis* (Steps 1 and 2 of the scientific method), you then decide to use an experiment for your *research design* (Step 3 of the scientific method).

Now carefully study the very simple experimental set up in the **Process Diagram 1.2.** You start by developing your *hypothesis*, which we defined earlier as a tentative and testable explanation (or "educated guess") about the relationship between two or more variables. Then you assign your research participants to either the **experimental group**, who receive the drug or treatment under study, or the **control group**, participants who do NOT receive the drug or treatment under study. Note that having at least two groups allows the performance of one group to be compared with that of another.

Next, you arrange the factors, or *variables*, you manipulate, which are called **independent variables (IV)**. You also decide which variables you plan to measure and examine for possible change, known as **dependent variables (DV)**. Participants who are assigned to the *experimental group* receive the IV, which is the treatment under study, and the variable being manipulated by you—the experimenter. Those assigned to the *control group* will be treated in every way just like the experimental group. The only difference is that they would NOT text while driving.

You, the experimenter, would then ask all participants to drive for a given amount of time (e.g., 30 minutes in the driving simulator). While they're driving, you would record the number of simulated traffic accidents (the DV). [Note: The goal of any experiment is to learn how the dependent variable is *affected by* (depends on) the independent variable.]

As a final step, you'll compare the results from both groups, and report your findings to a peer-reviewed scientific journal like the ones found in the reference list at the end of this book. Keep in mind that because the control group was treated exactly like the experimental group, any significant difference in the number of traffic accidents (the DV) between the two groups would be the result of the IV. In contrast, if you found little or no difference between the groups, you would conclude that texting does not affect traffic accidents.

Before going on, it's important to note that actual research does find that cell phone use, particularly texting, while driving definitely leads to increased accidents, and potentially serious or fatal consequences (e.g., Klauer et al., 2014; Rumschlag et al., 2015). In other words: "Let's all just put down the phone and drive."

See **TUTORIAL VIDEO, The Experiment: Understanding IV and DV**

Experimental Safeguards

As you've seen, every experiment is designed to answer essentially the same question: Does the independent variable (IV) *cause* the predicted change in the dependent variable (DV)? To answer this question, the experimenter must establish several safeguards. In addition to the previously mentioned controls within the experiment itself, a good scientific experiment also protects against potential sources of error from both the researcher and the participant.

Let's start with **sample bias**, which occurs when the researcher recruits and/or selects participants who do not accurately reflect the composition of the larger population from which they are drawn. For example, some critics suggest that psychological literature is biased because it too often uses college students as participants. We can counteract potential sample bias by selecting participants who constitute a *representative sample* of the entire population of interest.

The next step that involves potential bias comes when assigning participants to either the experimental or the control group. As you can see in Step 2 of **Process Diagram 1.2**, using a chance or random system, such as a coin toss, avoids bias because each participant is equally likely to be assigned to any particular group, a technique called **random assignment**.

Process Diagram 1.2

Experimental Research Design

When designing an experiment, researchers must follow certain steps to ensure that their results are scientifically meaningful. In this example, researchers want to test whether texting on a cell phone while driving causes more traffic accidents.

Step 1 The experimenter begins by identifying the hypothesis.

Step 2 In order to avoid sample bias, the experimenter first selects research participants who constitute a representative sample of the entire population of interest. Next, the experimenter randomly assigns these participants to two different groups.

Step 3 Having an experimental group, who receives the treatment, and a control group, who does not receive the treatment, allows a baseline comparison of responses between the two groups.

Step 4 Both the experimental and control groups are assigned to a driving simulator. The experimental group then texts while driving, whereas the control group does not text. Texting or not texting is the independent variable (IV). And the number of simulated traffic accidents is the dependent variable (DV). It's called "dependent" because the behavior (or outcome) exhibited by the participants is assumed to depend on manipulations of the IV.

Step 5 The experimenter then counts the number of simulated traffic accidents for each group, and then analyzes the data.

Step 6 The experimenter writes up his or her report, and submits it to scientific journals for possible publication.

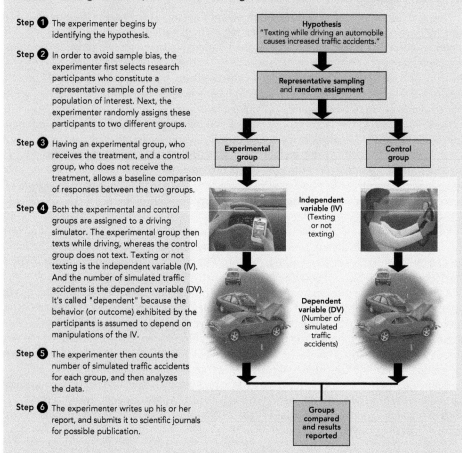

To help you remember the independent and dependent variables (IV and DV), carefully study these drawings, and create a visual picture in your own mind of how:

The experimenter "manipulates" the IV to determine its causal effect on the DV.

The experimenter "measures" the DV, which "depends" on the IV.

It's also critical to control for extraneous, *confounding variables* (such as time of day, lighting conditions, and room temperature). These variables must be held constant across both the experimental and control groups. Otherwise, if not controlled, these variables might contaminate your research results (■ **Figure 1.9**).

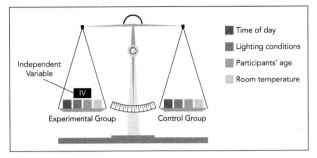

See **FIGURE 1.9, Controlling for confounding variables**

- Time of day
- Lighting conditions
- Participants' age
- Room temperature

Independent Variable
IV
Experimental Group
Control Group

In addition, if experimenters' beliefs and expectations are not controlled for, they can affect participants' responses, producing flawed results. Imagine what might happen if an experimenter breathed a sigh of relief when a participant gave a response that supported the researcher's hypothesis. A good example of this comes from the case of *Clever Hans*, the famous mathematical "wonder horse" (■ **Figure 1.10**). One way to prevent such **experimenter bias** from destroying the validity of participants' responses is to establish objective methods for collecting and recording data, such as using computers to present stimuli and record responses.

See **FIGURE 1.10, Can a horse add, multiply, and divide?**

Experimenters also can skew their results if they assume that behaviors typical in their own culture are typical in all cultures—a bias known as **ethnocentrism**. One way to avoid this problem is to have researchers from two cultures each conduct the same study twice, once with their own culture and once with at least one other culture. This kind of *cross-cultural sampling* isolates group differences in behavior that might stem from any researchers' ethnocentrism.

Martha Lazar/The Image Bank/Getty Images

We've seen that researchers can inadvertently introduce error (the experimenter bias). Unfortunately, participants also can produce a similar error, called **participant bias**. For example, research measuring accuracy of reports of alcohol consumption demonstrates that heavy drinkers tend to under-report how much alcohol they are consuming (Northcote & Livingston, 2011; Wetterling et al., 2014). In this case, participants obviously tried to present themselves in a good light—called the *social desirability response*. As you'll discover in the next section on ethical guidelines, one of the most effective (and controversial) ways of preventing this type of participant bias is to temporarily deceive participants about the true nature of the research project. For example, in studies examining when and how people help others, participants may not be told the true goal of the study because they might try to present themselves as more helpful than they actually would be in real life. In addition to the use of *deception*, researchers also attempt to control for both *experimenter* and *participant bias* by offering anonymous participation, along with guarantees of privacy and confidentiality.

See **FIGURE 1.11, A single- or double-blind experimental design**

Perhaps the most common techniques to minimize bias are **single-blind** and **double-blind studies**. As you can see in ■ **Figure 1.11**, this approach requires experimenters to keep participants and/or experimenters blind (unaware) of the treatment or condition to which the participants have been assigned.

Imagine yourself in an experiment and being told that the pill you're taking for 8 weeks will stop your headaches. Can you see how it's critical that you, as a participant, and possibly the experimenter who collects your data, should be blind as to whether you are in the control or experimental group? In this example, that means if you were a participant you wouldn't know if you are being given the actual, experimental drug, or a harmless, **placebo** pill that has no physiological effect. Researchers do this because your expectations or beliefs, rather than the experimental treatment, can produce a particular outcome, called a **placebo effect**. Giving members of the control group a placebo, while giving the experimental group a pill with the active ingredients, allows researchers to determine whether changes are due to the pill that's being tested, or simply to the participants' expectations.

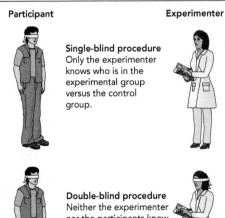

Participant
Experimenter

Single-blind procedure
Only the experimenter knows who is in the experimental group versus the control group.

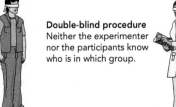

Double-blind procedure
Neither the experimenter nor the participants know who is in which group.

Research Methods—Final Take Home Message

Recognizing that we've offered a large number of research problems and safeguards associated with the various research methods (descriptive, correlational, and experimental), we've

See **FIGURE 1.12, Potential research problems and solutions**

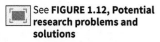

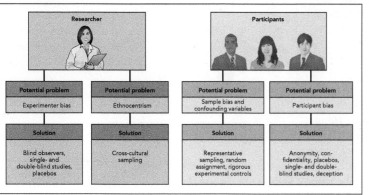

gathered them all into ■ **Figure 1.12**. Be sure to study it carefully. In addition, note that the problems and safeguards connected with descriptive and correlational research described earlier also apply to experimental research.

If you'd like additional information about research methods and statistical analyses.

Ethical Guidelines

So far, we've discussed how psychologists conduct research and analyze their data. Now we need to examine the general ethics that guide their research. The two largest professional organizations of psychologists, the American Psychological Association (APA) and the Association for Psychological Science (APS), both recognize the importance of maintaining high ethical standards in research, therapy, and all other areas of professional psychology. The preamble to the APA's publication *Ethical Principles of Psychologists and Code of Conduct* (2002, 2010) requires psychologists to maintain their competence, to retain objectivity in applying their skills, and to preserve the dignity and best interests of their clients, colleagues, students, research participants, and society. In addition, colleges and universities today have institutional review boards (IRBs) that carefully evaluate the ethics and methods of research conducted at their institutions.

Respecting the Rights of Human Participants

The APA and APS have developed rigorous guidelines regulating research with human participants, including:

- **Informed consent** Researchers must obtain an **informed consent** agreement from all participants *before* initiating an experiment. Participants are made aware of the nature of the study, what to expect, and significant factors that might influence their willingness to participate, including all physical risks, discomfort, and possibly unpleasant emotional experiences.

- **Voluntary participation** Participants must be told that they're free to decline to participate, or to withdraw from the research at any time.

- **Restricted use of deception, followed by debriefing** If participants knew the true purpose behind certain studies, they might not respond naturally. In one of psychology's most famous, and controversial, studies (Milgram, 1963), researchers ordered participants to give electric shocks to another participant (who was really a confederate of the researcher, and was not actually receiving any shocks). Although this study was testing participants' willingness to follow orders, they were told that the study was examining the use of shocks to assist with learning. Obviously in this case, participants' behavior could not be accurately measured if they were told the real focus of the study. Therefore, researchers occasionally need to temporarily deceive participants about the actual reason for the experiment.

  However, when deception is necessary, important guidelines and restrictions still apply. One of the most important is **debriefing**, which is conducted once the data collection has been completed. The researchers provide a full explanation of the research, including its design and purpose, any deceptions used, and then clarify participants' misconceptions, questions, or concerns.

- **Confidentiality** All information acquired about people during a study must be kept private, and not published in such a way that an individual's right to privacy is compromised.

Respecting the Rights of Nonhuman Animals

Although they are used in only 7 to 8% of psychological research (APA, 2009; ILAR, 2009; MORI, 2005), nonhuman animals—mostly rats and mice—have made significant contributions to almost every area of psychology (■ **Figure 1.13**). Without nonhuman animals in *medical research*, how would we test new drugs, surgical procedures, and methods for relieving pain? *Psychological research* with nonhuman animals has led to significant advances in virtually every area of psychology—the brain and nervous system, health and stress, sensation and perception, sleep, learning, memory, motivation, and emotion.

See **FIGURE 1.13, Research with nonhuman animals**

Jonathan Selig/Getty Images

Nonhuman animal research has also produced significant gains for some animal populations, such as the development of more natural environments for zoo animals, and more successful breeding techniques for endangered species.

Despite the advantages, using nonhuman animals in psychological research remains controversial (Baker & Serdikoff, 2013; Binder, 2014; Grimm, 2014). While debate continues about ethical issues in such research, psychologists take great care in handling research animals. Researchers also actively search for new and better ways to minimize any harm to the animals (APA Congressional Briefing, 2015; Morling, 2015; Pope & Vasquez, 2011).

Respecting the Rights of Psychotherapy Clients

Professional organizations, such as the APA and APS, as well as academic institutions, and state and local agencies, all may require that therapists, like researchers, maintain the highest ethical standards. Therapists must also uphold their clients' trust. All personal information and therapy records must be kept confidential. Furthermore, client records are only made available to authorized persons, and with the client's permission. However, therapists are legally required to break confidentiality if a client threatens violence to him- or herself or to others, if a client is suspected of abusing a child or an elderly person, and in other limited situations (Fisher, 2013; Kress et al., 2013; Tyson et al., 2011).

A Final Note on Ethical Issues

What about ethics and beginning psychology students? Once friends and acquaintances know you're taking a course in psychology, they may ask you to interpret their dreams, help them discipline their children, or even ask your opinion on whether they should start or end their relationships. Although you will learn a great deal about psychological functioning in this text, and in your psychology class, take care that you do not overestimate your expertise. Also remember that the theories and findings of psychological science are circular and cumulative—and continually being revised.

David L. Cole, a recipient of the APA Distinguished Teaching in Psychology Award, reminds us that, "Undergraduate psychology can, and I believe should, seek to liberate the student from ignorance, but also the arrogance of believing we know more about ourselves and others than we really do" (Cole, 1982, p. 24).

Q | Answer the **Concept Check** questions.

1.4 STRATEGIES FOR STUDENT SUCCESS

Congratulations! At this very moment, you're demonstrating one of the most important traits of a critical thinker and successful college student—your willingness to accept *suggestions for improvement*. Many students think they already know how to be a student, but would they similarly assume they could become top-notch musicians, athletes, or plumbers without mastering the tools of those trades? Trying to compete in a college environment with minimal, or even average, study skills is like trying to ride a bicycle on a high-speed freeway. *All students* (even those who seem to get A's without much effort) can improve their student tools.

In this section, you will find several important, well-documented study tips and techniques guaranteed to make you a more efficient and successful college student. Before we begin, be sure to complete the following *skills checklist*.

Skills for Student Success Checklist

(Answer true or false to each item. Then for each answer that you answered "True," pay particular attention to the corresponding headings in this Strategies for Student Success section.)

Study Habits

_____ 1.   While reading, I often get lost in all the details, and can't pick out the most important points.

_____ 2.   When I finish studying a chapter, I frequently can't remember what I've just read.

_____ 3.   I generally study with either the TV or music playing in the background.

_____ 4.   I tend to read each section of a chapter at the same speed, instead of slowing down on the difficult sections.

Time Management

_____ 5.   I can't keep up with my reading assignments given all the other demands on my time.

_____ 6.   I typically wait to study, and then "cram" right before a test.

_____ 7.   I go to almost all my classes, but I generally don't take notes, and often find myself texting, playing games on my computer, or daydreaming.

Grade Improvement

_____ 8.   I study and read ahead of time, but during a test I frequently find that my mind goes blank.

_____ 9.   Although I study and read before tests, and think I'll do well, I often find that the exam questions are much harder than I expected.

_____ 10. I wish I could perform better on tests, and read faster or more efficiently.

Study Habits

If you sometimes read a paragraph many times, yet remember nothing from it, try these six ways to successfully read (and remember) information in this and most other texts:

1. Familiarization The first step to good study habits is to familiarize yourself with the general text so that you can take full advantage of its contents. Scanning through the Table of Contents will help give you a bird's-eye view of the overall text. In addition, as you're familiarizing yourself with these features, be sure to also note the various tables, figures, photographs, and special feature boxes, all of which will enhance your understanding of the subject.

2. Active Reading The next step is to make a conscious decision to *actively* read and learn the material. Reading a text is *not* like reading a novel, or fun articles on the Internet! You must tell your brain to slow down, focus on details, and save the material for future recall.

Another way to read actively is to use the **SQ4R method**, which was developed by Francis Robinson (1970). The initials stand for six steps in effective reading: **S**urvey, **Q**uestion, **R**ead, **R**ecite, **R**eview, and w**R**ite. As you might have guessed, this text was designed to incorporate each of these steps (**Process Diagram 1.3**).

3. Avoid highlighting and rereading Marking with a yellow highlighter, or underlining potentially important portions of material, as well as rereading text material after initial reading, are common techniques students use while studying. Unfortunately, they are almost always a waste of time! Research clearly shows that they are among the

Process Diagram 1.3

Using the SQ$_4$R Method

Follow these steps to improve your reading effi ciency.

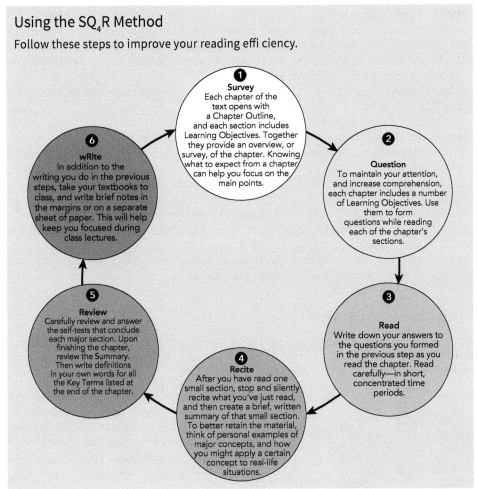

LEAST effective of all the major study techniques (Dunlosky et al., 2013). As previously discussed, you need to actively focus on your reading. Highlighting and rereading generally encourage passive reading.

4. Take notes While reading and listening to classroom lectures, ask yourself, "What is the main idea?" Write down key ideas and supporting details and examples. Effective note taking depends on active reading and active listening. Also, pay attention to the amount of pages and lecture time your text or instructors spend on various topics. It's generally a good indication of what they consider important for you to know.

5. Distributed practice Spreading your study sessions out over time (distributed practice) is far more efficient than waiting until right before an exam, and cramming in all the information at once (massed practice) (Chapter 7). If you were a basketball player, you wouldn't wait until the night before a big play-off game to practice. The same is true for exam preparation. Keep in mind that distributed practice is NOT simply rereading the chapter several times over a few days. As mentioned earlier, you need to actively focus and study what you're reading.

6. Practice test taking For most of us, taking tests is NOT one of our favorite activities. However, it's important to note that, as we mentioned earlier in the chapter, research has clearly shown that practice test taking and distributed practice are two of the most efficient ways to study and learn (Aziz et al., 2014; Bourne & Healy, 2014; Dunlosky et al., 2013). Just as you need to repeatedly practice your free throw shot to become a good basketball player, you need to repeatedly practice your test-taking skills.

If you think practice testing only applies to sports, consider this study conducted in actual introduction to psychology courses! Researchers (who were psych professors doing this study in their own psych classes) compared students who took brief, multiple-choice quizzes at the start of each class, which made up the bulk of their final grade, to those in a typical class with final grades being based entirely on four big exams (Pennebaker et al., 2013). As predicted by the previously mentioned value of practice test taking, the researchers found that this type of frequent testing led to higher grades (on average a half a letter grade increase) not only in this psychology class, but also in the students' subsequent college classes. Beyond the value of practice testing itself, how would you explain this grade increase? Students in the daily quiz condition showed higher class attendance, which included lectures over the material included on quizzes and exams. In addition, the researchers discovered that frequent testing required students to diligently keep up with the material, which led to better study skills and time management.

Based on this growing body of research, and our own teaching success with frequent testing, we've designed this text to include numerous opportunities for practice testing sprinkled throughout each chapter. As you're actively reading and studying each chapter, be sure to complete all these self-tests. When you miss a question, it's very helpful, and important, to immediately go back and reread the sections of the text that correspond to your incorrect response. You also can easily access the free flashcards, and other forms of self-testing in your Learning Space course.

Time Management

If you find that you can't always strike a good balance between work, study, and social activities, or that you aren't always good at budgeting your time, here are five basic time-management strategies:

See **FIGURE 1.14, Sample record of daily activities**

	Sunday	Monday	Tuesday	Wednesday	Thursday	Friday	Saturday
7:00		Breakfast		Breakfast		Breakfast	
8:00		History	Breakfast	History	Breakfast	History	
9:00		Psychology	Statistics	Psychology	Statistics	Psychology	
10:00		Review History & Psychology	Campus Job	Review History & Psychology	Statistics Lab	Review History & Psychology	
11:00		Biology		Biology		Biology	
12:00		Lunch / Study		Exercise	Lunch	Exercise	
1:00		Bio Lab	Lunch	Lunch	Study	Lunch	
2:00			Study	Study			

- **Establish a baseline.** Before attempting any changes, simply record your day-to-day activities for one to two weeks (see the sample in ▦ **Figure 1.14**). You may be surprised at how you spend your time.
- **Set up a realistic schedule.** Make a daily and weekly "to do" list, including all required activities, basic maintenance tasks (like laundry, cooking, child care, and eating), and a reasonable amount of down time. Then create a daily schedule of activities that includes time for each of these. To make permanent time-management changes, shape your behavior, starting with small changes and building on them.
- **Reward yourself.** Give yourself immediate, tangible rewards for sticking with your daily schedule.
- **Maximize your time.** To increase your efficiency, begin by paying close attention to the amount of time you spend on true, focused studying versus the time you waste worrying, complaining, and/or fiddling around getting ready to study ("fretting and prepping").

Time experts also point out that people often overlook important *time opportunities*—spare moments that normally go to waste that you might use productively. When you use public transportation, review notes, or read your textbook. While waiting for doctor or dental appointments, or to pick up your kids after school, take out your text and study for 10 to 20 minutes. Hidden moments count! (Can you see how this also is a form of distributed practice?)

In addition, time management is equally important during class lectures. Students often mistakenly believe that they can absorb information, and do well on exams, by simply going to class. They fail to realize that casually listening to the professor, while also texting, playing computer games, or talking to other classmates, is largely a waste of their valuable time. As you just discovered in the Study Habits section, it's very important to stay focused while studying AND during class lectures. Given that the professor generally lectures on what he or she considers most important (and will include on exams), paying full attention, and taking detailed notes during each class session, is the most efficient and profitable use of your time—and far better than what you can accomplish while studying on your own.

Grade Improvement

Here are three additional strategies for grade improvement and test taking that have a direct impact on your overall grade point average (GPA) in all your college classes, and your mastery of the material.

- **Improve your general test-taking skills.** Expect a bit of stress but don't panic. Pace yourself but don't rush. Focus on what you know. Skip over questions when you don't know the answers, and then go back if time allows. On multiple-choice exams, carefully read each question, and all the alternative answers, before responding. Answer all questions and make sure you have recorded your answers correctly.

  Also, bear in mind that information relevant to one question is often found in another test question. Do not hesitate to change an answer if you get more information—or even if you simply have a better guess about an answer. Contrary to the popular myth widely held by many students (and faculty) that "your first hunch is your best guess," research suggests this is NOT the case (Benjamin et al., 1984; Lilienfeld et al., 2010, 2015). Changing answers is far more likely to result in a higher score (■ **Figure 1.15**).

See **FIGURE 1.15, Should you change your answers?**

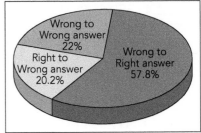

- **Overlearn.** Many people tend to study new material just to the point where they can recite the information, but they do not attempt to understand it more deeply. For best results, however, you should *overlearn*. In other words, be sure to fully understand how key terms and concepts are related to one another, and also be able to generate examples other than the ones in the text. In addition, you should repeatedly review the material (by visualizing the phenomena that are described, and explained in the text, and by rehearsing what you have learned) until the information is firmly locked in place. Overlearning is particularly important if you suffer from test anxiety.

  Would you like a quick demonstration of why it's so important to overlearn? See ■ **Figure 1.16**. You'll undoubtedly have trouble picking out the true penny, even though you've handled, and undoubtedly bought many things involving pennies your entire life. The point is that if you're going to take a test over pennies, or any material in this text, you need to carefully study to the point of overlearning in order to do well.

See **FIGURE 1.16, Which one of these 10 pennies is an exact duplicate of a real U.S. penny?**

- **Take study skills courses.** Improve your reading speed and comprehension, and your word-processing/typing skills, by taking additional courses designed to develop these specific abilities. Also, don't overlook important human resources. Your instructors, roommates, classmates, friends, and family members often provide useful tips and encouragement.

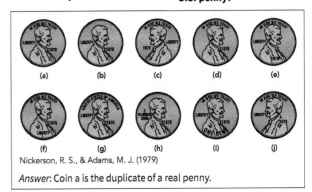

(a) (b) (c) (d) (e)
(f) (g) (h) (i) (j)

Nickerson, R. S., & Adams, M. J. (1979)

Answer: Coin a is the duplicate of a real penny.

Attitude Adjustment

Do you recall the admittedly corny joke we mentioned earlier regarding how many psychologists it takes to change a light bulb? The answer was none because the light bulb has to want to change itself. Although it's a simple joke, it does reflect an important truth—*change comes from within*. Although we're all heavily influenced by interacting biological, psychological, and social forces (the *biopsychosocial model*), if we want to change things (like becoming a more successful student), psychological forces are where we have the greatest personal control.

This explains why *attitude adjustment* is so important. You have the power to decide that you can, and will, improve your academic skills! Instead of focusing on negative thoughts like, "I can't go to the party because I have to study," or "Going to class feels like a waste of time," try counter statements such as, "I'm going to learn how to study and make better use of my class time, so that I can have more free time." Similarly, rather than thinking or saying, "I never do well on tests," do something constructive like taking study skills, and/or test preparation courses, at your college. In addition to the interactive tutorials and quizzes mentioned earlier, we also offer numerous Internet links for student success and time management.

A final word about attitude adjustment. Imagine for a moment that the toilet in your bathroom is overflowing, and creating a horrible, smelly mess. Whom should you reward? The plumber who quickly and efficiently solves the problem? Or someone who "tries very hard" but does little to solve the problem?

Q | Answer the **Concept Check** questions.

Some students may believe they can pass college courses by simply attending class and doing the assignments. This may have worked for *some* students in *some* classes in high school. But many college professors don't assign homework, and may not notice if you skip class. They assume students are independent, self-motivated adult learners, and that grades for their course should generally reflect knowledge and performance—not effort. Although hard work and perseverance are the true keys to college and life success (Chapter 10), your college grades are ultimately based on your final output. Did you fix the toilet or not?

WP LS Go to your WileyPLUS Learning Space course for video episodes, examples, art, tables, Concept Checks, practice, and other pedagogical resources that will help you succeed in this course.

NEUROSCIENCE AND BIOLOGICAL FOUNDATIONS

2

WP LS Go to your WileyPLUS Learning Space course for video episodes, examples, art, tables, Concept Checks, practice, and other pedagogical resources that will help you succeed in this course.

2.1 OUR GENETIC INHERITANCE

Why should we study genetics? Millions of years of evolution have contributed to what we are today. Our ancestors foraged for food, fought for survival, and passed on traits that were selected and transmitted down through the generations. How do these transmitted traits affect us today? For answers, psychologists often turn to **behavioral genetics**, the study of how heredity and environment affect us, and **evolutionary psychology**, the application of the principles of evolution to explain behavior and mental processes.

See multipart **FIGURE 2.1, Hereditary code**

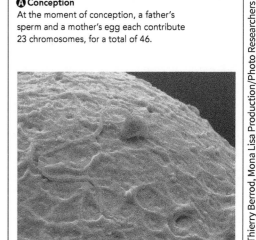

A Conception
At the moment of conception, a father's sperm and a mother's egg each contribute 23 chromosomes, for a total of 46.

Thierry Berrod, Mona Lisa Production/Photo Researchers

Behavioral Genetics

With a good heredity, nature deals you a fine hand at cards; and with a good environment, you learn to play the hand well.

WALTER C. ALVAREZ

In earlier times, people thought inherited characteristics were passed along through the blood—"He's got his family's bad blood." We now know it's a lot more complicated than that. Let's start at the beginning.

At the moment of your conception, your biological mother and father each contributed 23 **chromosomes**, which are threadlike, linear strands of **DNA** (deoxyribonucleic acid) encoded with the **genes** from your parents (▦ **Figure 2.1**). Interestingly, DNA of all humans (except identical twins) has unique, distinguishing features, much like the details on our fingerprints. This uniqueness allows DNA analysis to be used for genetic testing, which helps identify existing or potential future disorders, as well as in forensics to exclude or identify criminal suspects (▦ **Figure 2.2**).

▶ See **VIRTUAL FIELD TRIP: Reading Your DNA**

See **FIGURE 2.2, DNA Testing–Changing Lives, Saving Lives**

Note that our *genes* are the basic building blocks of our entire biological inheritance (Garrett, 2015; Scherman, 2014). Each of our human characteristics and behaviors is related to the presence or absence of particular genes that control the transmission of traits. In some traits, such as blood type, a single pair of genes (one from each parent) determines what characteristics we will possess. When two genes for a given trait conflict, the outcome depends on whether the gene is *dominant* or *recessive*. A dominant gene reveals its trait whenever the gene is present. In contrast, the gene for a recessive trait is normally expressed only if the other gene in the pair is also recessive.

Unfortunately, there are numerous myths and misconceptions about traits supposedly genetically determined by dominant genes. For example, we once assumed that characteristics such as eye color, hair color, and height were the result of either one dominant gene, or two paired recessive genes. But modern geneticists now believe that these characteristics are *polygenic*, meaning they are controlled by multiple genes. Most polygenic traits, like height and intelligence, also are affected by environmental and social factors. As you might imagine, children who are malnourished or abused may not reach their

Michael Biesecker/AP

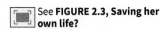
See **FIGURE 2.3, Saving her own life?**

Daniel Zuchnik/Getty Images

See multipart **FIGURE 2.4, Identical versus fraternal twins**

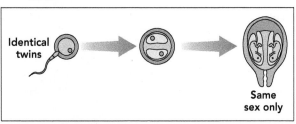

Identical twins

Same sex only

See **FIGURE 2.5, Adoption studies**

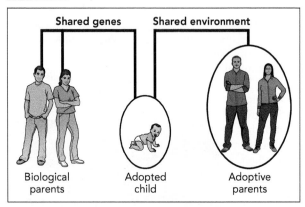

Shared genes   Shared environment

Biological parents   Adopted child   Adoptive parents

See **TUTORIAL VIDEO: The Interaction of Genes and Environment**

full potential genetic height or maximum genetic intelligence (e.g., von Stumm & Plomin, 2015).

Research Methods for Human Inheritance

Why and how do scientists study human inheritance? As you'll see throughout this text, researchers are finding increasing evidence that our personality traits, cognitive abilities, behavioral habits, sexual orientation, and psychological disorders are all determined, at least in part, by genetic factors. In addition, genetic or chromosomal abnormalities contribute to Alzheimer's disease and schizophrenia, as well as diabetes, heart disease, and cancer (■ **Figure 2.3**). Surprisingly, researchers have also discovered that our genes can influence which particular political party we identify with, as well as specific political views, including beliefs about the death penalty, unemployment, and abortion (Butkovic et al., 2015; Funk et al., 2013; Hatemi & McDermott, 2012). Furthermore, genes may even affect the age at which we first initiate sexual activity (Carlson et al., 2014).

In addition to studying how genes affect us as individuals, behavioral geneticists also examine how and when they're transmitted from one generation to the next. Can you see why this makes the study of identical and fraternal twins a top priority? Identical twins share all their genes, whereas fraternal twins share about half of their genes (■ **Figure 2.4**). Furthermore, when both sets of twins share the same parents and develop in relatively the same environment, they provide an invaluable "natural experiment." If heredity influences a trait or behavior to some degree, identical twins should be more alike than fraternal twins. For example, studies of intelligence show that identical twins have more similar intelligence test scores than fraternal twins (Trzaskowski et al., 2014; Woodley of Menie & Madison, 2015), which suggests a genetic influence on intelligence.

As you can see in ■ **Figure 2.5**, behavioral geneticists also study families with biological and adopted children to determine whether adopted children are more like their adoptive parents (who controlled the home environment) or their biological parents (who controlled the genetic inheritance).

Research using these family studies has found that many traits or mental disorders, such as intelligence, sociability, and depression, are strongly influenced by genetics (Gelernter, 2015; Plomin & Deary, 2015; Stein et al., 2012). Twin and adoption studies also have allowed behavioral geneticists to estimate the **heritability** of various traits, which is the percentage of variation in a population attributable to heredity. If genetics contributed nothing to a trait, it would have a heritability estimate of 0%. If a trait is completely due to genetics, it would have a heritability estimate of 100%. The estimates of heritability for most human behaviors fall in the range of 30% to 60%. But, as emphasized in Chapter 8, it's essential to recognize that *heritability estimates for intelligence, or any trait, apply to groups—NOT to individuals!*

Height, for example, has one of the highest heritability estimates—around 90% (Plomin, 1990). However, your own personal height may be very different from that of your parents or other blood relatives because we each inherit a unique combination of genes. Therefore, it is impossible to predict your exact individual height from a heritability estimate. You can only estimate for the group as a whole (■ **Figure 2.6**).

Evolutionary Psychology

As we've seen, behavioral genetics studies help explain the role of heredity (nature) and the environment (nurture) in our individual behavior. To increase our understanding of genetic dispositions, we also need to look at universal behaviors transmitted from our evolutionary past.

Evolutionary psychology suggests that many behavioral commonalities, from eating to fighting with our enemies, emerged and remain in human populations because they helped our ancestors survive (Buss, 2015; Goldfinch, 2015; Workman & Reader, 2014). This perspective stems from the writings of Charles Darwin (1859), who suggested that **natural selection** occurs when a particular genetic trait gives an organism a reproductive advantage over others. Because of natural selection, the fastest, strongest, smartest, or otherwise most fit organisms are most likely to live long enough to reproduce and thereby pass on their genes to the next generation.

Genetic mutations also help explain behavior. Everyone likely carries at least one gene that has mutated, or changed, from the original. Very rarely, a mutated gene will be significant enough to change an individual's behavior. It might cause someone to be more or less social, risk taking, shy, or careful. If the gene then gives the person a reproductive advantage, he or she will be more likely to pass on the gene to future generations. However, neither genetic mutations, nor natural selection, guarantee long-term survival. Even a well-adapted population can perish if its environment changes too rapidly.

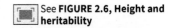

See **FIGURE 2.6, Height and heritability**

Image Source/Getty Images

Q Answer the **Concept Check** questions.

2.2 NEURAL BASES OF BEHAVIOR

You've undoubtedly heard the expression "information is power," and nowhere is this truer than in our human body. Without information, we literally could not survive! Cells within our nervous system must take in sensory information from the outside world through our eyes, ears, and other sensory receptors, and then decide what to do with it. Just as the circulatory system handles *blood*, which conveys chemicals and oxygen, our nervous system uses chemicals and electrical processes that convey *information*.

Given the vital importance of this type of communication, we need to discuss the mechanics of how it occurs. Our brain and the rest of our nervous system all contain microscopic cells called **neurons**. Each neuron is a tiny information-processing system with thousands of connections for receiving and sending electrochemical signals to other neurons. Each human body may have as many as 1 *trillion* neurons.

Neurons are held in place and supported by **glial cells**, which make up about 90% of the brain's total cells. Research indicates that these cells play a direct role in nervous system communication (Barry et al., 2014; Garrett, 2015; McCarthy et al., 2015). They also supply nutrients and oxygen, perform cleanup tasks, and insulate one neuron from another so that their neural messages are not scrambled. However, the "star" of the communication show is still the neuron.

See multipart **FIGURE 2.7, Key parts of a neuron**

Just as no two people are alike, no two neurons are the same. However, most neurons do share three basic features: *dendrites, cell body,* and *axon* (■ **Figure 2.7**). **Dendrites** look like leafless branches of a tree. In fact, the word *dendrite* means "little tree" in Greek. Each neuron may have hundreds or thousands of dendrites, which act like antennas to receive electrochemical information from other nearby neurons. The information then flows into the **cell body**, or *soma* (Greek for "body"). If the cell body receives enough information/stimulation from its dendrites, it will pass the message on to a long, tube-like structure, called the **axon** (from the Greek word for "axle"). Like a miniature cable, the axon carries information away from the cell body to the *terminal buttons*.

Dendrites receive information from other cells.

Cell body receives information from dendrites, and if enough stimulation is received, the message is passed on to the axon.

Axon carries neuron's message to other body cells.

Myelin sheath covers the axon of some neurons to insulate and help speed neural impulses.

Terminal buttons of axon form junctions with other cells and release chemicals called neurotransmitters.

Alfred Pasieka/Photo Researchers

The **myelin sheath**, a white, fatty coating around the axons of some neurons, is not considered one of the three key features of a neuron, but it does insulate and speed neural impulses. Its importance becomes readily apparent in certain diseases, such as *multiple sclerosis*, in which the myelin progressively deteriorates and the person gradually loses muscular coordination. Thankfully, the disease often goes into remission, but it can be fatal if it strikes the neurons that control basic life-support processes, such as breathing or heartbeat.

How Do Neurons Communicate?

A neuron's basic function is to transmit information throughout the nervous system. Neurons "speak" in a type of electrical and chemical language. The process of neural communication begins within the neuron itself, when the dendrites and cell body initially receive electrical messages. As you can see in **Process Diagram 2.1**, communication *within* our neurons is admittedly somewhat confusing. Therefore, we'll also add a brief, written summary after the diagram.

In short, neurons are normally at rest and ready to be activated, which explains why this resting stage is called the "resting potential." However, if the neuron receives a combined signal that exceeds the minimum threshold, the neuron will be activated and "fire," thus transmitting an electrical impulse (called an **action potential**) that travels down the axon via a chemical and electrical process. Note that this firing of an action potential is an *all-or-nothing* process. It's similar to a light switch, where once you apply the minimum amount of pressure needed to flip the switch, the light comes on. There is no "partial firing" of a neuron. It's either on or off. But if this is true, how do we detect the intensity of a stimulus, such as the difference between a rock landing on our hand or a butterfly? A strong stimulus (like the rock) causes more neurons to fire and to fire more often.

Isn't this amazing? Stop for a moment and appreciate how every second of our lives billions of neurons in our body are sending messages (action potentials) that keep us alive and functioning. Even more mind boggling is the process of how neurons communicate with each other and with our body (see **Process Diagram 2.2**).

Keep in mind that communication within the neuron is through action potentials. However, once the action potential reaches the *terminal buttons* at the end of the axon, it triggers the release of chemical messengers, called **neurotransmitters**. These chemicals then travel across the **synapse**, or gap, to bind to receptor sites on the nearby receiving neurons. Like a key fitting into a lock, the neurotransmitters unlock tiny channels in the receiving neuron, and send either excitatory ("fire") or inhibitory ("don't fire") messages.

How Do Neurotransmitters Affect Us?

Having studied the complex and exciting processes of neural communication within and between neurons, we can now explore how poisons from the bites of black widow spiders and snakes, and mind-altering drugs, like nicotine, alcohol, and cocaine, act at the synapse by replacing, decreasing, or enhancing the amount of different neurotransmitters. **Agonist drugs** enhance or "mimic" the action of particular neurotransmitters, whereas **antagonist drugs** block or inhibit the effects (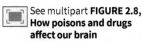 **Figure 2.8**).

In addition to explaining how drugs and poisons affect us, studying the brain and its neurotransmitters also helps us understand some common medical problems (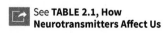 **Table 2.1**). For example, we know that decreased levels of the neurotransmitter dopamine are associated with Parkinson's disease (PD), whereas excessively high levels of dopamine appear to contribute to some forms of schizophrenia.

Perhaps the best-known neurotransmitters are the endogenous opioid peptides, commonly known as **endorphins** (a contraction of *endogenous* [self-produced] and *morphine*). These chemicals mimic the effects of opium-based drugs such as morphine: They elevate mood and reduce pain. For example, drinking alcohol causes endorphins to be released in parts of the brain that are responsible for feelings of reward and pleasure (Charbogne et al., 2014; Sdrulla et al., 2015).

Endorphins also affect memory, learning, blood pressure, appetite, and sexual activity. For example, rats that are injected in the brain with an endorphin-like chemical eat considerably more M&Ms than they would under normal conditions, even consuming as much

See multipart **FIGURE 2.8, How poisons and drugs affect our brain**

Altrendo Images/ Stockbyte/Getty Images

See **TABLE 2.1, How Neurotransmitters Affect Us**

Process Diagram 2.1

Communication *Within* the Neuron

The process of neural communication begins within the neuron itself, when the dendrites and cell body receive information, and conduct it toward the axon. From there, the information moves down the entire length of the axon via a brief, traveling electrical charge called an action potential, which can be described in the following three steps:

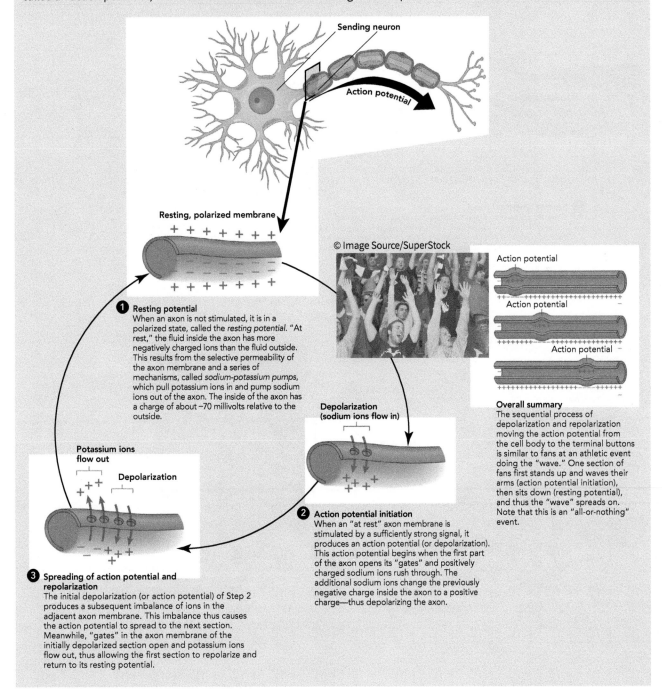

© Image Source/SuperStock

Resting, polarized membrane

❶ Resting potential
When an axon is not stimulated, it is in a polarized state, called the *resting potential*. "At rest," the fluid inside the axon has more negatively charged ions than the fluid outside. This results from the selective permeability of the axon membrane and a series of mechanisms, called *sodium-potassium pumps*, which pull potassium ions in and pump sodium ions out of the axon. The inside of the axon has a charge of about −70 millivolts relative to the outside.

Depolarization (sodium ions flow in)

❷ Action potential initiation
When an "at rest" axon membrane is stimulated by a sufficiently strong signal, it produces an action potential (or depolarization). This action potential begins when the first part of the axon opens its "gates" and positively charged sodium ions rush through. The additional sodium ions change the previously negative charge inside the axon to a positive charge—thus depolarizing the axon.

Potassium ions flow out

Depolarization

❸ Spreading of action potential and repolarization
The initial depolarization (or action potential) of Step 2 produces a subsequent imbalance of ions in the adjacent axon membrane. This imbalance thus causes the action potential to spread to the next section. Meanwhile, "gates" in the axon membrane of the initially depolarized section open and potassium ions flow out, thus allowing the first section to repolarize and return to its resting potential.

Action potential
Action potential
Action potential

Overall summary
The sequential process of depolarization and repolarization moving the action potential from the cell body to the terminal buttons is similar to fans at an athletic event doing the "wave." One section of fans first stands up and waves their arms (action potential initiation), then sits down (resting potential), and thus the "wave" spreads on. Note that this is an "all-or-nothing" event.

Process Diagram 2.2

Communication *Between* Neurons

Within the neuron, messages travel electrically (see **Process Diagram 2.1**). Between neurons, messages are transmitted chemically. The three steps shown here summarize this chemical transmission.

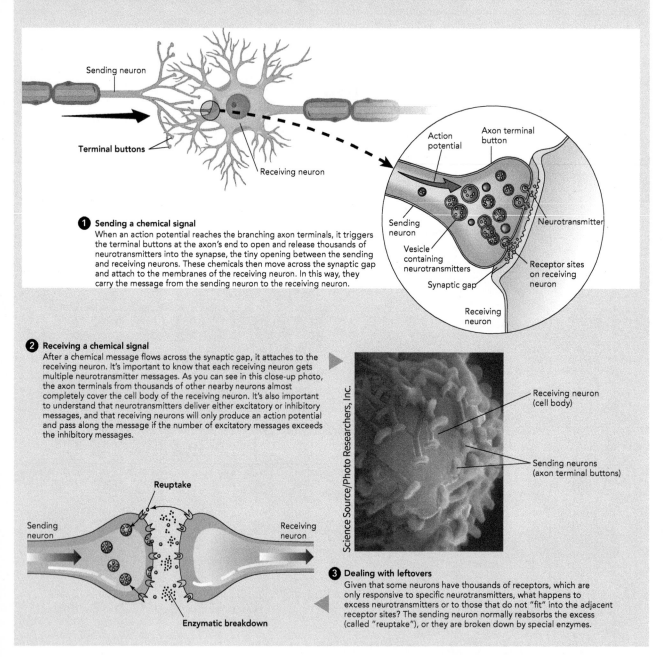

1 Sending a chemical signal

When an action potential reaches the branching axon terminals, it triggers the terminal buttons at the axon's end to open and release thousands of neurotransmitters into the synapse, the tiny opening between the sending and receiving neurons. These chemicals then move across the synaptic gap and attach to the membranes of the receiving neuron. In this way, they carry the message from the sending neuron to the receiving neuron.

2 Receiving a chemical signal

After a chemical message flows across the synaptic gap, it attaches to the receiving neuron. It's important to know that each receiving neuron gets multiple neurotransmitter messages. As you can see in this close-up photo, the axon terminals from thousands of other nearby neurons almost completely cover the cell body of the receiving neuron. It's also important to understand that neurotransmitters deliver either excitatory or inhibitory messages, and that receiving neurons will only produce an action potential and pass along the message if the number of excitatory messages exceeds the inhibitory messages.

3 Dealing with leftovers

Given that some neurons have thousands of receptors, which are only responsive to specific neurotransmitters, what happens to excess neurotransmitters or to those that do not "fit" into the adjacent receptor sites? The sending neuron normally reabsorbs the excess (called "reuptake"), or they are broken down by special enzymes.

Science Source/Photo Researchers, Inc.

as 17 grams (more than 5% of their body weight (DiFeliceantonio et al., 2012). Although this may not seem like a lot of chocolate to you, it is the equivalent of a normal-sized adult eating 7.5 pounds of M&Ms in a single session!

Hormones and the Endocrine System

We've just seen how the nervous system uses neurotransmitters to transmit messages throughout the body. A second type of communication system also exists. This second system is made up of a network of glands, called the **endocrine system**, which uses **hormones** to carry its messages (■ **Figure 2.9**).

See **FIGURE 2.9, The endocrine system**

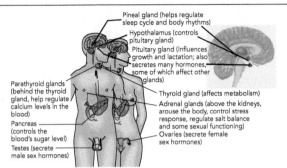

Why do we need two communication systems? Like e-mails we send to particular people, neurotransmitters deliver messages to specific receptors, which other neurons nearby probably don't "overhear." Hormones, in contrast, are like a global e-mail message that we send to everyone in our address book.

Another important difference between neurotransmitters and hormones is that neurotransmitters are released from a neuron's terminal buttons into the small open space (the *synapse*). Conversely, hormones are released from endocrine glands directly into our bloodstream. The messages are then carried by the blood throughout our bodies to any cell that will listen. Like our global e-mail recipients who may forward our message on to other people, messages from hormones are often forwarded on to other parts of the body. For example, a small part of the brain called the hypothalamus releases hormones that signal the pituitary (another small brain structure), which in turn stimulates or inhibits the release of other hormones.

What is the overall importance of our endocrine system? It helps regulate long-term bodily processes, such as growth and the development of sexual characteristics. It also maintains ongoing bodily processes, such as digestion and elimination. Finally, hormones control the body's response to emergencies. In times of crisis, the hypothalamus sends messages through two pathways—the neural system and the endocrine system (primarily the pituitary gland). The pituitary sends hormonal messages to the adrenal glands (located right above the kidneys). As you'll see in Chapter 3, the adrenal glands then release *cortisol*, a stress hormone that boosts energy and blood sugar levels, *epinephrine* (commonly called adrenaline), and *norepinephrine* (also called noradrenaline). (Remember that these same chemicals also can serve as neurotransmitters.)

Q Answer the **Concept Check** questions.

2.3 OUR NERVOUS SYSTEM'S ORGANIZATION

See **FIGURE 2.10, Why do we need helmets?**

Now that we've completed our brief journey inside the structure of the neuron, explored how neurons communicate, and discovered the importance of the endocrine system, we can discuss broader questions, such as: "Where do neural messages come from and why are they important?"

Have you ever wondered how we see and/or hear incoming text messages and then immediately type a response? In short, we have two basic types of neurons—sensory and motor. Our *sensory neurons* respond to physical stimuli, such as the light and sound waves connected to the text message, by sending neural messages to our brain and nervous system regarding the environment. Our *motor neurons* respond by transmitting signals that activate our muscles and glands, thereby allowing our fingers to type reply messages. Sadly, repeated head trauma, particularly when associated with loss of consciousness, can lead to serious illnesses due to the death of motor and other neurons (see ■ **Figure 2.10**).

For a better understanding of sensory and motor neurons, we now need to step back and visualize the overall organization of our entire nervous system. Look closely at **Study Organizer 2.1**. It provides a graphic "road map" representation of how our nervous system is divided into two separate, but interrelated, parts—the **central nervous system (CNS)** and the **peripheral nervous system (PNS)**.

The first part, the CNS, consists of our brain and a bundle of nerves (our *spinal cord*) that runs through our spinal column. Because it is located in the *center* of our body (within the

MLB Photos/Getty Images

See **STUDY ORGANIZER 2.1: The Nervous System**

skull and spine), it is called the *central* nervous system (CNS). Our CNS is primarily responsible for processing and organizing information.

Now, picture the many nerves that lie outside our skull and spine. This is the second major part of the nervous system—the PNS. It carries messages (action potentials) between the central nervous system and the *periphery* of our body, and is therefore known as the *peripheral* nervous system (PNS). The PNS links the CNS to our body's sense receptors (skin, eyes, ears, etc.), muscles, organs, and glands.

> *Note:* When attempting to learn and memorize a large set of new terms and concepts, like those in this chapter, organization is the best way to master the material and get it "permanently" stored in long-term memory (Chapter 7). A broad overview showing the "big picture" helps you organize and file specific details. Just as you need to see a globe of the world showing all the continents to easily place individual countries, you need a "map" of the entire nervous system to effectively study the individual parts.

Given this quick overview, let's go inside each of these two divisions for a closer look, beginning with the central nervous system (CNS).

Central Nervous System (CNS)

The central nervous system (CNS) is the branch of the nervous system that makes us unique. Most other animals can smell, run, see, and hear far better than we can. But thanks to our CNS, we can process information and adapt to our environment in ways that no other animal can. Unfortunately, our brain and spinal cord (the two components of the CNS) are incredibly fragile. Unlike neurons in the PNS that can regenerate and require less protection, neurons in the CNS can suffer serious and permanent damage.

Interestingly, our CNS may not be as "hardwired" as we once thought. Although scientists once believed that it was impossible to repair or replace damaged neurons in the brain, we now know that the brain is capable of lifelong **neuroplasticity** and **neurogenesis**.

Neuroplasticity

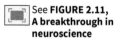
See **FIGURE 2.11,
A breakthrough in
neuroscience**

Courtesy Taub Therapy Clinc/
UAB Media Relations

Rather than being a fixed, solid organ, the brain is capable of changing its structure and function as a result of usage and experience (Ben-Soussan et al., 2015; Bracht et al., 2015; Starkey & Schwab, 2014). This "rewiring" is what makes our brains so wonderfully adaptive. For example, it makes it possible for us to learn a new sport or a foreign language. Remarkably, this rewiring has even helped "remodel" the brain following strokes (**Figure 2.11**).

However, some neuroscientists are concerned that the current publicity surrounding neuroplasticity is "overblown" and misleading the public into expecting more than is actually possible. This caution is an important reminder that we all still need to take the standard precautions to protect our CNS, such as wearing helmets and buckling our seat belts.

Neurogenesis

Our brains continually replace lost cells with new cells that originate deep within the brain and migrate to become part of its circuitry. The source of these newly created cells is neural **stem cells**—rare, immature cells that can grow and develop into any type of cell. Their fate depends on the chemical signals they receive. Experiments and clinical trials on both human and nonhuman animals have used stem cells to repopulate or replace cells, such as bone marrow or brain cells, which have been devastated by injury or disease. This research offers hope to patients suffering from strokes, Alzheimer's, Parkinson's, epilepsy, and depression (Alenina & Klempin, 2015; Kohl et al., 2015; Zhang et al., 2014). In addition, stem cell injections into the eyes of patients with untreatable eye diseases and severe visual problems have led to dramatic improvements in vision (Chucair-Elliott et al., 2015; Stanzel et al., 2014; Tzameret et al., 2014).

Now that we have discussed neuroplasticity and neurogenesis within the CNS, let's take a closer look at the spinal cord. Because of its central importance for psychology and behavior, we'll discuss the brain in more detail in the next major section.

Spinal Cord

Beginning at the base of our brains and continuing down our backs, the spinal cord carries vital information from the rest of the body into and out of the brain. The critical importance of the spinal cord is dramatically illustrated when the flow of information is disrupted by injuries or accidents and the body becomes paralyzed, as well as incapable of noticing touch or pain sensations on the body. When the damage occurs high on the spinal cord (close to the brain), the individual will most likely be *quadriplegic*, paralyzed everywhere but the head and neck. If damage occurs lower down the back, the person could be unable to voluntarily move any muscles in the lower half of the body—a condition known as *paraplegia*.

Are you wondering if stem cell transplants could help people paralyzed from spinal cord injuries walk again? Scientists have had some success transplanting stem cells into spinal cord–injured nonhuman animals (Granger et al., 2014; Sandner et al., 2015; Yamamoto et al., 2014). When the damaged spinal cord was viewed several weeks later, the implanted cells had survived and spread throughout the injured area. More important, the transplant animals also showed some improvement in previously paralyzed parts of their bodies. Medical researchers have recently begun testing the safety of embryonic stem cell therapy for human paralysis patients, and future trials may determine whether these cells will repair damaged spinal cords and/or improve sensation and movement in paralyzed areas (Granger et al., 2014; Robbins, 2013).

In addition to transmitting messages into and out of our brain, our spinal cord also initiates some automatic behaviors on its own. We call these involuntary, automatic behaviors **reflexes**, or *reflex arcs*, because the response to the incoming stimuli is automatically "reflected" back to the spinal cord. As you can see in **Process Diagram 2.3**, this reflex arc allows an immediate action response without the delay of routing signals directly to the brain.

This Process Diagram contains essential information NOT found elsewhere in the text, which is likely to appear on quizzes and exams. Be sure to study it CAREFULLY!

Process Diagram 2.3

How the Spinal Reflex Operates

In a simple reflex arc, a sensory receptor responds to stimulation and initiates a neural impulse that travels to the spinal cord. This signal then travels back to the appropriate muscle, which then contracts. The response is automatic and immediate in a reflex because the signal only travels as far as the spinal cord before action is initiated, not all the way to the brain. The brain is later "notified" of the action when the spinal cord sends along the message. What might be the evolutionary advantages of the reflex arc?

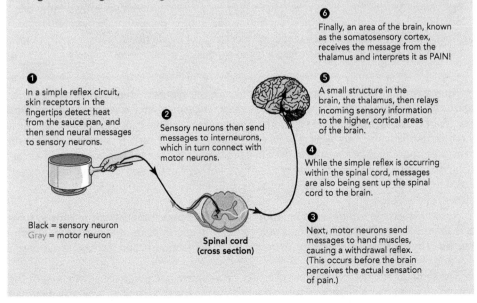

❻ Finally, an area of the brain, known as the somatosensory cortex, receives the message from the thalamus and interprets it as PAIN!

❶ In a simple reflex circuit, skin receptors in the fingertips detect heat from the sauce pan, and then send neural messages to sensory neurons.

❷ Sensory neurons then send messages to interneurons, which in turn connect with motor neurons.

❺ A small structure in the brain, the thalamus, then relays incoming sensory information to the higher, cortical areas of the brain.

❹ While the simple reflex is occurring within the spinal cord, messages are also being sent up the spinal cord to the brain.

Black = sensory neuron
Gray = motor neuron

Spinal cord
(cross section)

❸ Next, motor neurons send messages to hand muscles, causing a withdrawal reflex. (This occurs before the brain perceives the actual sensation of pain.)

We're all born with numerous reflexes, many of which fade over time. But even as adults, we still blink in response to a puff of air in our eyes, gag when something touches the back of the throat, and urinate and defecate in response to pressure in the bladder and rectum. Reflexes even influence our sexual responses. Certain stimuli, such as the stroking of the genitals, can lead to arousal and the reflexive muscle contractions of orgasm in both men and women. However, in order for us to have the passion, thoughts, and emotion we normally associate with sex, the sensory information from the stroking and orgasm must be carried by the spinal cord to the appropriate areas of the brain that receive and interpret these specific sensory messages.

Peripheral Nervous System (PNS)

The peripheral nervous system (PNS) is just what it sounds like—the part that involves neurons *peripheral* to (or outside) the brain and spinal cord. The chief function of the PNS is to carry information to and from the CNS. It links the brain and spinal cord to the body's sense receptors, muscles, and glands.

Looking back at Study Organizer 2.1, note that the PNS also is subdivided into the somatic nervous system and the autonomic nervous system. The **somatic nervous system (SNS)** consists of all the nerves that connect to sensory receptors and skeletal muscles. The name comes from the term *soma*, which means "body," and the somatic nervous system plays a key role in communication throughout the entire body. In a kind of two-way street, the SNS first carries sensory information to the brain and spinal cord (CNS) and then carries messages from the CNS to skeletal muscles.

The other subdivision of the PNS is the **autonomic nervous system (ANS)**. The ANS is responsible for involuntary tasks, such as heart rate, digestion, pupil dilation, and breathing. Like an automatic pilot, the ANS can sometimes be consciously overridden. But as its name implies, the autonomic system normally operates on its own (*autonomously*).

The ANS is further divided into two branches, the sympathetic and parasympathetic (see again Study Organizer 2.1). These two branches tend to work in opposition to each other to regulate the functioning of such target organs as the heart, the intestines, and the lungs (■ **Figure 2.12**). Like two children on a teeter-totter, one will be up while the other is down, but they essentially balance each other out.

See **FIGURE 2.12, Actions of the autonomic nervous system (ANS)**

Sympathetic dominance

Parasympathetic dominance

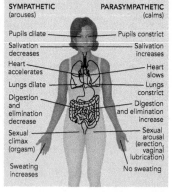

SYMPATHETIC (arouses)

PARASYMPATHETIC (calms)

SYMPATHETIC (arouses)	PARASYMPATHETIC (calms)
Pupils dilate	Pupils constrict
Salivation decreases	Salivation increases
Heart accelerates	Heart slows
Lungs dilate	Lungs constrict
Digestion and elimination decrease	Digestion and elimination increase
Sexual climax (orgasm)	Sexual arousal (erection, vaginal lubrication)
Sweating increases	No sweating

Geri Lavrov/ Photographer's Choice/ Beto Hacker/Getty Getty Images, Inc. Images, Inc.

Note: One way to differentiate the two subdivisions of the ANS is to imagine skydiving out of an airplane. When you initially jump, your sympathetic nervous system has "sympathy" for your stressful situation. It alerts and prepares you for immediate action. Once your "para" chute opens, your "para" sympathetic nervous system takes over, and you can relax as you float safely to earth.

During stressful times, either mental or physical, the **sympathetic nervous system** mobilizes bodily resources to respond to the stressor. This emergency response is often called the "fight or flight" response. If you noticed a dangerous snake coiled and ready to strike, your sympathetic nervous system would increase your heart rate, respiration, and blood pressure; stop your digestive and eliminative processes; and release hormones, such as cortisol, into the bloodstream. The net result of sympathetic activation is to get more oxygenated blood and energy to the skeletal muscles, thus allowing you to cope with the stress—to either fight or flee.

In contrast to the sympathetic nervous system, the **parasympathetic nervous system** is responsible for calming our bodies and conserving energy. It returns our normal bodily functions by slowing our heart rate, lowering our blood pressure, and increasing our digestive and eliminative processes.

The sympathetic nervous system provides an adaptive, evolutionary advantage. At the beginning of human evolution, when we faced a dangerous bear or an aggressive human

Q Answer the **Concept Check** questions.

intruder, there were only two reasonable responses—fight or flight. The automatic mobilization of bodily resources can still be critical, even in modern times. However, less life-threatening events, such as traffic jams, also activate our sympathetic nervous system. As the next chapter discusses, chronic sympathetic nervous system arousal to ongoing daily stressors can become detrimental to our health. For a look at how the autonomic nervous system affects our sexual lives, see ▣ **Figure 2.13**.

See multipart **FIGURE 2.13, Autonomic nervous system and sexual arousal**

Piotr Marcinski/Shutterstock

2.4 A TOUR THROUGH OUR BRAIN

We begin our exploration of the brain with a discussion of the tools that neuroscientists use to study it. Then we offer a quick tour of the brain, beginning at its lower end, where the spinal cord joins the base of the brain, and then move upward, all the way to the top of the skull. Note that as we move from bottom to top, "lower," basic processes, such as breathing, generally give way to more complex mental processes (▣ **Figure 2.14**).

Biological Tools for Research

How do we know how the brain and nervous system work? From early times, scientists have *dissected* the brains and other body parts of deceased human and nonhuman animals. They've also used *lesioning* techniques (systematically destroying bodily tissue to study the effects on behavior and mental processes). By the mid-1800s, this research had produced a basic map of the nervous system, including some areas of the brain. Early researchers also relied on clinical observations and case studies of living people who had experienced injuries, diseases, and disorders that affected brain functioning.

Modern researchers still use such methods, but they also employ other techniques to examine biological events that underlie our behavior and mental processes (▣ **Table 2.2**). For example, advances in brain science have led to various types of brain-imaging scans, which can be used in both clinical and laboratory settings. Most of these methods are relatively *noninvasive*—that is, their use does not involve breaking the skin or entering the body.

⊳ See **TUTORIAL VIDEO: Dissecting the Brain**

⊳ See **VIRTUAL FIELD TRIP: Neuroimaging**

⊳ See **TABLE 2.1: Sample Tools for Biological Research**

⊳ See **TUTORIAL VIDEO: Build a Brain—A Learning Activity**

▣ See **FIGURE 2.14, The human brain**

Our Brain's Organization

Having studied the tools scientists use for exploring the brain, we can now begin our tour. Let's talk first about brain size and complexity, which vary significantly from species to species. For example, fish and reptiles have smaller, less complex brains than do cats and dogs. The most complex brains belong to whales, dolphins, and higher primates, such as chimps, gorillas, and humans. The billions of neurons that make up the human brain control much of what we think, feel, and do. Certain brain structures are specialized to perform certain tasks, a process known as *localization of function*. However, most parts of the brain perform integrating, overlapping functions.

Just as our overall nervous system is divided into the CNS and PNS, our brain is typically divided into three major sections: the hindbrain, midbrain, and forebrain. And just as the CNS and PNS are overlapping, integrated systems, the same is true for these three sections of our brains.

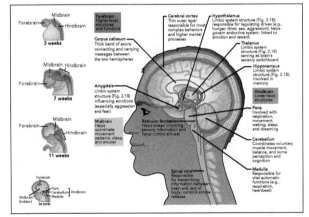

Hindbrain

Picture this: You're asleep and in the middle of a frightening nightmare. Your heart is racing, your breathing is rapid, and you're attempting to run away but find you can't move! Suddenly, your nightmare is shattered by a buzzing alarm clock. All your automatic behaviors and

See **FIGURE 2.15, Walk the line**

survival responses in this scenario are controlled or influenced by parts of your **hindbrain**, which includes your medulla, pons, and cerebellum.

The **medulla** is essentially an extension of the spinal cord, which has many neural fibers passing through it that carry information to and from the brain. The medulla also controls many essential automatic bodily functions, such as respiration and heart rate.

The **pons** is involved in respiration, movement, sleeping, waking, and dreaming (among other things). In addition, the pons contains axons that cross from one side of the brain to the other (*pons* is Latin for "bridge").

It's important to note that the medulla, pons, and much of the midbrain (discussed in the next section) are generally grouped together into an area called the **brain stem**, aptly named for its stem like appearance. It connects with the spinal cord, and plays an essential role in maintaining consciousness, regulating the sleep cycle, and managing basic survival functions, such as respiration and heartbeat.

The cauliflower-shaped **cerebellum** (Latin for "little brain") is, evolutionarily, a very old structure. It coordinates fine muscle movement and balance (◫ **Figure 2.15**). Researchers using functional magnetic resonance imaging (fMRI) have shown that parts of the cerebellum also are important for memory, sensation, perception, cognition, language, learning, and even "multitasking" (Durán et al., 2014; Garrett, 2015; Ulupinar et al., 2015).

Midbrain

The **midbrain** helps us orient our eye and body movements to visual and auditory stimuli, and it works with the pons to help control sleep and level of arousal. It also contains a small structure, the *substantia nigra*, that secretes the neurotransmitter dopamine. Parkinson's disease, an age-related degenerative condition, is related to the deterioration of neurons in the substantia nigra and the subsequent loss of dopamine.

Running through the core of the hindbrain, midbrain, and brainstem is the **reticular formation (RF)**. This diffuse, finger-shaped network of neurons helps screen incoming sensory information, as well helping to alert the higher brain centers to important events. Without our reticular formation, we would not be alert or perhaps even conscious.

Forebrain

See **FIGURE 2.16, Structures of the forebrain**

The **forebrain** is the largest and most prominent part of the human brain. It includes the cerebral cortex, hypothalamus, limbic system, and thalamus (◫ **Figure 2.16**). The last three structures are located near the top of the brainstem. The cerebral cortex (discussed separately, in the next section) is wrapped above and around them. (*Cerebrum* is Latin for "brain," and *cortex* is Latin for "covering" or "bark.")

The **thalamus** integrates input from the senses, and it may also function in learning and memory (Cordova et al., 2014; Edelstyn et al., 2014; McCormick et al., 2015). It receives input from nearly all sensory systems, except smell, and then directs the information to the appropriate cortical areas. The thalamus also transmits some higher brain information to the cerebellum and medulla. Think of the thalamus as the switchboard in an air traffic control center that receives information from all aircraft and directs them to landing or takeoff areas.

Because the thalamus is the brain's major sensory relay center to the cerebral cortex, damage or abnormalities in the thalamus might cause the cortex to misinterpret or not receive vital sensory information. As you'll discover in Chapter 12, brain-imaging research links thalamus abnormalities to schizophrenia, a serious psychological disorder characterized by problems with sensory filtering and perception (Argyelan et al., 2014; Cooper et al., 2014; Kim et al., 2015).

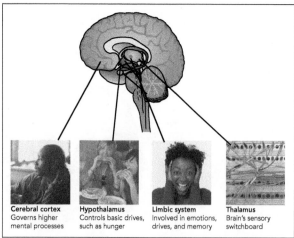

Cerebral cortex
Governs higher mental processes

Hypothalamus
Controls basic drives, such as hunger

Limbic system
Involved in emotions, drives, and memory

Thalamus
Brain's sensory switchboard

Beneath the thalamus lies the kidney bean–sized **hypothalamus** ("hypo" means "under"). This organ has been called the "master control center" for emotions and for many basic motives such as hunger, thirst, sex, and aggression (Ahmed et al., 2014; Maggi et al., 2015; Zhang et al., 2015). The hypothalamus also controls the body's internal environment, including temperature, which it accomplishes by regulating the endocrine system (■ **Figure 2.17**).

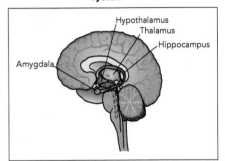

See **FIGURE 2.17, Diet and the hypothalamus**

© donmedia/iStockphoto

Hanging down from the hypothalamus, the *pituitary gland* is usually considered the master endocrine gland because it releases hormones that activate the other endocrine glands. The hypothalamus influences the pituitary through direct neural connections and through release of its own hormones into the blood supply of the pituitary. The hypothalamus also directly influences some important aspects of behavior, such as eating and drinking patterns.

An interconnected group of forebrain structures, known as the **limbic system**, is located roughly along the border between the cerebral cortex and the lower-level brain structures (■ **Figure 2.18**). The limbic system is generally responsible for emotions, drives, and memory. In Chapter 7, you'll discover how the **hippocampus**, a key part of the limbic system, is involved in forming and retrieving our memories. However, the limbic system's major focus of interest is the **amygdala**, which is linked to the production and regulation of emotions–especially aggression and fear (Denny et al., 2015; LeDoux, 1998, 2007; Negrón-Oyarzo et al., 2014).

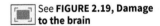

See **FIGURE 2.18, Major brain structures commonly associated with the limbic system**

Hypothalamus
Thalamus
Hippocampus
Amygdala

Another well-known function of the limbic system is its role as part of the so-called "pleasure center," a set of brain structures whose stimulation leads to highly enjoyable feelings (Bjork et al., 2012; Kolb, 2013; Olds & Milner, 1954; Stopper & Floresco, 2015). Even though limbic system structures and neurotransmitters are instrumental in emotions, the frontal lobes of the cerebral cortex also play an important role.

The Cerebral Cortex

The gray, wrinkled **cerebral cortex**, the surface layer of the cerebral hemispheres, is responsible for most complex behaviors and higher mental processes. It plays such a vital role that many consider it the essence of life itself.

Although the cerebral cortex is only about one-eighth of an inch thick, it's made up of approximately 30 billion neurons and nine times as many glial cells. Its numerous wrinkles, called *convolutions*, significantly increase its surface area. Damage to the cerebral cortex is linked to numerous problems, including suicide, substance abuse, and dementia (Maller et al., 2014; Masel & DeWitt, 2014; Sharma et al., 2015). Evidence suggests that brain injuries due to concussions are particularly common in athletes who experience head injuries in sports like football, ice hockey, boxing, and soccer (■ **Figure 2.19**).

See **FIGURE 2.19, Damage to the brain**

The full cerebral cortex and the two cerebral hemispheres beneath it closely resemble an oversized walnut. The division, or *fissure*, down the center marks the separation between the left and right *hemispheres* of the brain, which make up about 80% of the brain's weight. The hemispheres are mostly filled with axon connections between the cortex and the other brain structures. Each hemisphere controls the opposite side of the body.

NICHOLAS KAMM/AFP/Getty Images

The cerebral hemispheres are each divided into four distinct areas, or lobes (■ **Figure 2.20**). Like the lower-level brain structures, each lobe specializes in somewhat different tasks, another example of *localization of function*. However, some functions overlap two or more lobes.

Frontal Lobes

By far the largest of the cortical lobes, the two **frontal lobes** are located at the top front portion of the two brain hemispheres—right behind your forehead. The frontal lobes receive

See **FIGURE 2.20, Lobes of the brain**

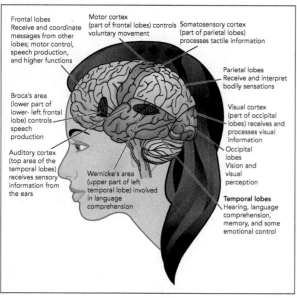

Frontal lobes
Receive and coordinate messages from other lobes; motor control, speech production, and higher functions

Motor cortex (part of frontal lobes) controls voluntary movement

Somatosensory cortex (part of parietal lobes) processes tactile information

Parietal lobes
Receive and interpret bodily sensations

Broca's area (lower part of lower-left frontal lobe) controls speech production

Visual cortex (part of occipital lobes) receives and processes visual information

Occipital lobes
Vision and visual perception

Auditory cortex (top area of the temporal lobes) receives sensory information from the ears

Wernicke's area (upper part of left temporal lobe) involved in language comprehension

Temporal lobes
Hearing, language comprehension, memory, and some emotional control

and coordinate messages from all other lobes of the cortex, while also being responsible for at least three additional functions:

1. *Speech production* Broca's area, located in the *left* frontal lobe near the bottom of the motor control area, plays a crucial role in speech production. In 1865, French physician Paul Broca discovered that damage to this area causes difficulty in speech, but not language comprehension. This type of impaired language ability is known as *Broca's aphasia*.

2. *Motor control* At the very back of the frontal lobes lies the *motor cortex*, which sends messages to the various muscles that instigate voluntary movement. When you want to call your friend on your cell phone, the motor control area of your frontal lobes guides your fingers to press the desired sequence of numbers.

3. *Higher functions* Most complex, higher functions, such as thinking, personality, emotion, and memory, are controlled primarily by the frontal lobes. Damage to the frontal lobe also affects motivation, drives, creativity, self-awareness, initiative, and reasoning. Abnormalities in the frontal lobes are often observed in patients with schizophrenia (Chapter 12). For example, patients with schizophrenia often show overall loss of gray matter, as well as increases in cerebrospinal fluid in the frontal lobes (DeRosse et al., 2015; Kubera et al., 2014; Schnack et al., 2014). However, one recent study using fMRI data found that the functional changes with the strongest links to schizophrenia involved the parietal lobes, NOT the frontal lobes (Guo et al., 2014).

At this point in the chapter, you may be feeling inundated with the large number of terms and functions of the various parts of your brain and nervous system. If so, please understand that mastering this information is essential to every chapter of this text and to the entire field of psychology. More importantly, recognizing that your brain is YOU—all your personality, thoughts, and your very life—will undoubtedly increase your motivation to study and protect it. Just as we routinely wear shoes to protect our feet, we need seat belts and helmets to protect our far more fragile brains! On a more encouraging note, updated information on the famous case of Phineas Gage, indicates that damage to the brain may not be as permanent as we once thought, thanks to the two processes we discussed earlier—*neuroplasticity* and *neurogenesis*.

Parietal Lobes

At the top of the brain, just behind the frontal lobes, are the two **parietal lobes**, which process information about touch, pressure, vibration, and pain. When you step on a sharp nail, you quickly (and reflexively) withdraw your foot because the messages travel directly to and from your spinal cord. The actual experience of "pain" occurs when the neural messages reach the *motor cortex* of the parietal lobes of your brain. The front part of the parietal lobes also contains the *somatosensory cortex*, a band of tissue that processes tactile information about different body parts.

Note in **Study Organizer 2.2** how certain areas of your body, like your hands and face, are represented as proportionally quite large compared to other body parts. This is meant to graphically demonstrate how the outer layer (cortex) of your brain dedicates more space for body parts that require more sensitivity, precision, and control. Thus, your hands and face have far more nerve endings and overall sensitivity than does your arm, which explains why a paper cut on your thumb is so disproportionately painful, and why you can easily feel the lightest caress on your cheek. Interestingly, an increase in the use of certain body parts, such as learning to play the piano, will lead to an increase in the amount of somatosensory cortex dedicated to your fingers—another example of *neuroplasticity*.

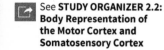

See **STUDY ORGANIZER 2.2: Body Representation of the Motor Cortex and Somatosensory Cortex**

Temporal Lobes

The **temporal lobes** are for hearing, language comprehension, memory, and some emotional control. The *auditory cortex*, which processes sound, is located at the top front of each temporal lobe. This area is responsible for receiving incoming sensory information and sending it on to the parietal lobes, where it is combined with other sensory information.

A part of the left temporal lobe called *Wernicke's area* aids in language comprehension. About a decade after Broca's discovery, German neurologist Carl Wernicke noted that patients with damage in this area could not understand what they read or heard, but they could speak quickly and easily. However, their speech was often unintelligible because it contained made-up words, sound substitutions, and word substitutions. This syndrome is now referred to as *Wernicke's aphasia*.

Occipital Lobes

The **occipital lobes** are responsible for, among other things, vision and visual perception. Damage to the occipital lobes can produce blindness, even if the eyes and their neural connection to the brain are perfectly healthy.

Association Areas

One of the most popular myths in psychology is that we use only 10% of our brain. This myth might have begun with early research which showed that approximately three-fourths of the cortex is "quiet" (with no precise, specific function responsive to electrical brain stimulation). These areas are not dormant, however. They are clearly engaged in interpreting, integrating, and acting on information processed by other parts of the brain. They are called **association areas** because they *associate*, or connect, various areas and functions of the brain. The association areas in the frontal lobes, for example, help in decision making and planning. Similarly, the association area right in front of the motor cortex aids in the planning of voluntary movement.

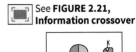
See **FIGURE 2.21, Information crossover**

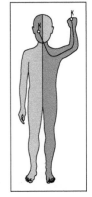

Two Brains in One?

We mentioned earlier that the brain's left and right cerebral hemispheres control opposite sides of the body (see ▣ **Figure 2.21**). Each hemisphere also has separate areas of specialization. This is another example of localization of function, technically referred to as *lateralization*.

Although complex activities typically involve both hemispheres, for most adults the left hemisphere is specialized not only for language functions (speaking, reading, writing, and understanding language), but also for analytical functions, such as mathematics (▣ **Figure 2.22**). The right hemisphere also contributes to complex word and language comprehension (Granger et al., 2015; Ivanitskii et al., 2015; Rapp & Lipka, 2011). Interestingly, the right hemisphere can take over additional language function when damage occurs to the left hemisphere—another example of our brain's remarkable plasticity (Staudt, 2010).

Note that right hemisphere does have some language functions, but it's primarily specialized for nonverbal functions. This includes art and musical abilities, as well as perceptual and spatiomanipulative skills, such as maneuvering through space, drawing or building geometric designs, working jigsaw puzzles, building model cars, painting pictures, and recognizing metaphors, faces and facial expressions (Argyriou et al., 2015; Gazzaniga, 1970, 2009; Ono et al., 2015).

Is this left- and right-brain specialization reversed in left-handed people? About 68% of left-handers (people who use their left hands to write, hammer a nail, and throw a ball) and 97% of right-handers have their major language areas on the left hemisphere. This suggests that even though the right side of the brain is dominant for movement in left-handers, other types of skills are often localized in the same brain areas as for right-handers.

Before going on, it's important to note that the popular media has made numerous exaggerated claims about the differences between the left and right hemispheres.

See **FIGURE 2.22, Functions of the left and right hemispheres**

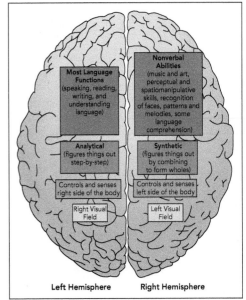

Split-Brain Surgery

How do we know that our two hemispheres perform somewhat differently, but overlapping functions?

Early researchers believed the right hemisphere was subordinate or nondominant to the left, with few special functions or abilities. In the 1960s, landmark **split-brain surgeries** began to change this view.

 See **FIGURE 2.23, Views of the corpus callosum**

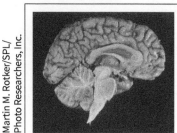
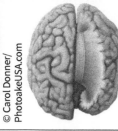

The primary connection between the two cerebral hemispheres is a thick, ribbon-like band of neural fibers under the cortex called the **corpus callosum** (**Figure 2.23**). In some rare cases of severe epilepsy, when other forms of treatment have failed, surgeons cut the corpus callosum to stop the spread of epileptic seizures from one hemisphere to the other. Because this operation cuts the only direct communication link between the two hemispheres, it reveals what each half of the brain can do in isolation from the other. The resulting research has profoundly improved our understanding of how the two halves of the brain function.

For example, when someone has a brain stroke and loses his or her ability to speak, we know this generally points to damage on the left hemisphere, because this is where *Broca's area*, which controls speech production, is located (refer back to **Figure 2.20**). However, keep in mind that when specific regions of the brain are injured or destroyed, their functions can sometimes be picked up by a neighboring region—even the opposite hemisphere.

Although most split-brain surgery patients generally show little outward change in their behavior, other than fewer epileptic seizures, the surgery does create some unusual responses. One split-brain patient reported that when he dressed himself, he sometimes pulled his pants down with his left hand and up with his right (Gazzaniga, 2009). The subtle changes in split-brain patients normally appear only with specialized testing. See **Study Organizer 2.3** for an illustration and description of this type of specialized test. Keep in mind that in actual split-brain surgery on live patients, only some fibers within the corpus callosum are cut (*not* the lower brain structures), and this surgery is performed only in rare cases of epilepsy.

See **STUDY ORGANIZER 2.3: Split-Brain Research**

Q Answer the **Concept Check** questions.

In our tour of the nervous system, the principles of localization of function, lateralization, and specialization recur: Dendrites receive information, the occipital lobes specialize in vision, and so on. Keep in mind, however, that all parts of the brain and nervous system also play overlapping and synchronized roles.

WP LS Go to your *WileyPLUS* Learning Space course for video episodes, examples, art, tables, Concept Checks, practice, and other pedagogical resources that will help you succeed in this course.

Reading for

STRESS AND HEALTH PSYCHOLOGY

3

3.1 UNDERSTANDING STRESS

Everyone experiences **stress**, and we generally know what a person means when he or she speaks of being "stressed." Thanks to the pioneering research of Hans Selye, the widely-acknowledged "father" of stress research, when we talk about stress we're generally focused on its biological effects—the wear and tear on our bodies. Most scientists define stress as *the interpretation of specific events, called **stressors**, as threatening or challenging*. The resulting physical and psychological reactions to stressors are known as the *stress response* (Friedman, 2015; Sanderson, 2013; Selye, 1936, 1983). Using these definitions, can you see how an upcoming exam on this material could be called a stressor if you're an unprepared student, and your physical and psychological reactions are your stress response? In the following section, we'll discuss the key sources of stress and how it affects us.

Sources of Stress

Although most of us can name numerous events, situations, and/or people that cause stress in our modern lives, psychological science has focused on seven major sources (■ **Figure 3.1**).

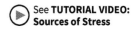

See **TUTORIAL VIDEO: Sources of Stress**

See **FIGURE 3.1, Seven major sources of stress**

Life Changes

Early stress researchers Thomas Holmes and Richard Rahe (1967) believed that any *life change* that required some adjustment in behavior or lifestyle could cause some degree of stress. They also believed that exposure to numerous stressful events in a short period could have a direct, detrimental effect on both psychological and physical health.

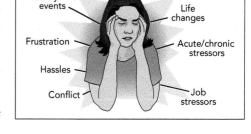

To investigate the relationship between change and stress, Holmes and Rahe created the Social Readjustment Rating Scale (SRRS), which asks people to check off all the life events they have experienced in the previous year.

The SRRS is an easy and popular tool for measuring stress, and cross-cultural studies have shown that most people rank the magnitude of their stressful events similarly (Consedine & Soto, 2014; Loving & Sbarra, 2015; Smith et al., 2014). But the SRRS is not foolproof. For example, it only shows a correlation between stress and illness; it does not prove that stress actually *causes* illnesses.

> *Future shock is the shattering stress and disorientation that we induce in individuals by subjecting them to too much change in too short a time.*
>
> —Alvin Toffler

Acute/Chronic Stressors

In addition to the stress caused by life-changing events, it's also important to note that the stressors can be either acute or chronic—and sometimes both. **Acute stress** is generally severe, but short term, with a definite endpoint, such as nearly missing a very important deadline. Surprisingly, repeated media coverage of tragedies, like the Boston Marathon bombings, can lead to more acute stress in viewers, than in those who had direct exposure to the tragedy (Holman et al., 2014).

See **FIGURE 3.2,**
Discrimination as a chronic
stressor

© JamieWilson/iStockphoto

War, a bad marriage, poor working conditions, poverty, and/or discrimination (discussed in ▣ **Figure 3.2**), can all be significant sources of **chronic stress** (Arbona & Jimenez, 2014; Chaby et al., 2015; Sotiropoulos et al., 2015). Surprisingly, chronic stress can even suppress sexual desire and damage testicular cells in male rats (Hou et al., 2014). In addition, persistent environmental noise is associated with measurable hormonal and brain changes (Fouladi et al., 2012). Our social lives also can be chronically stressful because making and maintaining friendships require considerable thought and energy (Fox & Moreland, 2015; Kushlev & Dunn, 2015; Vishwanath, 2015).

Job Stressors

For many people, one of their most pressing concerns is *job stress*, which can result from unemployment, job change, and/or worries about job performance. It can even lead to suicide (Adams, 2015; Cartwright & Cooper, 2014; Rees et al., 2015). One study found that leaders—including military officers and government officials—experience lower levels of stress than nonleaders, presumably because they have higher levels of control in their work environments (Sherman et al., 2012). The most stressful jobs are those that have little job security, and make great demands on performance and concentration, with little or no allowances for creativity or opportunity for advancement (Bauer & Hammig, 2014; Sanderson, 2013).

Other common sources of job stress come from *role conflict*, being forced to take on separate and incompatible roles, or *role ambiguity*, being uncertain about the expectations and demands of your role (Bauer & Hammig, 2014; Memili et al., 2015; Olson, 2014). Being a mid-level manager who reports to many supervisors, while also working among the people he or she is expected to supervise, is a prime example of both role conflict and role ambiguity.

Stress at work can also cause serious stress at home, not only for the worker but for other family members as well. In our private lives, divorce, child and spousal abuse, alcoholism, and money problems can all place severe stress on a family (Carpenter et al., 2015; Fan et al., 2015; Pederson, 2014).

Conflict

See **STUDY ORGANIZER 3.1:**
Types of Conflict

Stress also can arise when we experience **conflict**—that is, when we are forced to make a choice between at least two incompatible alternatives. There are three basic types of conflict: **approach-approach**, **approach-avoidance**, and **avoidance-avoidance**.

Generally, approach–approach conflicts are the easiest to resolve and produce the least stress. Avoidance–avoidance conflicts, on the other hand, are usually the most difficult and take the longest to resolve because either choice leads to unpleasant results. Furthermore, the longer any conflict exists, or the more important the decision, the more stress a person will experience.

Hassles

The minor **hassles** of daily living also can pile up and become a major source of stress. We all share many hassles, such as time pressures and financial concerns. But our reactions to them vary. Persistent hassles can lead to a form of physical, mental, and emotional exhaustion known as **burnout** (Guveli et al., 2015; Marchand et al., 2014; You et al., 2015). This is particularly true for some people in chronically stressful professions with little personal control, such as firefighters, police officers, doctors, and nurses. Their exhaustion and "burnout" then lead to more work absences, reduced productivity, and increased risk of illness.

Some authorities believe hassles can be more significant than major life events in creating stress (Kubiak et al., 2008; Stefanek et al., 2012). Divorce is extremely stressful, but it may be so because of the increased number of hassles it brings—changes in finances, child-care arrangements, new responsibilities, and so on.

Frustration

Like hassles, **frustration**, a negative emotional state resulting from a blocked goal, can cause stress. And the more motivated we are, the more frustrated we become when our goals are blocked. After getting stuck in traffic and missing an important job interview, we may become very frustrated. However, if the same traffic jam causes us to be five minutes late showing up to a casual party, we may experience little or no frustration.

Cataclysmic Events

The September 11th 2001 terrorist attacks in America, Hurricane Katrina in 2005, and the 2015 earthquakes in Nepal are what stress researchers call **cataclysmic events**. They occur suddenly and generally affect many people simultaneously. Politicians and the public often imagine that such catastrophes inevitably create huge numbers of seriously depressed and permanently scarred survivors.

Interestingly, relief agencies typically send large numbers of counselors to help with the psychological aftermath. However, researchers have found that because the catastrophe is shared by so many others, there is already a great deal of mutual social support from those with firsthand experience with the same disaster, which may help people cope (Aldrich & Meyer, 2015; Ginzburg & Bateman, 2008). On the other hand, these cataclysmic events are clearly devastating to all parts of the victims' lives (Alvarez, 2011; Gulliver et al., 2014; Joseph et al., 2014). And some survivors may develop a prolonged and severe stress reaction, known as *posttraumatic stress disorder (PTSD)*, which we discuss later in this chapter (Blanc et al., 2015; Ronan et al., 2015).

Effects of Stress

Have you ever experienced a near accident or some other sudden, frightening event? If so, you may have noticed how stress increased your heart rate, blood pressure, respiration, and muscle tension, while simultaneously decreasing your digestion and constricting your blood vessels. As you recall from Chapter 2, under stressful conditions, the *sympathetic* part of the autonomic nervous system is dominant, and your reactions to the stressor are part of the general "fight-or-flight" syndrome.

Once the danger passes, a longer sequence of important health-related reactions occurs. Selye's *general adaptation syndrome* and the SAM-HPA axis (discussed next) control the most significant of these changes.

Selye's General Adaptation Syndrome (GAS)

Canadian physician Hans Selye identified a generalized physiological reaction to chronic stress that he called the **general adaptation syndrome (GAS)** (Selye, 1936). The GAS occurs in three phases—*alarm*, *resistance*, and *exhaustion*—activated by efforts to adapt to any stressor, whether physical or psychological (**Process Diagram 3.1**).

Most of Selye's ideas about the GAS pattern of stress response have proven to be correct, but we now know that different stressors evoke different responses and that people vary widely in their reactions to them. One of the most interesting differences has to do with gender. Men more often choose "fight or flight." In contrast, women are more likely to "tend and befriend" (Cardoso et al., 2013; Taylor, 2006, 2012; von Dawans et al., 2012). This means that, when stressed, women more often take care of themselves and their children (tending), and form strong social bonds with others (befriending).

This Process Diagram contains essential information NOT found elsewhere in the text, which is likely to appear on quizzes and exams. Be sure to study it CAREFULLY!

Process Diagram 3.1

General Adaptation Syndrome (GAS)

The three phases of Selye's syndrome (*alarm*, *resistance*, and *exhaustion*) focus on the biological response to stress—particularly the "wear and tear" on the body that results from prolonged stress.

1 Alarm phase
When surprised or threatened, your body enters an alarm phase during which your resistance to stress is temporarily suppressed, while your arousal is high (e.g., increased heart rate and blood pressure) and blood is diverted to your skeletal muscles to prepare for "fight-or-flight" (Chapter 2).

2 Resistance phase
If the stress continues, your body rebounds to a phase of increased resistance. Physiological arousal remains higher than normal, and there is an outpouring of stress hormones. During this resistance stage, people use a variety of coping methods. For example, if your job is threatened, you may work longer hours and give up your vacation days.

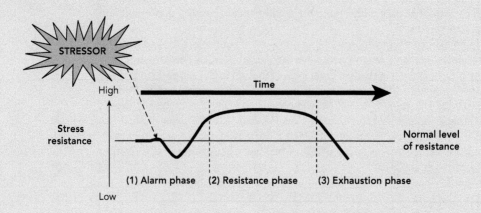

3 Exhaustion phase
Your body's resistance to stress can only last so long before exhaustion sets in. During this final phase, you become more susceptible to serious illnesses, and to experience possibly irreversible damage to your body. Selye maintained that one outcome of this exhaustion phase for some people is the development of *diseases of adaptation*, including asthma, ulcers, and high blood pressure. Unless a way of relieving stress is found, the eventual result may be complete collapse and death.

Some researchers believe these differences are hormonal in nature. Although oxytocin is released during stress in both men and women, the women's higher level of estrogen tends to enhance oxytocin, which results in more calming and nurturing feelings. In contrast, the hormone testosterone, which men produce in higher levels during stress, reduces the effects of oxytocin.

What is Selye's most important take-home message? *Our bodies are relatively well designed for temporary stress but poorly equipped for prolonged stress.* The same biological processes that are adaptive in the short run, such as the fight-or-flight response, can be hazardous in the long run (Papathanasiou et al., 2015; Russell et al., 2014).

The SAM System and HPA Axis

To understand these dangers, we need to first describe how our bodies (ideally) respond to stress. As you can see in **Process Diagram 3.2**, once our brains identify a stressor, our **SAM** (sympatho–adreno–medullary) **system** and **HPA** (hypothalamic–pituitary–adreno-cortical) **axis** then work together to increase our arousal and energy levels to deal with the stress (Garrett, 2015; Nicolaides et al., 2015; Sanderson, 2013). Once the stress is resolved, these systems turn off, and our bodies return to normal, baseline functioning, known as **homeostasis**.

Unfortunately, given our increasingly stressful modern lifestyle, our bodies are far too often in a state of elevated, chronic arousal, which can wreak havoc on our health. Some of the most damaging effects of stress are on our *immune system* and our *cognitive functioning*.

Stress and Our Immune System

The discovery of the relationship between stress and our immune system has been very important. When people are under stress, the immune system is less able to regulate the normal inflammation system, which makes us more susceptible to diseases, such as bursitis,

Process Diagram 3.2

The Stress Response—an Interrelated System

Faced with stress, our sympathetic nervous system prepares us for immediate action—to "fight or flee." Our slower-acting HPA axis maintains our arousal. Here's how it happens:

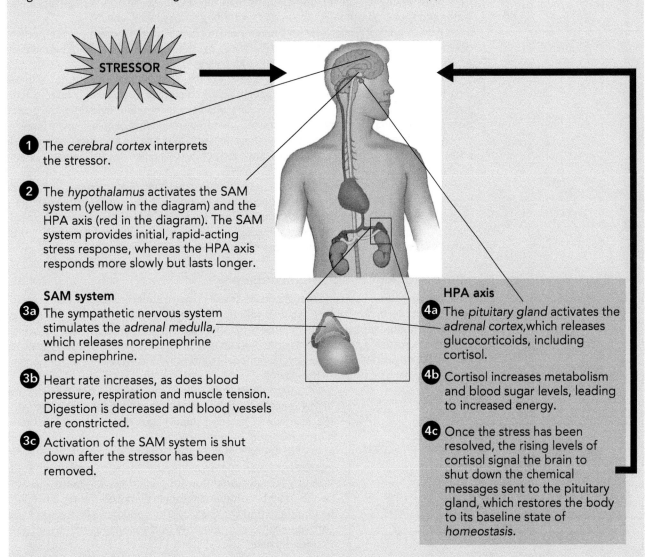

STRESSOR

1 The *cerebral cortex* interprets the stressor.

2 The *hypothalamus* activates the SAM system (yellow in the diagram) and the HPA axis (red in the diagram). The SAM system provides initial, rapid-acting stress response, whereas the HPA axis responds more slowly but lasts longer.

SAM system

3a The sympathetic nervous system stimulates the *adrenal medulla*, which releases norepinephrine and epinephrine.

3b Heart rate increases, as does blood pressure, respiration and muscle tension. Digestion is decreased and blood vessels are constricted.

3c Activation of the SAM system is shut down after the stressor has been removed.

HPA axis

4a The *pituitary gland* activates the *adrenal cortex*, which releases glucocorticoids, including cortisol.

4b Cortisol increases metabolism and blood sugar levels, leading to increased energy.

4c Once the stress has been resolved, the rising levels of cortisol signal the brain to shut down the chemical messages sent to the pituitary gland, which restores the body to its baseline state of *homeostasis*.

Test Yourself

Reviewing the SAM System and HPA Axis

Arrange the following 8 key terms in their proper order as part of the SAM system or the HPA axis, and then compare your answers with those provided.

Key terms: *adrenal cortex, adrenal medulla, glucocorticoids, cortisol, norepinephrine, epinephrine, pituitary gland, homeostasis* colitis, Alzheimer's disease, rheumatoid arthritis, periodontal disease, and even the common cold (Campbell et al., 2015; Cohen et al., 2003, 2012; Sotiropoulos et al., 2015).

Prolonged, excessive, and/or chronic stress also contributes to hypertension, depression, posttraumatic stress disorder (PTSD), drug and alcohol abuse, and even low birth weight (Brannon et al., 2014; Nicolaides et al., 2015; Zhou & Kreek, 2014).

Interestingly, the relationship between stress and depression has even been shown in rats. For example, researchers in one study caused stress in rats by restricting their food and play time, isolating them from other rats, and switching around their sleep and awake times for three weeks (Son et al., 2012). After experiencing this stress, the rats showed clear signs of depression, including having little interest in eating or drinking tasty sugar water and showing immobility instead of swimming when placed in water.

As if the long list of ill effects for both human and nonhuman animals weren't enough, severe or prolonged stress can also produce premature aging, faster growth of cancer cells, and even death (Germain, 2014; Lamkin et al., 2015; Prenderville et al., 2015).

How does this happen? *Cortisol*, a key element of the HPA axis, plays a critical role in the long-term negative effects of stress. Although increased cortisol levels initially help us fight stressors, if these levels stay high, which occurs when stress continues over time, the body's disease-fighting immune system is suppressed. One study found that people who are lonely—which is another type of chronic stressor—have an impaired immune response, leaving their bodies vulnerable to infections, allergies, and many of the other illnesses cited above (Jaremka et al., 2013).

Knowledge that psychological factors have considerable control over these and other illnesses has upset the long-held assumption in biology and medicine that diseases are strictly physical. The clinical and theoretical implications are so important that a new interdisciplinary field, called **psychoneuroimmunology**, has emerged (Kusnecov & Anisman, 2014). It studies the effects of psychological and other factors on the immune system.

Stress and Our Cognitive Functioning

What happens to our brain and thought processes when we're under immediate, *acute* stress? As we've just seen, cortisol helps us deal with immediate dangers by mobilizing our energy resources. It also helps us create memories for short-term, highly emotional events, which explains the so-called *flashbulb memories* discussed in Chapter 7. Can you see why scientists believe that our increased memories for emotional events may have evolved to help us remember what to avoid or protect in the future?

See **FIGURE 3.3, Our brain under chronic stress**

Stress and our prefrontal cortex
Chronic stress results in a reduction in the size of neurons in the prefrontal cortex and a diminished performance during cognitive tasks.

Stress and our hippocampus
Cortisol released in response to immediate stress can be beneficial. However, under chronic stress it can produce a vicious cycle leading to permanent damage of the hippocampus.

Although short-term stress can improve our memory for highly emotional, flashbulb events, it can also interfere with the retrieval of existing memories, the laying down of new memories, and general information processing (Aggarwal et al., 2014; Olver et al., 2014; Wingenfeld & Wolf, 2015). This interference with cognitive functioning helps explain why you may forget important information during a big exam and why people may become dangerously confused during a fire and be unable to find the fire exit. The good news is that once the cortisol washes out, memory performance generally returns to normal levels.

What happens during prolonged stress? Chronic stress may lead to permanent damage to the prefrontal cortex. In addition, long-term exposure to cortisol can permanently damage cells in the hippocampus, a key part of the brain involved in memory (Chapter 7). Furthermore, once the hippocampus has been damaged, it cannot provide proper feedback to the hypothalamus, so cortisol continues to be secreted, and a vicious cycle can develop (**Figure 3.3**).

The Benefits of Stress

So far in our discussion, we've focused primarily on the harmful, negative side of stress, but there are also some positive aspects. Our bodies are nearly always in some state of stress, whether pleasant or unpleasant, mild or severe. *Anything* placing a demand on the body can cause stress. When stress is pleasant or perceived as a manageable challenge, it can be beneficial. As seen in athletes, business tycoons, entertainers, or great leaders, this type of desirable stress, called **eustress**, helps arouse and motivate us to persevere and accomplish challenging goals. Stress that is unpleasant and threatening is called **distress** (Selye, 1974).

Consider large life events like graduating from college, securing a highly desirable job, and/or getting married. Each of these occasions involves enormous changes in our lives and inevitable conflicts, frustration and other sources of stress, yet for most of us they are incredibly positive events. Rather than being the source of discomfort and distress, *eustress* is pleasant and motivating. It encourages us to overcome obstacles and even enjoy the effort and work we expend toward achieving our goals. Physical exercise is an even clearer example of the benefits of eustress. When we're working out at a gym, or even just walking in a park, we're placing some level of stress on our bodies. However, this stress encourages the development and strengthening of all parts of our body, particularly our muscles, heart, lungs, and bones. Exercise also releases endorphins (Chapter 2), which helps lift depression and overall mood.

See **FIGURE 3.4, Stress and task complexity**

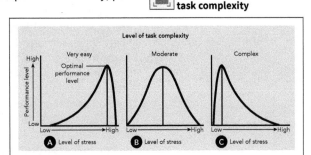

Keep in mind that all the achievements related to eustress do require considerable effort. As you well know, going through college requires long hours of study, self-discipline, and delayed gratification. It's also important to note that the optimal level of stress depends on the task complexity (▣ **Figure 3.4**). For example, during well learned, easy tasks, you need a higher level of stress to perform at your best. This is why athletes, who are performing their well-learned tasks, perform better during high stakes competition—their stress levels are higher. In contrast, you need a really low level of stress during a hard, complex exam in your psychology class (unless you've taken the time to do a lot of practice testing, like those provided within this text).

Answer the **Concept Check** questions.

3.2 STRESS AND ILLNESS

As we've just seen, stress has dramatic effects on our bodies. This section explores how stress is related to four serious illnesses—*gastric ulcers, cancer, cardiovascular disorders, and posttraumatic stress disorder (PTSD).*

Gastric Ulcers

Gastric ulcers are lesions to the lining of the stomach (and duodenum—the upper section of the small intestine) that can be quite painful. In extreme cases, they may even be life-threatening. Beginning in the 1950s, psychologists reported strong evidence that stress can lead to ulcers. Studies found that people who live in stressful situations have a higher incidence of ulcers than people who don't. And numerous experiments with laboratory animals have shown that stressors, such as shock or confinement to a very small space, can produce ulcers in a few hours in some laboratory animals (e.g., Fomenko et al., 2014; Landeira-Fernandez, 2015).

The relationship between stress and ulcers seemed well established until researchers identified a bacterium (*Helicobacter pylori*, or *H. pylori*) that appears to be associated with ulcers. Later studies confirmed that this bacterium clearly damages the stomach wall, and that antibiotic treatment helps many patients. However, approximately 75% of normal, healthy people's stomachs also have the bacterium. This suggests that the bacterium may cause the ulcer, but only in people who are compromised by stress. Furthermore, behavior modification and other psychological treatments, used alongside antibiotics, can also help ease ulcers. In other words, although stress *by itself* does not cause ulcers, it is a contributing factor, along with biological factors (Fink, 2011; Lemogne et al., 2015; Southwick & Watson, 2015).

Before going on, note that many believe ulcers are "psychosomatic," and this means they're imaginary. However, psychosomatic (*psyche* means "mind" and *soma* means "body") refers to symptoms or illnesses that are caused or aggravated by psychological factors, especially stress. Most researchers and health practitioners believe that almost all illnesses are partly psychosomatic in this sense.

Cancer

Cancer is among the leading causes of death for adults in the United States. It occurs when a particular type of primitive body cell begins rapidly dividing, and then forms a tumor that invades healthy tissue. Unless destroyed or removed, the tumor eventually damages organs and causes death. More than 100 types of cancer have been identified. They appear to be caused by an interaction between environmental factors (such as diet, smoking, and pollutants) and inherited predispositions.

Note that research does *not* support the popular myths that stress *causes* cancer, or that positive attitudes can cure it (Chang et al., 2015; Coyne & Tennen, 2010; Lilienfeld et al., 2010, 2015). However, stress may increase the spread of cancer cells to other organs, including the bones, which decreases the likelihood of survival (Campbell et al., 2012; Klink, 2014).

But this is not to say that developing a positive attitude and reducing our stress levels aren't worthy health goals (Hays, 2014; Quick et al., 2013; Tamagawa et al., 2015). In a healthy person, whenever cancer cells start to multiply, the immune system checks the uncontrolled growth by attacking the abnormal cells. Unfortunately, as you read earlier, prolonged stress causes the adrenal glands to release hormones that negatively affect the immune system, and a compromised immune system is less able to resist infection or to fight off cancer cells (Bick et al., 2015; Jung et al., 2015; Kokolus et al., 2014). For example, when researchers disrupted the sleep of 21 healthy people over a period of six weeks, they found increases in blood sugar and decreases in metabolism, which may lead to obesity and diabetes (Buxton et al., 2012).

Cardiovascular Disorders

Cardiovascular disorders contribute to over half of all deaths in the United States (American Heart Association, 2013). Understandably, health psychologists are concerned because stress is a major contributor to these deaths (Go et al., 2014; Orth-Gomér et al., 2015; Steptoe & Kivimäki, 2013).

Heart disease is a general term for all disorders that eventually affect the heart muscle and lead to heart failure. *Coronary heart disease* occurs when the walls of the coronary arteries thicken, reducing or blocking the blood supply to the heart. Symptoms of such disease include *angina* (chest pain due to insufficient blood supply to the heart) and *heart attack* (death of heart muscle tissue).

How does stress contribute to heart disease? Recall that one of the major brain and nervous system "fight or flight" reactions is the release of epinephrine (adrenaline) and cortisol into the bloodstream. These hormones increase heart rate and release fat and glucose from the body's stores to give muscles a readily available source of energy. If no physical fight-or-flight action is taken (and this is most likely the case in our modern lives), the fat that was released into the bloodstream is not burned as fuel. Instead, it may adhere to the walls of blood vessels. These fatty deposits are a major cause of blood-supply blockage, which causes heart attacks. A large, meta-analysis of the correlation between job strain and coronary heart disease found that people with stressful jobs are 23% more likely to experience a heart attack than those without stressful jobs (Kivimäki et al., 2012).

Of course, the stress-related buildup of fat in our arteries is not the only risk factor associated with heart disease. Other factors include smoking, obesity, a high-fat diet, and lack of exercise (Bharmal et al., 2014; Christian et al., 2015; Orth-Gomér et al., 2015).

Posttraumatic Stress Disorder (PTSD)

One of the most powerful examples of the effects of severe stress is **posttraumatic stress disorder (PTSD),** which is a long-lasting, trauma and stressor-related disorder that overwhelms an individual's ability to cope (American Psychiatric Association, 2013; Brewin, 2014; Levine, 2015). Have you ever been in a serious car accident or been the victim of a violent crime? According to the National Institute on Mental Health (NIMH) (2014), it's natural to feel afraid when in danger. But for some people in certain situations, the normal "fight or flight" response is modified or damaged. This change helps explain why people with PTSD continue to experience extreme stress and fear, even when they're no longer in danger.

Research shows that approximately 40% of all children and teens will experience a traumatic stressor, and that the lifetime prevalence for trauma is between 50 to 90% (Brown et al., 2013; Cohen et al., 2014). However, the vast majority will not go on to develop PTSD. The core symptoms of PTSD, shown in  **Table 3.1**, may continue for months or even years after the event. Unfortunately, some victims of PTSD turn to alcohol and other drugs to cope, which generally compounds the problem (Currier et al., 2014; McLean et al., 2015; Ouimette & Read, 2014).

Sadly, one of the most dangerous problems associated with PTSD is the increased risk for suicide. Did you know that the number of suicides committed by young male veterans under the age of 30 jumped by 44% between 2009 and 2011, while the rate for female vets increased by 11% in the same time period? The precise cause for this astronomically high and climbing jump in suicides is unknown, but experts point to PTSD, along with combat injuries, and the difficulties of readjusting to civilian life (Finley et al., 2015; Legarreta et al., 2015; Zoroya, 2014).

See **TABLE 3.1: Key Characteristics of PTSD**

See **FIGURE 3.5, Stress and PTSD**

Lest you think PTSD only develops from military experiences, it's important to note that victims of natural disasters, physical or sexual assault, terrorist attacks, and rescue workers facing overwhelming situations also may develop PTSD. Sadly, research shows that simply watching television coverage of major natural disasters, such as hurricanes, earthquakes and tornados, can increase the number of PTSD symptoms, especially in children who are already experiencing some symptoms (Holman et al., 2014; Weems et al., 2012).

PTSD is not a new problem. During the Industrial Revolution, workers who survived horrific railroad accidents sometimes developed a condition very similar to PTSD. It was called "railway spine" because experts thought the problem resulted from a twisting or concussion of the spine. Later, doctors working with combat veterans referred to the disorder as "shell shock" because they believed it was a response to the physical concussion caused by exploding artillery. Today, we know that PTSD is caused by exposure to extraordinary stress ( **Figure 3.5**).

Elaine Thompson/AP Photo

See **TABLE 3.2: Seven Important Tips for Coping with Crisis**

Answer the **Concept Check** questions.

What can we do to help? Professionals have had success with various forms of therapy and medication for PTSD. In addition to the tips for helping someone with PTSD in the **Applying Psychology**, professionals also have offered seven general tips to help all of us cope with stress and crisis situations (**Table 3.2**).

3.3 STRESS MANAGEMENT

As noted at the beginning of this chapter, stress is a normal, and necessary, part of our life. Therefore, *stress management* is the goal—not stress elimination. Although our initial, bodily responses to stress are largely controlled by nonconscious, autonomic processes, our higher brain functions can help us avoid the serious damage from chronic overarousal. The key is to consciously recognize when we are overstressed and then to choose resources that activate our parasympathetic, relaxation response. In this section, we'll first discuss general patterns often used for coping with stress. Then we'll explore how individual differences in personality affect our coping responses. Finally, we'll present six important resources for healthy living and stress management.

Coping with Stress

Because we can't escape stress, we need to learn how to effectively cope with it. Simply defined, *coping* is an attempt to manage stress in some effective way. It is not one single act, but a process that allows us to deal with various stressors (**Process Diagram 3.3**).

See **TUTORIAL VIDEO: Coping with Stress**

Process Diagram 3.3

Cognitive Appraisal and Coping

Research suggests that our emotional response to an event depends largely on how we interpret the event. First, we go through a *primary appraisal* process to evaluate the threat, and decide whether it's harmful or potentially harmful. Next, during *secondary appraisal*, we assess our available and potential resources for coping with the stress. Then, we generally choose either *emotion-* or *problem-focused* methods of coping, or a combination of the two.

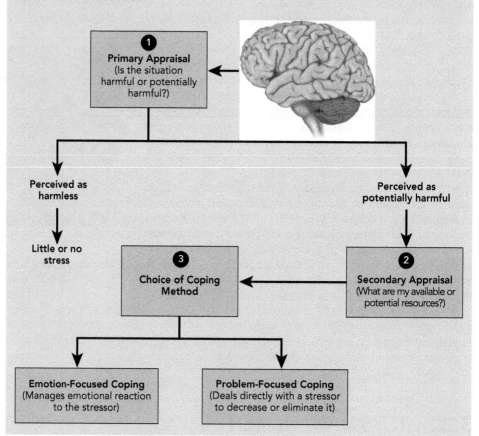

One of the biggest challenges during the process of coping is deciding whether to try to change the stressor itself or our emotional reactions to it. **Problem-focused coping** strategies work to deal directly with a stressor in order to eventually decrease or eliminate it (Arends et al., 2014; Gil & Weinberg, 2015; Mayordomo-Rodriquez et al., 2015). We tend to choose this approach, and find it most effective, when we have some control over a stressful situation.

Many times, however, it seems that nothing can be done to alter the stressful situation, so we turn to **emotion-focused coping**, in which we attempt to relieve or regulate our emotional reactions. Imagine you were refused a highly desirable job. You might initially feel disappointed and/or rejected. But you could successfully cope with these feelings by reappraising the situation, and deciding the job wasn't the right match for you, or that you weren't really qualified or ready for it.

When faced with unavoidable stress, Freud believed that we often resort to **defense mechanisms**, in which we unconsciously distort reality to protect our egos and to avoid anxiety (see Chapter 11). These defense mechanisms can sometimes act as a beneficial type of emotion-focused coping, like the example above. But if taken too far they can be destructive. If we fail to get the promotion, and then resort to elaborate excuses (rationalizations) for our failure, it may block us from seeing a situation more clearly and realistically, which in turn can prevent us from developing valuable skills. In short, emotion-focused coping can be beneficial, as long as it's not overused, and does not distort reality (Hertel et al., 2015; Levine, 2015).

Keep in mind that emotion-focused forms of coping can't change the problem, but they do make us feel better about the stressful situation. For example, teenagers who are asked to think about the benefits of a recent stressful event—such as having a traffic accident or losing a valued relationship—show increases in positive mood and decreases in negative mood (Rood et al., 2012). Interestingly, instant messaging (IM) also helps distressed teenagers share their emotions, and receive immediate social support and advice (Dolev-Cohen & Barak, 2013).

Personality and Coping

A person's ability to cope effectively with stress depends in large part on his or her own personality and internal resources. Researchers have found that people who see stress as enhancing, not debilitating, show more adaptive cortisol responses (Crum et al., 2013; Dunn et al., 2014; Purc-Stephenson, 2014).

In addition, hope can sustain a person in the face of severe odds, as is often documented in news reports of people who have triumphed over seemingly unbeatable circumstances. Similarly, having a good sense of humor is one of the best ways to reduce stress. The ability to laugh at oneself and at life's inevitable ups and downs allows us to relax and gain a broader perspective. In short: "Don't sweat the small stuff."

Internal versus External Locus of Control

Perhaps one of the most important personal resources for stress management is a sense of personal control. People who believe they are the "masters of their own destiny" have what is known as an **internal locus of control**. Believing they control their own fate, they tend to make more effective decisions and healthier lifestyle changes, are more likely to follow treatment programs, and more often find ways to positively cope with a situation.

Conversely, people with an **external locus of control** believe that chance or outside forces beyond our control determine our fate. Therefore, they tend to feel powerless to change their circumstances, are less likely to make effective and positive changes, and are more likely to experience higher levels of stress (e.g., Au, 2015; Rotter, 1966; Zhang et al., 2014).

 See **FIGURE 3.6, Positive psychology in action**

Positive Affect (Emotion)

Have you ever wondered why some people survive in the face of great stress (personal tragedies, demanding jobs, or an abusive home life) while others do not? One answer may be that these "survivors" have a unique trait called **positive affect**, meaning they have a sense of pleasure in their environment, including feelings of happiness, joy, enthusiasm, and contentment (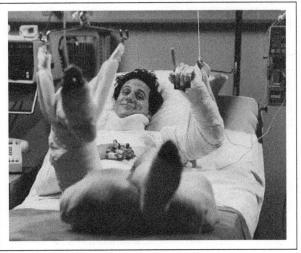 **Figure 3.6**). Interestingly, people who are high in positive affect also experience fewer colds and auto accidents, as well as better overall amount and quality of sleep, and an enhanced quality of life (Brannon et al., 2014; Eaton et al., 2014; von Känel, 2014).

Positive states are also associated with longer life expectancy. One study of 600 patients with heart disease found that those with more positive attitudes were less likely to die during a five-year follow-up period (Hoogwegt et al., 2013). Of those with positive attitudes, only 9.9% had died five years later, compared to 16.5% of those with less positive attitudes. This study provides powerful evidence for the mind-body link. Positive affect is also associated with a lower risk of mortality in patients with diabetes, which is a leading cause of death in the United States (Cohn et al., 2014; Moskowitz et al., 2008).

In addition, positive affect is closely associated with **optimism**, the expectation that good things will happen in the future and bad things will not. If you agree with statements such as, "I tend to expect the best of others," or "I generally think of several ways to get out of a difficult situation," you're probably an optimist. On the other hand, if you're a pessimist, you're far more likely to expect the worst of others, and to give up in the face of adversity.

As you might expect, optimists tend to be much better at coping with stress. Rather than seeing bad times as a constant threat and assuming personal blame for them, they generally assume that bad times are temporary and external to themselves. Optimists also tend to have better overall physical and psychological health, and to lead longer and overall happier lives (Dumitrache et al., 2015; Seligman, 2011; Sergeant & Mongrain, 2014).

See **TUTORIAL VIDEO:** **Positive Psychology**

The good news is that according to Martin Seligman, a leader in the field of positive psychology, optimism can be learned (Seligman, 2012). In short, he believes optimism requires careful monitoring and challenging of our thoughts, feelings, and self-talk. For example, if you don't get a promotion at work or you receive a low grade on an exam, don't focus on all the negative possible outcomes and unreasonably blame yourself. Instead, force yourself to think of alternate ways to meet your goals and develop specific plans to improve your performance. Chapter 13 offers additional help and details for overcoming faulty thought processes.

Resources for Healthy Living

In addition to the individual's personality and internal resources, effective coping also depends on the available external resources, such as the **mindfulness-based stress reduction (MBSR)** programs, which are based on developing a state of consciousness that attends to ongoing events in a receptive, non-judgmental way. The practice of MBSR has proven to be particularly effective in managing stress and treating mood disturbances, and its even been linked to positive, and perhaps permanent cell and brain changes (Chiesa et al., 2015; Esch, 2014; Schutte & Malouff, 2014).

One of the most effective and frequently overlooked external resources for stress management is *social support*. When we are faced with stressful circumstances, our friends and family often help us take care of our health, listen, hold our hands, make us feel important, and provide stability to offset the changes in our lives.

See **VIRTUAL FIELD TRIP:** **Biofeedback: Learning to Understand Your Body**

This support can help offset the stressful effects of chronic illness, pregnancy, physical abuse, job loss, and work overload. People who have greater social support also experience better health outcomes, including greater psychological well-being, greater physical well-being, faster recovery from illness, and a longer life expectancy (Cherry et al., 2015; Hughes et al., 2014; Kovács et al., 2015; Panagioti et al., 2014).

See **STUDY ORGANIZER 3.2:** **Six Additional External Stress Resources**

These findings may help explain why married people live longer than unmarried people (Liu, 2009), and why people who experience a marital separation or divorce are at greater risk of an early death (Sbarra & Nietert, 2009). So what is the important take-home message from this emphasis on social support? Don't be afraid to offer help and support to others—or to ask for the same for yourself!

Q Answer the **Concept Check** questions.

Six additional external resources for healthy living and stress management are exercise, social skills, behavior change, stressor control, material resources, and relaxation.

3.4 HEALTH PSYCHOLOGY

Health psychology emphasizes the interplay between our physical health and our psychological well being. As mentioned earlier in this chapter, ulcers, cancer, cardiovascular disorders, and PTSD significantly affect our physical well-being, as well as our cognitive, emotional, and behavioral responses. In this final section, we'll discuss the work of health psychologists, followed by an exploration of stress in the workplace.

What Does a Health Psychologist Do?

As researchers, health psychologists are particularly interested in how changes in behavior can improve health outcomes (Brannon et al., 2014; Straub, 2014). They also emphasize the relationship between stress and the immune system. As we discovered earlier, a normally functioning immune system helps defend against disease. On the other hand, a suppressed immune system leaves the body susceptible to a number of illnesses.

As practitioners, health psychologists can work as independent clinicians or as consultants alongside physicians, physical and occupational therapists, and other health care workers. The goal of health psychologists is to reduce psychological distress or unhealthy behaviors. They also help patients and families make critical decisions and prepare psychologically for surgery or other treatment. Health psychologists have become so involved with health and illness that medical centers are one of their major employers (Considering a Career, 2011).

Health psychologists also educate the public about illness *prevention* and health *maintenance*. For example, they provide public information about the effects of stress, smoking, alcohol, lack of exercise, and other health issues. Tobacco use endangers both smokers and those who breathe secondhand smoke, so it's not surprising that health psychologists are concerned with preventing smoking and getting those who already smoke to stop.

See **FIGURE 3.7, Understanding nicotine addiction**

Did you know that according to the U.S. Department of Health and Human Services, smoking has killed 10 times the number of Americans who died in all our nation's wars combined (Sebelius, 2014)? Thanks in large part to comprehensive mass media campaigns, smoke-free policies, restrictions on underage access to tobacco, and large price increases, adult smoking rates have fallen from about 43% in 1965 to about 18% in 2014. Unfortunately, cigarette smoking still remains the chief preventable killer in America (Koh, 2014).

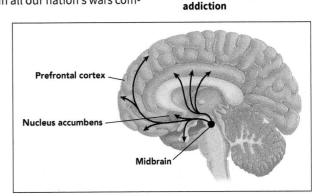

Given that almost everyone recognizes the serious consequences of smoking, and the fact that the first puff is rarely pleasant, why do people start smoking? The answer can be found in the biopsychosocial model (■ **Figure 3.7**). In addition to encouraging smokers to stop, or to never start, health psychologists also help people cope with conditions such as chronic pain, diabetes, and high blood pressure, as well as unhealthful behaviors such as inappropriate anger and/or lack of assertiveness. If you're interested in a career in this field, check with your college's counseling or career center.

Health Psychology at Work

Have you ever dragged yourself home from work so tired you feared you couldn't make it to your bed? Do you think your job may be killing you? You may be right! Some research suggests that job stress and overwork can greatly increase your risk of dying from heart disease and stroke (Bannai & Tamakoshi, 2014; Landsbergis et al., 2011; Nakao, 2010). And the Japanese even have a specific word for it, "karoshi" [KAH-roe-she], which is translated literally as "death from overwork."

Job Stress

Starting in the late 1970s, Japanese health officials began to notice serious, and potentially lethal, effects of working 10 or 12 hours a day, 6 and 7 days a week, year after year (Kanai, 2009; Kondo & Oh, 2010; Nakashima et al., 2011). Some research suggests that more than 10,000 workers die from work-related cardiovascular diseases in Japan each year, but few victims of karoshi are compensated under the Japanese workers' compensation system (Hsiu-Hui, 2007). Intense job stressors reportedly not only increase the risk for karoshi, but they also leave some workers disoriented and suffering from serious stress even when they're not working.

Things are not much better in the United States. The average number of work hours per week is among the highest in the developed world (Brown, 2011). Unfortunately, in our global economy, pressures to reduce costs and to increase productivity will undoubtedly continue, and job stress may prove to be a serious and growing health risk.

If you're not suffering from overwork, are you hassled and stressed by the ever-changing technology at your workplace? Do the expensive machines your employers install to "aid productivity" create stress-related problems instead? If so, you may be suffering from the well-documented, ill-effects of **technostress**, a feeling of anxiety or mental pressure from overexposure or involvement with technology (Maier et al., 2015; Sellberg & Susi, 2014; Tarafdar et al., 2015).

Although technology is often described as a way of bringing people together, how often have you noticed busy executives frantically checking their e-mail while on vacation? It's even more common to see families eating dinners at restaurants with their children playing video games or text messaging, and the parents loudly talking on separate cell phones. Furthermore, researchers have found that people who spend more time on Facebook experience lower levels of day-to-day happiness as well as lower overall feelings of life satisfaction (Kross et al., 2013). In fact, simply placing a cell phone on the table between two people - even if no one ever picks it up - leads to lower levels of closeness, connection, and meaning in their conversation (Przybylski & Weinstein, 2013)

Coping with Job Stress

Experts are suggesting that we can (and must) control technology and its impact on our lives. Admittedly, we all find the new technologies convenient and useful. But how can we control technostress? First, evaluate each new technology on its usefulness for you and your lifestyle. It isn't a black or white, "technophobe" or "technophile," choice. If something works for you, invest the energy to adopt it. Second, establish clear boundaries. Technology came into the world with an implied promise of a better and more productive life. But, for many, the servant has become the master. Like any healthy relationship, our technology interactions should be based on moderation and balance (Ashton, 2013).

In addition to potential death from overwork and increased stress and illness from technostress, health psychologists have identified several additional factors in job-related stress. Their findings suggest that one way to prevent these stresses is to gather lots of information before making a career decision.

If you would like to apply this to your own career plans, start by identifying what you like and don't like about your current (and past) jobs. With this information in hand, you'll be prepared to find jobs that will better suit your interests, needs, and abilities, which will likely reduce your stress. To start your analysis, answer *Yes* or *No* to these questions:

1. Is there a sufficient amount of laughter and sociability in my workplace?

2. Does my boss notice and appreciate my work?

3. Is my boss understanding and friendly?

4. Am I embarrassed by the physical conditions of my workplace?

5. Do I feel safe and comfortable in my place of work?

6. Do I like the location of my job?

7. If I won the lottery and were guaranteed a lifetime income, would I feel truly sad if I also had to quit my job?

8. Do I watch the clock, daydream, take long lunches, and leave work as soon as possible?

9. Do I frequently feel stressed and overwhelmed by the demands of my job?

10. Compared to others with my qualifications, am I being paid what I am worth?

11. Are promotions made in a fair and just manner where I work?

12. Given the demands of my job, am I fairly compensated for my work?

Now score your answers. Give yourself one point for each answer that matches the following: 1. No; 2. No; 3. No; 4. Yes; 5. No; 6. No; 7. No; 8. Yes; 9. Yes; 10. No; 11. No; 12. No.

The questions you just answered are based on four factors that research shows are conducive to increased job satisfaction and reduced stress: supportive colleagues, supportive working conditions, mentally challenging work, and equitable rewards (Robbins, 1996). Your total score reveals your overall level of dissatisfaction. A look at specific questions can help identify which of these four factors is most important to your job satisfaction—and most lacking in your current job.

1. Supportive colleagues (items 1, 2, 3): For most people, work fills important social needs. Therefore, having friendly and supportive colleagues and superiors leads to increased satisfaction.

2. Supportive working conditions (items 4, 5, 6): Not surprisingly, most employees prefer working in safe, clean, and relatively modern facilities. They also prefer jobs close to home.

3. Mentally challenging work (items 7, 8, 9): Jobs with too little challenge create boredom and apathy, whereas too much challenge creates frustration and feelings of failure.

4. Equitable rewards (items 10, 11, 12): Employees want pay and promotions based on job demands, individual skill levels, and community pay standards.

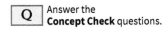 Answer the **Concept Check** questions.

WP LS Go to your WileyPLUS Learning Space course for video episodes, examples, art, tables, Concept Checks, practice, and other pedagogical resources that will help you succeed in this course.

Reading for

SENSATION AND PERCEPTION

4.1 HOW WE SENSE AND PERCEIVE OUR WORLD

Is all that we see or seem, but a dream within a dream?

—EDGAR ALLAN POE

Psychologists are keenly interested in our senses because they are our mind's window to the outside world. We're equally interested in how our mind perceives and interprets the information it receives from the senses.

Sensation begins with specialized receptor cells located in our sense organs (eyes, ears, nose, tongue, skin, and internal body tissues). When sense organs detect an appropriate stimulus (light, mechanical pressure, chemical molecules), they convert it into neural impulses (action potentials) that are transmitted to our brain. Through the process of **perception**, the brain then assigns meaning to this sensory information (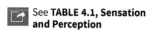 **Table 4.1**).

 See **TABLE 4.1, Sensation and Perception**

An easy way to understand the difference between sensation and perception is to study the lines of text in this paragraph. The waves of light from the black and white dots and lines are first received by our eyes, which convert this stimulus energy into neural messages that are sent on to our brain. This is technically what is called "sensation." When these sensory messages arrive in the appropriate area of our cerebral cortex, and our brain selects, organizes, and interprets the information, we call it "perception." Thanks to your early childhood teachers, you can successfully perceive and interpret the dots and lines as letters and words.

It's important to point out that there is no clear boundary between sensation and perception, and some researchers consider them as a single process. In our everyday experience, the two normally blend into one continuous process. However, psychologists distinguish between bottom-up and top-down processing. The perspective that our sensory receptors register information from the external environment, and send it up to the brain, is called **bottom-up processing**. In other words, we build up complete perceptions by starting "at the bottom" with a sensory analysis of smaller features. In contrast, **top-down processing** begins at the top, within our brain, and works down. Our higher, top-level, cognitive processes (such as prior knowledge and expectations) are used to recognize individual features as a unified whole (Johns & Jones, 2015; Megumi et al., 2015; Naccache et al., 2014). One way to understand the difference between bottom-up and top-down processing, is to think about what happens when we "see" a helicopter flying overhead in the sky. According to the *bottom-up processing* perspective, receptors in our eyes and ears record the sight and sound of this large, loud object, and send these sensory messages on to our brain for interpretation. The other, top-down processing, approach suggests that our brain quickly makes a "best guess," and interprets the large, loud object as a "helicopter," based on our previous knowledge and expectations.

As you can see, the processes of sensation and perception are complex, but also very interesting. Your brain normally floats around in a silent, dark, unfeeling world. Thanks to sensation, information from the external environment is brought into your isolated brain, which then performs a seemingly magical act of turning these sensory messages into perception. Sensation plus perception is essentially life itself! Now that you understand and appreciate the overall purpose of these two processes, let's dig deeper, starting with the first step of sensation—*processing*.

Processing

See **FIGURE 4.1, Sensory processing within the brain**

Looking again at Table 4.1, note that our eyes, ears, skin, and other sense organs all contain special cells called receptors, which receive and process sensory information from the environment. For each sense, these specialized cells respond to a distinct stimulus, such as sound waves or odor molecules. Next, during the process of **transduction**, the receptors convert the energy from the specific sensory stimulus into neural impulses that are then sent on to the brain. For example, in hearing, tiny receptor cells in the inner ear convert mechanical vibrations from sound waves into electrochemical signals. Neurons then carry these signals to the brain, where specific sensory receptors detect and interpret the information.

How does our brain differentiate between sensations, such as sounds and smells? Through a process known as **coding**, the brain interprets different physical stimuli as distinct sensations because their neural impulses travel by different routes and arrive at different parts of the brain (▣ **Figure 4.1**).

It's also important to recognize that species have evolved selective receptors that suppress or amplify information for survival. Humans, for example, cannot sense ultraviolet light, electric or magnetic fields, the ultrasonic sound of a dog whistle, or infrared heat patterns from warm-blooded animals, as some other animals can. In this process of *sensory reduction*, we analyze and then filter incoming sensations before sending neural impulses on for further processing in other parts of our brain. Without this natural filtering of stimuli, we would constantly hear blood rushing through our veins and feel our clothes brushing against our skin. Some level of filtering is needed to prevent our brain from being overwhelmed with unnecessary information.

Psychophysics

How can scientists measure the exact amount of stimulus energy it takes to trigger a conscious experience? The answer comes from the field of **psychophysics**, which studies and measures the link between the physical characteristics of stimuli and the sensory experience of them.

One of the most interesting insights from psychophysics is that what is out there is not directly reproduced in here—inside our own minds and bodies. At this moment, there are light waves, sound waves, odors, tastes, and microscopic particles touching us that we cannot see, hear, smell, taste, or feel. We are consciously aware of only a narrow range of stimuli in our environment. German scientist Ernst Weber (1795–1878) was one of the first to study the smallest difference between two weights that could be detected (Goldstein, 2014; Schwartz & Krantz, 2016). This **difference threshold**, also known as *Weber's Law of just noticeable differences (JND)*, is the minimum difference that is consciously detectable 50% of the time. Another scientist, Gustav Fechner (1801–1887), expanded on Weber's law to determine what is called the **absolute threshold**, the minimum stimulation necessary to consciously detect a stimulus 50% of the time. See ▣ **Table 4.2** for a list of absolute thresholds for our various senses.

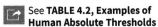See **TABLE 4.2, Examples of Human Absolute Thresholds**

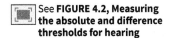See **FIGURE 4.2, Measuring the absolute and difference thresholds for hearing**

To measure your senses, an examiner presents a series of signals that vary in intensity and asks you to report which signals you can detect. In a hearing test, the softest level at which you can consistently hear a tone is your absolute threshold. The examiner then compares your threshold with those of people with normal hearing to determine whether you have hearing loss (▣ **Figure 4.2**).

Interestingly, many nonhuman animals have higher and lower thresholds than humans. For example, a dog's absolute and difference thresholds for smell are far more sensitive than

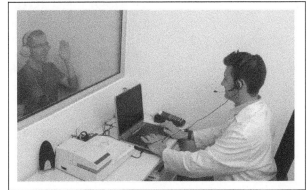

© Carmen Martínez Banús/iStockphoto

those of a human. For this reason, specially trained dogs provide invaluable help in sniffing out dangerous plants, animals, drugs, and explosives; tracking criminals; and assisting in search-and-rescue operations (Porritt et al., 2015). Some researchers believe dogs can even detect hidden corrosion, fecal contamination, chemical signs of certain illnesses (such as diabetes or cancer), and possibly even predict seizures in humans (Partyka et al., 2014; Schoon, 2014; Urbanová et al., 2015). Interestingly, they also can detect a human scent trace, even if the person did not touch an object (Vyplelová et al., 2014). Also, a recent study found that when dogs were presented with five different scents from humans and dogs, sensory receptors in the dogs' noses easily picked up all five scents (Berns et al., 2015). However, only the human scents activated a part of the dog's brain (the caudate nucleus), which has a well-known association with positive expectations. The researchers concluded that this brain activation, and the dog's positive association with human scents, point to the importance of humans in dogs' lives. A related study has shown that dogs can even discriminate among many emotional expressions on human faces (Muller et al., 2015).

Subliminal Stimuli

Have you heard some of the wild rumors about subliminal messages? During the 1970s, it was said that rock songs contained demonic messages, which could only be heard when the songs were played backwards! Similarly, in the 1990s, many suggested that some Disney films contained obscene subliminal messages. For example, in the film *Aladdin*, the lead character supposedly whispers, "all good teenagers take off your clothes," and *The Lion King* reportedly showed close ups of the dust with a secret spelling out of the word "sex." In addition, at one time movie theaters were reportedly flashing messages like "Eat popcorn" and "Drink Cola-Cola" on the screen. Even though the messages were so brief that viewers weren't aware of seeing them, it was believed they increased consumption of these products (Bargh, 2014; Blecha, 2004; Vokey & Read, 1985).

Can unconscious stimuli affect our behavior? To answer this question, it's first important to acknowledge that it's clearly possible to perceive something without conscious awareness (Atas et al., 2014; Farooqui & Manly, 2015; Ito et al., 2015; Peremen & Lamy, 2014). Experimental studies on **subliminal perception** commonly use an instrument, called a *tachistoscope*, to flash images too quickly for conscious recognition, but slowly enough to be registered by the brain.

Despite clear evidence that subliminal perception does occur, it doesn't necessarily mean that such processes lead to significant behavioral changes, known as *subliminal persuasion*. Subliminal stimuli are basically weak stimuli. At best, they have a modest (if any) effect on consumer behavior, the attitudes of youth listening to rock music, or on citizens' voting behavior (Fennis & Stroebe, 2010; Salpeter & Swirsky, 2012; Smarandescu & Shimp, 2014). However, they do seem to have an effect on indirect, more subtle reactions such as, our attitudes and feelings. For example, one study found that workers who were primed with positive affective cues, while performing work-related tasks, increased their task satisfaction and performance on certain tasks (Hu & Kaplan, 2015). Can you see how these findings might be used to improve the overall work environment for employees, or how it could be abused to increase profits?

Sensory Adaptation

Imagine that friends have invited you to come visit their beautiful new baby. As they greet you at the door, you are overwhelmed by the odor of a wet diaper. Why don't your friends do something about that smell? The answer lies in the previously mentioned sensory reduction, as well as **sensory adaptation**. When a constant stimulus is presented for a length of time, sensation often fades or disappears. Receptors in our sensory system become less sensitive. They get "tired" and actually fire less frequently.

Sensory adaptation can be understood from an evolutionary perspective. We can't afford to waste attention and time on unchanging, normally unimportant stimuli. "Turning down the volume" on repetitive information helps the brain cope with an overwhelming amount of sensory stimuli and enables us to pay attention to change. Sometimes, however, adaptation can be dangerous, as when people stop paying attention to a small gas leak in the kitchen.

Although some senses, like smell and touch, adapt quickly, we never completely adapt to visual stimuli or to extremely intense stimuli, such as the odor of ammonia or the pain of a bad burn. From an evolutionary perspective, these limitations on sensory adaptation aid survival by reminding us, for example, to keep a watch out for dangerous predators, avoid strong odors and heat, and take care of that burn.

If we don't adapt to pain, how do athletes keep playing despite painful injuries? In certain situations, including times of physical exertion, the body releases natural painkillers called *endorphins* (Chapter 2), which inhibit pain perception. This is the so-called "runner's high," which may help explain why athletes have been found to have a higher pain tolerance than nonathletes (Tesarz et al., 2012). (As a critical thinker, is it possible that individuals with a naturally high pain tolerance are just more attracted to athletics? Or might the experience of playing sports change your pain tolerance?)

In addition to endorphin release, one of the most widely accepted explanations of pain perception is the **gate-control theory of pain**, first proposed by Ronald Melzack and Patrick Wall (1965). According to this theory, the experience of pain depends partly on whether the neural message gets past a "gatekeeper" in the spinal cord. Normally, the gate is kept shut, either by impulses coming down from the brain or by messages coming from large-diameter nerve fibers that conduct most sensory signals, such as touch and pressure. However, when body tissue is damaged, impulses from smaller pain fibers open the gate (Kumar & Rizvi, 2014; Price & Prescott, 2015; Zhao & Wood, 2015).

 See **FIGURE 4.3, Overcoming excruciating pain!**

Can you see how this gate-control theory helps explain why massaging an injury or scratching an itch can temporarily relieve discomfort? It's because pressure on large-diameter neurons interferes with pain signals.

Messages from our brain also can control the pain gate, which explains how athletes and soldiers can carry on despite excruciating pain. When we're soothed by endorphins or distracted by competition or fear, our experience of pain can be greatly diminished (**Figure 4.3**). Similarly, music therapy, and actively concentrating on specific musical cues in songs people know well, such as "Mary Had a Little Lamb"—can help reduce pain, particularly when we are very anxious (Bradshaw et al., 2012; Gardstrom & Sorel, 2015). Research also suggests that the pain gate may be chemically controlled. A neurotransmitter called *substance P* opens the pain gate, and endorphins close it (Krug et al., 2015; Luz et al., 2014; Wu et al., 2015).

MacNicol/Getty Images

Other research finds that when normal sensory input is disrupted, the brain can generate pain and other sensations on its own (Deer et al., 2015; Inui & Masumoto, 2015; McGeary et al., 2015; Melzack, 1999). For example, after an amputation, most people report detecting their missing limb as if it were still there with no differences at all. In fact, up to 80 percent of people who have had amputations sometimes "feel" pain (and itching, burning, or tickling sensations) in the missing limb, long after the amputation. Numerous theories attempt to explain this *phantom limb pain* (PLP), but one of the best researched suggests that there is a mismatch between the sensory messages sent and received in the brain. Can you see how this is an example of our earlier description of how *bottom-up processes* (such as the sensory messages sent from our limbs to our brain) combine with our *top-down processes* (our brain's interpretation of these messages)? In this case of PLP, areas of the brain that receive sensory information are still intact (and confused) even though the sensory receptors in the amputated limb are gone. Messages are no longer being transmitted from the missing limb to the brain (bottom up), but the brain believes the limb should be there (top down). The brain's confusion results in pain.

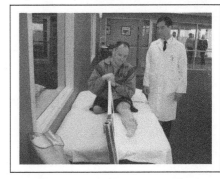 See **FIGURE 4.4, Treating amputee phantom limb pain**

In line with this idea of mismatched signals, when amputees wear prosthetic limbs, or when *mirror visual therapy* is used, phantom pain often disappears. In mirror therapy (**Figure 4.4**), pain relief apparently occurs because the brain is somehow tricked into believing there is no longer a missing limb (Foell et al., 2014; Hagenberg & Carter, 2014; Plumbe et al., 2013). Others state that it also helps the brain reorganize and incorporate this phantom limb into a new nervous system configuration. Interestingly, scientists who question this "brain-centric" explanation have reported significant relief from PLP after injecting patients with a local anesthetic near where the nerves from the amputated limb enter the spinal cord (Vaso et al., 2014).

Navy Mass Communication Specialist 2nd Class Jeff Hopkins

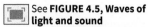

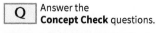

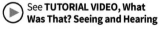

Q Answer the **Concept Check** questions.

▶ See **TUTORIAL VIDEO, What Was That? Seeing and Hearing**

See **FIGURE 4.5, Waves of light and sound**

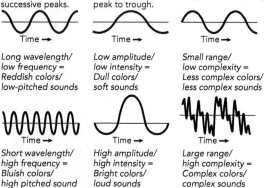

© S. Greg Panosian/iStockphoto

See **FIGURE 4.6, Properties of light and sound**

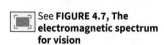

Wavelength/ frequency The distance between successive peaks.	Wave height/ amplitude The height from peak to trough.	Range or complexity The mixture of wave length and wave height.
Long wavelength/ low frequency = Reddish colors/ low-pitched sounds	Low amplitude/ low intensity = Dull colors/ soft sounds	Small range/ low complexity = Less complex colors/ less complex sounds
Short wavelength/ high frequency = Bluish colors/ high pitched sound	High amplitude/ high intensity = Bright colors/ loud sounds	Large range/ high complexity = Complex colors/ complex sounds

See **FIGURE 4.7, The electromagnetic spectrum for vision**

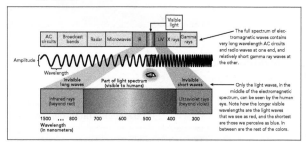

Now that we've studied how we perceive pain, how we might ignore or "play through" it, and how we might misperceive it with phantom limb pain, it's important to point out that when we get anxious or dwell on our pain, we can intensify it (Lin et al., 2013; Ray et al., 2015; Wertli et al., 2014). Interestingly, social and cultural factors, such as well-meaning friends or anxious parents who ask pain sufferers about their pain, may unintentionally reinforce and increase it (Esteve et al., 2014; Flor, 2013; Langer et al., 2014).

4.2 HOW WE SEE AND HEAR

Many people mistakenly believe that what they see and hear is a copy of the outside world. In fact, vision and hearing are the result of what our brains create in response to light and sound waves. What we see and hear is based on wave phenomena, similar to ocean waves (**Figure 4.5**).

In addition to wavelength/frequency shown in Figure 4.5, waves also vary in height (technically called *amplitude*). This wave height/amplitude determines the intensity of sights and sounds. Finally, waves also vary in range, or complexity, which mixes together waves of various wavelength/frequency and wave height/amplitude (**Figure 4.6**).

Vision

Did you know that Major League batters can routinely hit a 90-miles-per-hour fastball four-tenths of a second after it leaves the pitcher's hand? How can the human eye receive and process information that fast? To understand the marvels of vision, we need to start with the basics—that light waves are a form of electromagnetic energy and only a small part of the full *electromagnetic spectrum* (**Figure 4.7**).

To fully appreciate how our eyes turn these light waves into the experience we call *vision*, we need to first examine the various structures in our eyes that capture and focus the light waves. Then, we need to understand how these waves are transformed (transduced) into neural messages (action potentials) that our brain can process into images we consciously see. (Be sure to carefully study the **Process Diagram 4.1**.)

Color Vision

Our ability to perceive color is almost as remarkable and useful as vision itself. Humans may be able to discriminate among 7 million different hues, and research conducted in many cultures suggests that we all seem to see essentially the same colored world (Maule et al., 2014; Ozturk et al., 2013). Furthermore, studies of infants old enough to focus and move their eyes show that they are able to see color nearly as well as adults and have color preferences similar to those of adults (Bornstein et al., 2014; Yang et al., 2015).

This Process Diagram contains essential information NOT found elsewhere in the text, which is likely to appear on quizzes and exams. Be sure to study it CAREFULLY!

Process Diagram 4.1

How Our Eyes See

Various structures of your eye work together to capture and focus the light waves from the outside world. Receptor cells in your retina (rods and cones) then convert these waves into messages that are sent along the optic nerve to be interpreted by your brain.

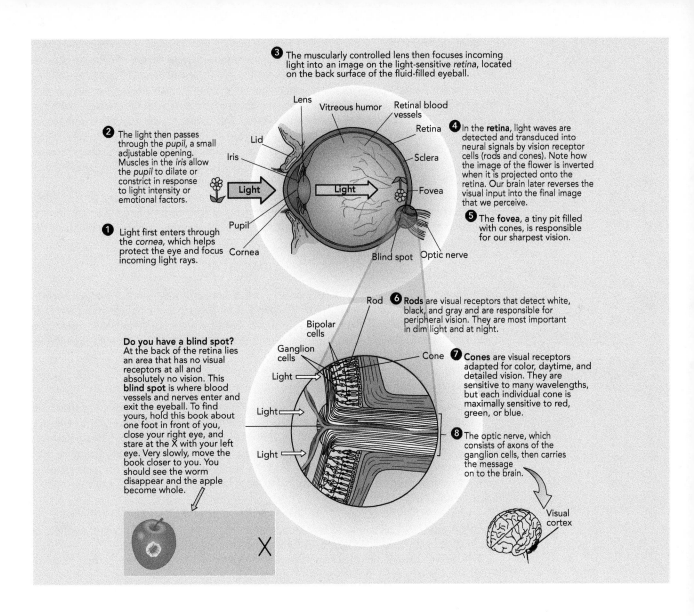

❸ The muscularly controlled lens then focuses incoming light into an image on the light-sensitive *retina*, located on the back surface of the fluid-filled eyeball.

❷ The light then passes through the *pupil*, a small adjustable opening. Muscles in the *iris* allow the *pupil* to dilate or constrict in response to light intensity or emotional factors.

❶ Light first enters through the *cornea*, which helps protect the eye and focus incoming light rays.

❹ In the **retina**, light waves are detected and transduced into neural signals by vision receptor cells (rods and cones). Note how the image of the flower is inverted when it is projected onto the retina. Our brain later reverses the visual input into the final image that we perceive.

❺ The **fovea**, a tiny pit filled with cones, is responsible for our sharpest vision.

Do you have a blind spot?
At the back of the retina lies an area that has no visual receptors at all and absolutely no vision. This **blind spot** is where blood vessels and nerves enter and exit the eyeball. To find yours, hold this book about one foot in front of you, close your right eye, and stare at the X with your left eye. Very slowly, move the book closer to you. You should see the worm disappear and the apple become whole.

❻ **Rods** are visual receptors that detect white, black, and gray and are responsible for peripheral vision. They are most important in dim light and at night.

❼ **Cones** are visual receptors adapted for color, daytime, and detailed vision. They are sensitive to many wavelengths, but each individual cone is maximally sensitive to red, green, or blue.

❽ The optic nerve, which consists of axons of the ganglion cells, then carries the message on to the brain.

Have you ever heard someone say: "I'm so angry that I'm seeing red"? Interestingly, a recent study lends support for this somewhat common expression. Researchers have found that the color red generally elicits more intense responses in both human and nonhuman animals. And, as interpersonal hostility rises, and during hostile social decision-making, participants are more likely to show a preference for the color red (Díaz-Román et al., 2015; Fetterman et al., 2015).

In addition, people pay higher bids on eBay for electronic items—such as a Nintendo Wii video game console—when those items appear on a red background, and waitresses wearing red get higher tips from male customers (Bagchi & Cheema, 2013; Gueguen & Jacob, 2014).

Although we know color is produced by different wavelengths of light, the actual way in which we perceive color is a matter of scientific debate. Traditionally, there have been two theories of color vision: the trichromatic (three-color) theory and the opponent-process theory. The **trichromatic theory of color** (from the Greek word *tri*, meaning "three," and *chroma*, meaning "color") suggests that we have three "color systems," each of which is maximally sensitive to red, green, or blue (Young, 1802). The proponents of this theory demonstrated that mixing lights of these three colors could yield the full spectrum of colors we perceive (▣ **Figure 4.8**).

See **FIGURE 4.8, Primary colors**

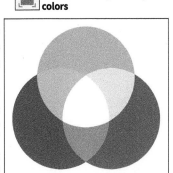

However, trichromatic theory doesn't fully explain color vision, and other researchers have proposed alternative theories. For example, the **opponent-process theory of color** agrees that we have three color systems, but it says that each system is sensitive to two opposing colors—blue and yellow, red and green, black and white—in an "on/off" fashion. In other words, each color receptor responds either to blue or yellow, or to red or green, with the black-or-white systems responding to differences in brightness levels. This theory makes a lot of sense because when different-colored lights are combined, people are unable to see reddish greens and bluish yellows. In fact, when red and green lights or blue and yellow lights are mixed in equal amounts, we see white. This opponent-process theory also explains *color afterimages*, a fun type of optical illusion in which an image briefly remains after the original image has faded.

Today we know that both trichromatic and opponent-process theories are correct—they just operate at different levels in visual processing. Color vision is processed in a trichromatic fashion in the retina. In contrast, color vision during opponent processing involves the retina, optic nerve, and brain.

Color-Deficient Vision

Most people perceive three different colors—red, green, and blue—and are called *trichromats*. However, a small percentage of the population has a genetic deficiency in the red–green system, the blue–yellow system, or both. Those who perceive only two colors are called *dichromats*. People who are sensitive to only the black–white system are called *monochromats*, and they are totally color blind.

Vision Problems and Peculiarities

Looking back at the process of vision detailed in Process Diagram 4.1 offers clues that help us understand several visual peculiarities. For example, small abnormalities in the eye sometimes cause images to be focused in front of the **retina** in the case of *nearsightedness (myopia)*, or behind it in the case of *farsightedness (hyperopia)*. In addition, during middle age, most people's lenses lose elasticity and the ability to accommodate for near vision, a condition known as *presbyopia*. Corrective lenses or laser surgery can often improve most of these visual acuity problems.

A visual peculiarity occurs where the optic nerve exits the eye. Because there are no receptor cells for visual stimuli in that area, we have a tiny hole, or **blind spot**, in our field of vision. (See again Process Diagram 4.1 for a demonstration.)

Two additional peculiarities happen when we go from a bright to dark setting and vice versa. Have you noticed that when you walk into a dark movie theater on a sunny afternoon, you're almost blind for a few seconds? The reason is that in bright light, the pigment inside the **rods** (see again Process Diagram 4.1) has been "bleached" by the bright light, making the rods temporarily nonfunctional. It takes a second or two for the rods to become functional enough again for you to see. This process of *dark adaptation* continues for 20 to 30 minutes.

In contrast, *light adaptation*, the adjustment that takes place when you go from darkness to a bright setting, takes about 7 to 10 minutes and is the work of the **cones**. Interestingly, a region in the center of the retina, called the **fovea**, has the greatest density of cones, which are most sensitive in brightly lit conditions. They're also responsible for color vision and fine detail.

Hearing

The sense or act of hearing, known as **audition**, has a number of important functions, ranging from alerting us to dangers to helping us communicate with others. In this section we talk first about sound waves, then about the ear's anatomy and function, and finally about problems with hearing.

Like the visual process, which transforms light waves into vision, the auditory system is designed to convert sound waves into hearing. Sound waves are produced by air molecules moving in a particular wave pattern. For example, vibrating objects like vocal cords or guitar strings create waves of compressed and expanded air resembling ripples on a lake that circle out from a tossed stone. Our ears detect and respond to these waves of small air pressure changes, our brain then interprets the neural messages resulting from these waves, and we hear!

To fully understand this process, pay close attention to **Process Diagram 4.2**.

Process Diagram 4.2

How Our Ears Hear

The **outer ear** captures and funnels sound waves into the eardrum. Next, three tiny bones in the **middle ear** pick up the eardrum's vibrations, and transmit them to the **inner ear**. Finally, the snail-shaped **cochlea** in the inner ear transforms (transduces) the sound waves into neural messages (action potentials) that our brain processes into what we consciously hear.

❶ The **outer ear** captures and funnels sound waves onto the tympanic membrane (ear drum).

❷ Vibrations of the tympanic membrane strike the **middle ear's** ossicles (hammer, anvil, and stirrup). Then the stirrup hits the oval window.

❸ Vibrations of the oval window create waves in the **inner ear's** cochlear fluid which deflects the basilar membrane. This movement bends the hair cells.

❹ The hair cells communicate with the auditory nerve, which sends neural impulses to the brain.

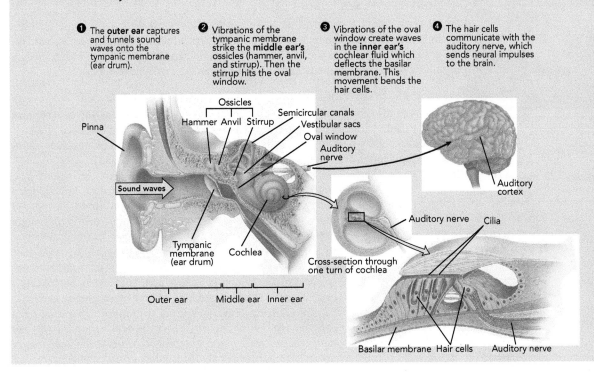

Pitch Perception

How do we determine that certain sounds are from a child's voice, and not from an adult's? We distinguish between high- and low-pitched sounds by the *frequency* of the sound waves. The higher the frequency, the higher the pitch. There are three main explanations for how we perceive *pitch*:

- According to the **place theory for hearing**, we hear different pitches because different sound waves stimulate different sections (or *places*) on our cochlea's basilar membrane (see again Process Diagram 4.2). Our brain figures out the pitch of a sound by detecting the position of the hair cells that sent the neural message. High frequencies produce large vibrations near the start of the basilar membrane—next to the oval window. However, this theory does not predict well for low frequencies, which tend to excite the entire basilar membrane.

- **Frequency theory for hearing** differs from place theory because it states that we hear pitch by the *frequency* of the sound waves traveling up the auditory nerve. High-frequency sounds trigger the auditory nerve to fire more often than do low-frequency sounds. The problem with this theory is that an individual neuron cannot fire faster than 1,000 times per second, which means that we could not hear many of the notes of a soprano singer.

See **FIGURE 4.9, Students exploiting age-related hearing loss**

© Monkey Business Images/iStockphoto

- The **volley principle for hearing** solves the problem of frequency theory, which can't account for the highest pitched sounds. It states that clusters of neurons take turns firing in a sequence of rhythmic *volleys*. Pitch perception depends upon the frequency of volleys, rather than the frequency carried by individual neurons.

Interestingly, as we age, we tend to lose our ability to hear high-pitched sounds, but are still able to hear low-pitched sounds. Given that young students can hear a special cell phone ringtone that sounds at 17 kilohertz—too high for most adult ears to detect—they can take advantage of this age-related hearing difference by texting, or even calling, one another during class (📺 **Figure 4.9**). Ironically, the cell phone's ringtone is an offshoot of another device, called the Mosquito, which was originally designed to help shopkeepers annoy and drive away loitering teens!

Softness versus Loudness

How we detect a sound as being soft or loud depends on its amplitude (or wave height). Waves with high peaks and low valleys produce loud sounds; waves with relatively low peaks and shallow valleys produce soft sounds. The relative loudness or softness of sounds is measured on a scale of *decibels* (📺 **Figure 4.10**).

Hearing Problems

What are the types, causes, and treatments of hearing loss? **Conduction hearing loss**, also called conduction deafness, results from problems with the mechanical system that conducts sound waves to the cochlea. Hearing aids that amplify the incoming sound waves, and some forms of surgery, can help with this type of hearing loss.

In contrast, **sensorineural hearing loss**, also known as nerve deafness, results from damage to the cochlea's receptor (hair) cells or to the auditory nerve. Disease and biological changes associated with aging can result in sensorineural hearing loss. But its most common (and preventable) cause is continuous exposure to loud noise, which can damage hair cells and lead to permanent hearing loss. Even brief exposure to really loud sounds, like a stereo or headphones at full blast, a jackhammer, or a jet airplane engine, can cause permanent nerve deafness (see again Figure 4.10). In fact, a high volume on earphones can reach the same noise level as a jet engine! All forms of high volume noise can damage the coating on nerve cells, making it harder for the nerve cells to send information from the ears to the brain (Eggermont, 2015; Fonseca et al., 2015; Johnson et al., 2014).

Although most hearing loss is temporary, damage to the auditory nerve or receptor cells is generally considered irreversible. The only treatment for auditory nerve damage is a small electronic device called a *cochlear implant*. If the auditory nerve is intact, the implant bypasses hair cells to stimulate the nerve directly. Currently, cochlear implants produce only a crude approximation of hearing, but the technology is improving.

Given the limited benefits of medicine or technology to help improve hearing following damage, it's important to protect our sense of hearing. We can do this by avoiding exceptionally loud noises, wearing earplugs when we cannot avoid such stimuli, and paying attention to bodily warnings of possible hearing loss, including a change in our normal hearing threshold and *tinnitus*, a whistling or ringing sensation in the ears. These relatively small changes can have lifelong benefits!

See **FIGURE 4.10, Beware of loud sounds**

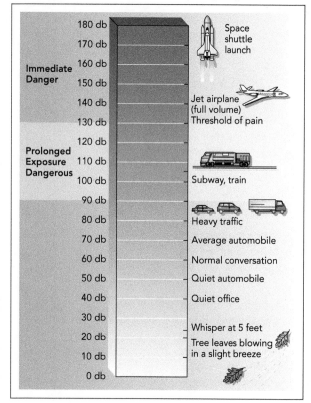

180 db	Space shuttle launch
170 db	
Immediate Danger 160 db	
150 db	
140 db	Jet airplane (full volume)
130 db	Threshold of pain
120 db	
Prolonged Exposure Dangerous 110 db	
100 db	Subway, train
90 db	
80 db	Heavy traffic
70 db	Average automobile
60 db	Normal conversation
50 db	Quiet automobile
40 db	Quiet office
30 db	
20 db	Whisper at 5 feet
10 db	Tree leaves blowing in a slight breeze
0 db	

Q Answer the **Concept Check** questions.

4.3 OUR OTHER IMPORTANT SENSES

Vision and audition may be the most prominent of our senses, but the others—smell, taste, and the body senses—are also important for gathering information about our environment.

Smell and Taste

See **STUDY ORGANIZER 4.1, Why Do We Enjoy Eating Pizza? Olfaction Plus Gustation**

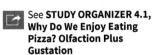

Smell and taste are sometimes called the *chemical senses* because they both rely on chemo-receptors that are sensitive to certain chemical molecules (**Study Organizer 4.1**). Have you noticed that you often have trouble separating smell and taste?

Our sense of smell, **olfaction**, is remarkably useful and sensitive. We possess more than 1,000 types of olfactory receptors, which allow us to detect more than 10,000 distinct smells. The nose is more sensitive to smoke than any electronic detector, and—through practice—blind people can quickly recognize others by their unique odors.

Have you noticed how certain smells seem to evoke strong memories and emotions? This is because the neural pathways that carry smell sensations from our nose to our cortex first run through our limbic center, which as you recall from Chapter 2 is responsible for emotions and memory.

What about sexual attraction? Some research on **pheromones** —chemicals found in natu-ral body scents that affect various behaviors—supports the idea that certain chemical odors increase sexual behaviors—even in humans (Baum & Cherry, 2015; Jouhanneau et al., 2014; Ottaviano et al., 2015). However, others suggest that human sexuality is far more complex than that of other animals—and more so than perfume advertisements would have you believe.

Today, the sense of taste, **gustation**, may be the least critical of our senses. In the past, however, it probably contributed significantly to our survival. For example, humans and other animals have a preference for sweet foods, which are generally nonpoisonous and are good sources of energy. However, the major function of taste, aided by smell, is to help us avoid eating or drinking harmful substances. Because many plants that taste bitter contain toxic chemicals, an animal is more likely to survive if it avoids bitter-tasting plants (French et al., 2015; Sagong et al., 2014; Schwartz & Krantz, 2016).

Interestingly, taste and smell receptors normally die and are replaced every few days, which probably reflects the fact that these receptors are directly exposed to the environment, whereas our vision receptors are protected by our eyeball and hearing receptors are protected by the ear-drum. However, as we grow older, the number of taste cells diminishes, which helps explain why adults enjoy spicier foods than infants. Scientists are particularly excited about the regenerative capabilities of the taste and olfactory cells because they hope to learn how to transfer this regen-eration to other types of cells that are currently unable to self-replace when damaged.

Learning and Culture

Many food and taste preferences also are learned from an early age and from personal expe-riences (Fildes et al., 2014; Mennella, 2014; Tan et al., 2015). For example, adults who are told a bottle of wine costs $90 (rather than its real price of $10) report that it tastes better than the supposedly cheaper brand. Ironically, these false expectations actually trigger areas of the brain that respond to pleasant experiences (Plassmann et al., 2008). This means that in a neurochemical sense, the wine we believe is better does, in fact, taste better!

See **FIGURE 4.11, Our body senses--skin, vestibular, and kinesthesis**

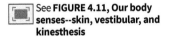

The culture we live in also affects our taste preferences. Many Japanese children eat raw fish, and some Chinese children eat chicken feet as part of their normal diet. Although most U.S. children might consider these foods "yucky," they tend to love cheese, which children in many other cultures find repulsive.

Before going on, we need to update you on recent research find-ings on taste perception. It was long believed that we had only four distinct tastes: sweet, sour, salty, and bitter. However, we now know that we also have a fifth taste sense, *umami*, a word that means "deli-cious" or "savory" and refers to sensitivity to an essential, excitatory neurotransmitter and amino acid, called glutamate (Bredie et al., 2014; Jinap & Hajeb, 2010; Lanfer et al., 2013). Glutamate is found in meats, meat broths, and monosodium glutamate (MSG).

The Body Senses

In addition to smell and taste, we have three important body senses that help us navigate our world—skin, vestibular, and kinesthesis (■ **Figure 4.11**).

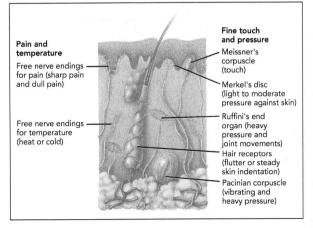

Pain and temperature

Free nerve endings for pain (sharp pain and dull pain)

Free nerve endings for temperature (heat or cold)

Fine touch and pressure

Meissner's corpuscle (touch)

Merkel's disc (light to moderate pressure against skin)

Ruffini's end organ (heavy pressure and joint movements)

Hair receptors (flutter or steady skin indentation)

Pacinian corpuscle (vibrating and heavy pressure)

Skin Senses

Our skin is uniquely designed for the detection of touch (or pressure), temperature, and pain (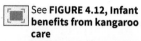 **Figure 4.11a**). The concentration and depth of the receptors for each of these stimuli vary (Fitzpatrick & Mooney, 2012; Hsiao & Gomez-Ramirez, 2013; Ruzzoli & Soto-Faraco, 2014). For example, touch receptors are most concentrated on the face and fingers and least concentrated in the back and legs. Getting a paper cut can feel so painful because we have many receptors on our fingertips. Some receptors respond to more than one type of stimulation. For example, itching, tickling, and vibrating sensations seem to be produced by light stimulation of both pressure and pain receptors.

The benefits of touch are so significant for human growth and development that the American Academy of Pediatrics recommends that all parents and babies have skin-to-skin contact in the first hours after birth. This type of contact, which is called *kangaroo care*, is especially beneficial for preterm and low-birth-weight infants, who then experience greater weight gain, fewer infections, and improved cognitive and motor development. How does kangaroo care lead to these improvements in infant health? (See  **Figure 4.12**).

See **FIGURE 4.12, Infant benefits from kangaroo care**

Vestibular Sense

Our sense of balance, the **vestibular sense**, informs our brain of how our body (particularly our head) is oriented with respect to gravity and three-dimensional space (**Figure 4.11b**). When our head tilts, liquid in the *semicircular canals*, located in our inner ear, moves and bends hair cell receptors. In addition, at the end of the semicircular canals are *vestibular sacs*, which contain hair cells sensitive to our whole body's movement (linear acceleration) relative to gravity (as shown in the amusement park ride example in Figure 4.11b). Information from the semicircular canals and the vestibular sacs is converted to neural impulses that are then carried to our brain.

Kinesthesis

The sense that provides the brain with information about bodily posture, orientation, and movement of individual body parts is called **kinesthesis** (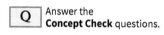 **Figure 4.11c**). Kinesthetic receptors are found throughout the muscles, joints, and tendons of our body. As we sit, walk, bend, lift, and/or turn, our kinesthetic receptors respond by sending vital messages to our brain regarding which muscles are being contracted or relaxed, how our body weight is distributed, where our arms and legs are in relation to the rest of our body, and so on. Without our kinesthetic sense, we would literally have to watch every step or movement we made.

Answer the **Concept Check** questions.

4.4 THE THREE PROCESSES OF PERCEPTION

We are ready to move from *sensation* and the major senses to *perception*, the process of selecting, organizing, and interpreting incoming sensations into useful mental representations of the world.

Normally, our perceptions agree with our sensations. When they do not, the result is called an **illusion**, a false or misleading impression produced by errors in the perceptual process or by actual physical distortions, as in desert mirages. Illusions provide psychologists with a tool for studying the normal process of perception. (See **Study Organizer 4.2.**)

See **STUDY ORGANIZER 4.2: Understanding Perceptual Illusions**

Selection

In almost every situation, we confront more sensory information than we can reasonably pay attention to. Three major factors help us focus on some stimuli and ignore others: *selective attention*, *feature detectors*, and *habituation*.

Mike Kemp/Getty Images

Certain basic mechanisms for perceptual selection are built into the brain. For example, through the process of **selective attention** (◾ **Figure 4.13**), we selectively pick out one message to attend to from a variety of messages coming to us simultaneously (Burham et al., 2014; Lehmann & Schönwiesner, 2014; Rosner et al., 2015).

See **FIGURE 4.13, Selective attention**

In addition to selective attention, the brains of humans and other animals contain specialized cells, called **feature detectors**, which respond only to certain stimuli. For example, studies with humans have found feature detectors in the temporal and occipital lobes that respond maximally to faces (◾ **Figure 4.14**). Problems in these areas can produce a condition called *prosopagnosia* (*prosopon* means "face," and *agnosia* means "failure to know"). Interestingly, people with prosopagnosia can recognize that they are looking at a face. But they cannot say whose face is reflected in a mirror, even if it is their own or that of a friend or relative (Rivolta, 2014; Tanzer et al., 2014; Van Belle, 2015).

Other examples of our brain's ability to filter experience occurs with **habituation**, a decrease in responding due to repeated stimulation by the same stimulus. Apparently, the brain is "prewired" to pay more attention to changes in the environment than to stimuli that remain constant. As you'll discover in Chapter 9, developmental psychologists often use measurements of habituation to tell when a stimulus can be detected and discriminated by infants who are too young to speak. When presented with a new stimulus, infants pay attention, but with repetition their responses weaken.

See **FIGURE 4.14, Location of feature detectors in the brain**

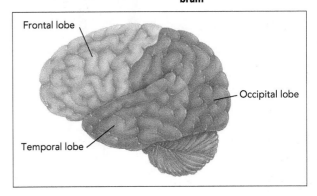

Frontal lobe

Occipital lobe

Temporal lobe

How does habituation differ from *sensory adaptation*, which we discussed earlier? Habituation is a perceptual process that occurs in the brain, and it involves a reduced responsiveness to unchanging (boring) stimuli. In contrast, sensory adaptation Zoccurs within the sensory receptors. The receptor cells in our eyes, ears, skin, and so on actually slow down their rate of firing if a stimulus doesn't change much. A smoke-filled bar might smell very smoky when you first arrive, but you hardly notice it after a few minutes.

To combat both habituation and sensory adaptation, parents and teachers often select stimuli that are *intense, novel,* and *contrasting*. Similarly, advertisers and politicians spend millions of dollars developing these same attention-getting principles into a fine art. The next time you're watching television, note the commercial and political ads. Are they flashier or brighter than the regular program (*intensity*)? Do they use talking cows to promote California cheese (*novelty*)? Is the promoted product or candidate set in *favorable contrast* to the competition?

Have you ever wondered why attention and compliments from a complete stranger seem more exciting and valuable to you than similar actions and words from your long-term romantic partner? Does this make you wonder if you're with the right person? Think again. Remember that we all habituate to unchanging stimuli. If we move on to other relationships, they too will soon fall victim to habituation.

What can we do to keep romance alive? If we're the person being complimented, we can be grateful that we've learned about habituation, and then remind ourselves to not be overly influenced by a stranger's attention. As the long-term romantic partner hoping to offset the dangers of habituation, we can take a note from advertisers by using more *intensity*, *novelty*, and *contrast* in our compliments and interactions with our loved ones.

Organization

In the previous section, we discussed how we select certain stimuli in our environment to pay attention to and not others. The next step in perception is to organize this selected information into useful mental representations of the world around us. Raw sensory data are like the parts of a watch—the parts must be assembled in a meaningful way before they are useful. We organize sensory data in terms of *form*, *depth*, and *constancy*.

See **FIGURE 4.15, Form perception and "impossible figures"**

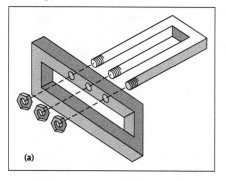

(a)

See **FIGURE 4.16, Understanding Gestalt principles of organization**

Figure–Ground:
Objects (the *figure*) are seen as distinct from the surroundings (the *gound*). (Here the red objects are the figure and the yellow backgound is the ground).

Proximity:
Objects that are physically close together are grouped together. (In this figure, we see 3 groups of 6 hearts, not 18 separate hearts.)

Continuity:
Objects that continue a pattern are grouped together. (When we see line **a.**, we normally see a combination of lines **b.** and **c.** — not **d.**)

When we see this,

a.

we normally see this

b.

plus this.

c.

Not this.

d.

Closure:
The tendency to see a finished unit (triangle, square, or circle) from an incomplete stimulus.

Similarity:
Similar objects are grouped together (the green colored dots are grouped together and perceived as the number 5).

See **FIGURE 4.17, Understanding reversible figures**

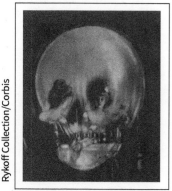

Rykoff Collection/Corbis

Form Perception

Look at the first drawing in ◾ **Figure 4.15a**. What do you see? Can you draw a similar object on a piece of paper? This is known as an "impossible figure." Now look at ◾ **Figure 4.15b**, which shows a painting by M. C. Escher, a Dutch painter who created striking examples of perceptual distortion. Although drawn to represent three-dimensional objects or situations, the parts don't assemble into logical wholes. Like the illusions studied earlier, impossible figures and distorted paintings help us understand perceptual principles—in this case, the principle of *form perception*.

Gestalt psychologists were among the first to study how our brain organizes sensory impressions into a *gestalt*—a German word meaning "form" or "whole." They emphasized the importance of organization and patterning in enabling us to perceive the whole stimulus rather than perceive its discrete parts as separate entities. The Gestaltists proposed several laws of organization that specify how people perceive form (◾ **Figure 4.16**).

The most fundamental Gestalt principle of organization is our tendency to distinguish between the *figure* (our main focus of attention) and *ground* (the background or surroundings). Your sense of figure and ground is at work in what you are doing right now—reading. Your brain is receiving sensations of black lines and white paper, but your brain is organizing these sensations into black letters and words on a white background. You perceive the letters as the figure and the white as the ground. If you make a great effort, you might be able to force yourself to see the page reversed, as though a black background were showing through letter-shaped-holes in a white foreground. There are times, however, when it is very hard to distinguish the figure from the ground, as you can see in ◾ **Figure 4.17**. This is known as a *reversible figure*. Your brain alternates between seeing the light areas as the figure and seeing them as the ground.

Culture and the Gestalt Laws

Gestalt psychologists conducted most of their work with formally educated people from urban European cultures. A.R. Luria (1976) was one of the first to question whether their laws held true for all participants, regardless of their education and cultural setting. Luria recruited a wide range of participants living in what was then the U.S.S.R. He included Ichkeri women from remote villages (with no formal education), collective farm activists (with some formal education), and female students in a teachers' school (with years of formal education).

Luria found that when presented with the stimuli shown in ◾ **Figure 4.18**, the formally trained female students were the only ones who identified the first three shapes by their categorical name of "circle." Whether circles were made of solid lines, incomplete lines, or solid colors, they called them all circles. However, participants with no formal education named the shapes according to the objects they resembled. They called a circle a watch, plate, or moon, and referred to the square as a mirror, house, or apricot-drying board. When asked if items 12 and 13 from Figure 4.18 were alike, one woman answered, "No, they're not alike. This one's not like a watch, but that one's a watch because there are dots."

Apparently, the Gestalt laws of perceptual organization are valid only for people who have been schooled in geometrical concepts. But an alternative explanation for Luria's findings has also been suggested. Luria's study, as well as most research on visual perception and optical illusions, relies on two-dimensional presentations—either on a piece of paper or projected on a screen. It may be that experience with pictures and photographs (not formal education in geometrical concepts) is necessary for learning to interpret two-dimensional figures as portraying three-dimensional forms. Westerners who have had years of practice learning to interpret two-dimensional drawings of three-dimensional objects may not remember how much practice it took to learn the cultural conventions about judging the size and shape of objects drawn on paper (Berry et al., 2011; Lim et al., 2015).

Depth Perception

In our three-dimensional world, the ability to perceive the depth and distance of objects—as well as their height and width—is essential. **Depth perception** is learned primarily through experience. Research using an apparatus called the *visual cliff* (■ **Figure 4.19**) suggests that very young infants can perceive depth, and will actively avoid it.

Some have suggested that this visual cliff research proves that depth perception, and an avoidance of heights, is inborn. However, the modern consensus is that infants are indeed able to perceive depth. But the idea that an infant's fear of heights causes their avoidance is not supported (Adolph et al., 2014). Instead, these researchers found that infants display a flexible and adaptive response at the edge of a drop-off. They pat the surface, attempt to reach through the glass, and even rock back and forth at the edge. They decide whether or not to cross or avoid a drop-off based on previous locomotor experiences, and gained knowledge of their own muscle strength, balance, and other criteria. This notion that fear of heights is NOT innate is supported by other research showing that infants and young children willingly approach, rather than withdraw from, photos, videos, and even live snakes and spiders (LoBue, 2013). Although they do show a heightened sensitivity to snakes, spiders, and heights, which may facilitate fear learning later in development, they do not innately fear them. In short, infants perceive depth, but their fear of heights apparently develops over time, like walking or language acquisition.

Although we do get some sense of distance based on hearing and even smell, most depth perception comes from several visual cues, which are summarized in ■ **Figure 4.20.** The first mechanism we use is the interaction of both of our eyes, which produces **binocular cues (Study Organizer 4.3)**. The binocular (two eyes) cues of **retinal disparity** and **convergence** are inadequate in judging distances longer than the length of a football field. Luckily, we have several **monocular cues (Study Organizer 4.4)**, which need only one eye to work.

Two additional monocular cues for depth perception, **accommodation** of the lens of the eye and *motion parallax*, cannot be used by artists. In *accommodation*, muscles that adjust the shape of the lens as it focuses on an object send neural messages to the brain, which interprets the signal to perceive distance. For near objects, the lens bulges; for far objects, it flattens. *Motion parallax* (also known as *relative motion*) refers to the fact that when we are moving, close objects appear to whiz by, whereas farther objects seem to move more slowly or remain stationary. This effect can easily be seen when traveling by car or train.

Constancy Perception

To organize our sensations into meaningful patterns, we develop **perceptual constancies**, the learned tendency to perceive the environment as stable, despite changes in an object's *size*, *shape*, *color*, and *brightness*. Without perceptual constancy, things would seem to grow larger as we get closer to them, change shape as our viewing angle changes, and change color as light levels change (Albright, 2015; Fleming, 2014; Howard et al., 2014).

Size Constancy

Regardless of the distance from us (or the size of the image it casts on our retina), *size constancy* allows us to interpret an object as always being the same size. For example,

See **FIGURE 4.18**, Luria's stimuli

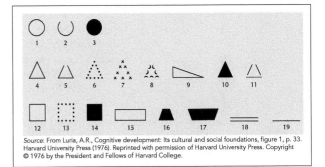

Source: From Luria, A.R., Cognitive development: Its cultural and social foundations, figure 1, p. 33. Harvard University Press (1976). Reprinted with permission of Harvard University Press. Copyright © 1976 by the President and Fellows of Harvard College.

See **FIGURE 4.19**, Visual cliff

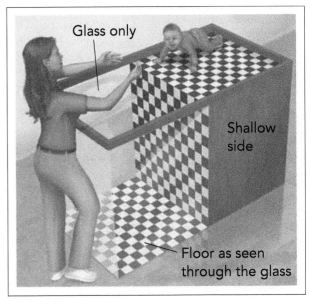

Glass only

Shallow side

Floor as seen through the glass

See **FIGURE 4.20**, Visual cues for depth perception

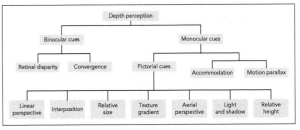

See **STUDY ORGANIZER 4.3**, Binocular Depth Cues

See **STUDY ORGANIZER 4.4**, Monocular Depth Cues

See **FIGURE 4.21, Size constancy**

Courtesy Karen Huffman

See **FIGURE 4.22, Color and brightness constancies**

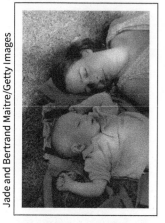

Jade and Bertrand Maitre/Getty Images

See **TUTORIAL VIDEO, A World Turned Upside-Down: Visual Perception**

See **FIGURE 4.23, Shape constancy**

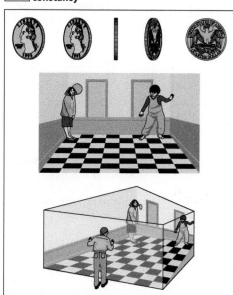

the image of the couple in the foreground of the photo (■ **Figure 4.21**) is much larger on our retina than the trees behind them. However, thanks to size constancy, we perceive them to be of normal size. Without this constancy, we would perceive people as "shrinking" when they move away from us, and "growing" when they move toward us. Although researchers have found evidence of size constancy in newborns, it also develops from learning and the environment. Case studies of people who have been blind since birth, and then have their sight restored, find that they initially have little or no size constancy (Sacks, 1995, 2015).

Color and Brightness Constancies

Our perception of color and brightness remain the same even when the light conditions change. Look at the actual variations in the skin tones of the mother and child in this photo (■ **Figure 4.22**). We perceive the color and brightness as constant despite the fact that the wavelength of light reaching our retina may vary as the light changes.

Shape Constancy

One additional perceptual constancy is the tendency to perceive an object's shape as staying constant even when the angle of our view changes (■ **Figure 4.23**).

Interpretation

After selectively sorting through incoming sensory information and organizing it, our brain uses this information to explain and make judgments about the external world. This final stage of perception—*interpretation*—is influenced by several factors, including *sensory adaptation*, *perceptual set*, and *frame of reference*.

Sensory Adaptation

Imagine that your visual field has been suddenly inverted and reversed. Things you normally expect to be on your right are now on your left—those above your head are now below. How would you ride a bike, read a book, or even walk through your home? Do you think you could ever adapt to this upside-down world? To answer that question, psychologist George Stratton (1896) invented, and for eight days wore, special prism goggles that flipped his view of the world from up to down and right to left. For the first few days, Stratton had a great deal of difficulty navigating in this environment and coping with everyday tasks. But by the third day, he noted:

Walking through the narrow spaces between pieces of furniture required much less care than hitherto. I could watch my hands as they wrote, without hesitating or becoming embarrassed thereby.

By the fifth day, Stratton had almost completely adjusted to his strange perceptual environment, and when he later removed the headgear, he quickly readapted.

What does this experiment have to do with everyday life? Stratton's experiment with the inverting goggles illustrates the critical role that *sensory adaptation* plays in the way we interpret the information that our brains gather. Without his ability to adapt his perceptions to a skewed environment, Stratton would not have been able to function. His brain's ability to retrain itself to interpret his new surroundings allowed him to create coherence from what would otherwise have been chaos.

Perceptual Set

As you can see in ■ **Figure 4.24**, our previous experiences, assumptions, and expectations also affect how we interpret and perceive the world, by creating a **perceptual set**, or a readiness to perceive in a particular manner (Dunning & Balcetis, 2013). In other words, we largely see what we expect to see!

Interestingly, factors in the environment, such as singing "Happy Birthday" before eating a birthday cake, have been found to actually change our perception of the food we eat. For example, in one set of studies, researchers investigated how the rituals we perform before eating might influence our perception and consumption of several foods (Vohs et al., 2013). Half the participants were asked to eat a piece of chocolate after following these specific instructions: "Without unwrapping the chocolate bar, break it in half. Unwrap half of the bar and eat it. Then, unwrap the other half and eat it." The second group of participants was instructed to simply relax for a short amount of time, and then eat the chocolate bar in whatever fashion they wished.

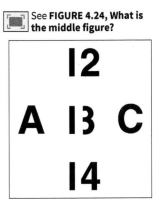

See **FIGURE 4.24, What is the middle figure?**

The results showed that those who had performed the "unwrapping ritual" rated the chocolate more highly, savored it more, and were willing to pay more for the chocolate than the other group. The findings suggest that a short, fabricated ritual can influence our perceptions of the food we eat. You can now see why singing "Happy Birthday" does improve the cake's taste, and why blowing out the candles is so important—given how they're such a large, ritualistic part of birthday celebrations in our Western culture. Using a positive psychology approach, can you think of ways to use this research in your own life? Perhaps you can try creating pleasant rituals to perform around specific healthy foods that you're trying to incorporate into your diet?

Unfortunately, in some cases, our *perceptual sets*, or expectations, can have hazardous effects. For example, researchers asked both White and Black participants to play a videogame in which they were told to shoot targets who were carrying a gun but not those who were unarmed (Correll et al., 2002). Participants of both races made the decision to shoot an armed target more quickly if the target was Black, rather than White. They also chose NOT to shoot an unarmed target more quickly if the target was White, rather than Black.

This study points to the influence of our expectations on real-life situations in which police officers must decide almost instantly whether to shoot a potential suspect—and may partially explain why Blacks are at greater risk than Whites of being wrongfully shot by police officers. Regardless of how you feel about the outcome of certain high-profile cases of police officer shootings, like the one in Ferguson, Missouri, can you see how perceptual set might have played a role?

Frame of Reference

In addition to possible problems with *perceptual set*, the way we perceive people, objects, or situations is also affected by the *frame of reference*, or context. An elephant is perceived as much larger when it is next to a mouse than when it stands next to a giraffe. This is the reason professional athletes who make huge amounts of money sometimes feel underpaid: They're comparing what they make to the pay of those around them, who also make huge sums, and not to the average person in the United States! The influence of frame of reference on perception also helps explain why people rate themselves as more athletic if they compare themselves to the Pope than to a professional basketball player (Mussweiler et al., 2004)!

Science and ESP

So far in this chapter, we have talked about sensations provided by our eyes, ears, nose, mouth, and skin. What about a so-called sixth sense? Can some people perceive things that cannot be sensed through the usual sensory channels, by using **extrasensory perception (ESP)**? Those who claim to have ESP profess to be able to read other people's minds (*telepathy*), to see remote events, such as a house on fire in another part of the country (*clairvoyance*), or the ability to see and predict the future (*precognition*). (*Psychokinesis*, the ability to move or change objects with mind power alone, such as bending a spoon or levitating a table, is generally not considered a type of ESP because, unlike the other three alleged abilities, it does not involve the senses, like "seeing the future.")

See **FIGURE 4.25, Zener cards and ESP research**

Scientists have studied ESP experiences and professed "psychics" since the late 19th century. For example, in the early 1900s, Joseph B. Rhine, a respected researcher, began experiments using Zener cards, a procedure spoofed in the popular 1984 movie, *Ghostbusters*. This deck of 25 cards included five different symbols (▣ **Figure 4.25**). "Senders" were asked to concentrate on the card, while "receivers" tried to "read the mind" of the sender. Although Rhine found a few people who appeared to score somewhat better than chance, his methodology was severely criticized and his findings were discredited.

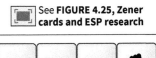

Like Rhine's research, virtually all later studies of ESP have produced similarly weak or controversial results (Alcock, 2011; Lilienfeld et al., 2015; Schick & Vaughn, 2014). Findings in ESP are notoriously "fragile," meaning that they do not hold up to scientific scrutiny under standard controls.

Perhaps the most serious weakness of ESP is its failure of replication by rivals in independent laboratories, which is a core requirement for scientific acceptance (Francis, 2012; Hyman, 1996; Rouder et al., 2013). (Recall also from Chapter 1 that magician James Randi and the MacArthur Foundation have offered $1 million to "anyone who proves a genuine psychic power under proper observing conditions." But even after many years, the money has never been collected!)

Despite the lack of credible, scientific evidence, 40 to 75% of all adults in the United States believe in ESP (Lamont, 2013; Moore, 2005). Why? One reason is that, as mentioned earlier in the chapter, our motivations and interests often influence our perceptions, driving us to selectively attend to things we want to see or hear. In addition, the subject of extrasensory perception often generates strong emotional responses. When individuals feel strongly about an issue, they sometimes fail to recognize the faulty reasoning underlying their beliefs.

Q | Answer the **Concept Check** questions.

WP LS Go to your WileyPLUS Learning Space course for video episodes, examples, art, tables, Concept Checks, practice, and other pedagogical resources that will help you succeed in this course.

Reading for

STATES OF CONSCIOUSNESS

5.1 CONSCIOUSNESS

What is **consciousness**? Most psychologists define it as a two-part awareness of both ourselves and our environment (Li, 2016; Thompson, 2015). This awareness explains how we can be deeply engrossed in studying or a conversation with others, and still hear the doorbell. However, we may not hear the doorbell when we're deeply asleep because we are in what is known as an **alternate state of consciousness (ASC)**. In this chapter, we will discuss the ASCs of sleep, dreaming, meditation, and hypnosis. But we first need to explore the general nature of consciousness in some depth.

No problem can be solved from the same level of consciousness that created it.

—ALBERT EINSTEIN

Selective Attention

William James, one of the early U.S. psychologists, likened *consciousness* to a stream that's constantly changing yet always the same. It meanders and flows, sometimes where the person wills and sometimes not. However, the process of **selective attention** (Chapter 4) allows us to control this *stream of consciousness* through deliberate concentration and full attention. For example, when listening to classroom lectures, your attention may unfortunately drift away to thoughts of a laptop computer you want to buy, or an attractive classmate. However, at this moment, you are (hopefully) fully awake and concentrating on the words in this text.

 See **FIGURE 5.1, Levels of awareness**

Interestingly, when we're fully focused and selectively attending to certain stimuli, we're sometimes less likely to pick up on other stimuli, such as not hearing someone ask us a question when we're fully concentrated on a book or movie. This is particularly true when the outside stimulus is unexpected, a phenomenon known as **inattentional blindness**. Have you seen any of the popular *YouTube* videos in which observers fail to notice a grown man dressed in a gorilla costume as he repeatedly passes through a group of people? These videos are based on a clever experiment that first asked participants to count the number of passes in a videotaped basketball game. Researchers then sent an assistant, dressed in a full gorilla suit, to walk through the middle of the ongoing game.

Can you predict what happened? The research participants were so focused on their pass-counting task that they failed to notice the gorilla (Simons & Chabris, 1999). This type of "blindness" is a common tool of magicians who ask us to focus on a distracting element, such as the deck of cards, while they manipulate the real object of their magic, such as removing an unsuspecting volunteer's wallet or watch. On a more important note, this type of inattentional blindness (or divided attention) can lead to serious problems for pilots, car and truck drivers, and pedestrians.

Levels of Awareness

As this example of inattentional blindness indicates, our *stream of consciousness* also varies in its level of awareness. Consciousness is not an all-or-nothing phenomenon—conscious or unconscious. Instead, it exists along a continuum, ranging from high awareness and sharp, focused alertness at one extreme, to middle levels of awareness, to low awareness or even nonconsciousness and coma at the other extreme ( **Figure 5.1**).

Note that this continuum also involves both *controlled* and *automatic* processes. When you're working at a demanding task or learning something new, such as how to drive a car, your consciousness is at the high end of the continuum. These **controlled processes** demand focused attention and generally interfere with other ongoing activities (Cohen & Israel, 2015; Vidal et al., 2015). Have you ever been so absorbed during an exam that you completely forgot your surroundings until the instructor announced, "Time is up," and asked for your paper? This type of focused attention is the hallmark of controlled processes—and good test takers!

In sharp contrast to the high awareness and focused attention required for controlled processes, **automatic processes** require minimal attention, and generally do not interfere with other ongoing activities. Think back to your teen years when you were first learning how to drive a car, and it took all of your attention (controlled processing). The fact that you can now effortlessly steer a car, and work the brakes all at one time (with little or no focused attention), is thanks to automatic processing. In short, learning a new task requires complete concentration and *controlled processing*. Once that task is well learned, you can switch to *automatic processing*.

Keep in mind that although automatic processes are generally helpful, such as changing gears and steering a car, there are times when we should be using controlled processing. For example, during class lectures, you need focused attention and controlled processing, in order to listen carefully enough to take notes and really learn the material—and to do well on exams.

Before going on, it's also important to address the apparently never-ending philosophical debates over the *mind–body problem*. Is the "mind" (consciousness and other mental functions) fundamentally different from matter (the body)? How can a supposedly nonmaterial mind influence a physical body and vice versa? Most psychologists today believe the mind *is* the brain and *consciousness* involves an activation and integration of several parts of the brain (■ **Figure 5.2**). However, awareness is generally limited to the *cerebral cortex*, particularly the frontal lobes. In contrast, arousal generally results from *brain stem activation* (Pinel, 2014; Raj et al., 2015).

▶ See **TUTORIAL VIDEO,**
Automatic Processing and
Multitasking

⬛ See **FIGURE 5.2,**
Consciousness and our
brain

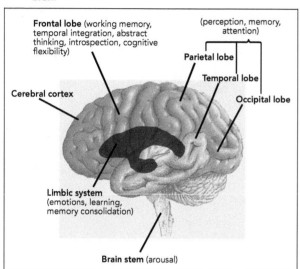

Frontal lobe (working memory, temporal integration, abstract thinking, introspection, cognitive flexibility)

(perception, memory, attention)

Parietal lobe

Temporal lobe

Occipital lobe

Cerebral cortex

Limbic system (emotions, learning, memory consolidation)

Brain stem (arousal)

Q Answer the
Concept Check questions.

5.2 SLEEP AND DREAMS

Having explored the definition and description of everyday, waking consciousness and its properties, we now turn to two of our most common *alternate states of consciousness* (ASCs)—sleep and dreaming. These ASCs are fascinating to both scientists and the general public. Why are we born with a mechanism that forces us to sleep and dream for approximately a third of our lives? How can an ASC that requires reduced awareness and responsiveness to our environment be beneficial in an evolutionary sense? What are the functions and causes of sleep and dreams? To understand sleep and dreaming, we need to first explore the topic of circadian rhythms.

Circadian Rhythms and Sleep

Most animals have adapted to our planet's cycle of days and nights by developing a pattern of bodily functions that wax and wane over each 24-hour period. For humans, our alertness, core body temperature, moods, learning efficiency, blood pressure, metabolism, immune responses, and pulse rate all follow these **circadian rhythms** (Crowley & Eastman, 2015; Du Pré et al., 2014; Sharma et al., 2015). Usually, these activities reach their peak during the day and their low point at night. (See **Study Organizer 5.1.**)

One of the biggest challenges for circadian rhythms happens during the teen years. Do you remember having trouble going to bed at a "reasonable hour" when you were a teenager and then having a really difficult time getting up each morning? This common pattern of staying up late at night and then sleeping longer in the morning appears to be a result of the natural shift

↗ See **STUDY ORGANIZER 5.1,**
Understanding Circadian
Rhythms

in the timing of circadian rhythms that occurs during puberty (McGlinchey, 2015; Paiva et al., 2015). This shift is caused by a delay in the release of the hormone melatonin. In adults, this hormone is typically released around 10 p.m., signaling the body that it is time to go to sleep. But in teenagers, melatonin isn't released until around 1 a.m.—thus explaining why it's more difficult for teenagers to fall asleep as early as adults or younger children do.

Recognition of this unique biological shift in circadian rhythms among teenagers has led some school districts to delay the start of school in the morning. Research shows that even a 25- to 30-minute delay allows teenagers to be more alert and focused during class, and contributes to improvements in their moods and overall health (Boergers et al., 2014; Bryant & Gómez, 2015).

Disruptions in circadian rhythms are not just a problem for teenagers (D'Ambrosio & Redline, 2014). We're all at risk of serious health issues and personal concerns, including increased risk of cancer, heart disease, autoimmune disorders, obesity, sleep disorders, and accidents, as well as decreased concentration and productivity (Du Pré et al., 2014; Tsimakouridze et al., 2015; Zuurbier et al., 2015).

The most serious disruptions and ill effects of these sleep and circadian disturbances tend to be with physicians, nurses, police, and others—about 20% of employees in the United States—whose occupations require rotating "shift work" schedules. Typically divided into a first shift (8 a.m. to 4 p.m), second shift (4 p.m. to midnight) and a third shift (midnight to 8 a.m.), these work shifts often change from week to week, and clearly disrupt the workers' circadian rhythms.

Like shift work, flying across several time zones can also cause fatigue, irritability, decreased alertness and mental agility, and worsening of psychiatric disorders (Chiesa et al., 2015; Selvi et al., 2015; Srinivasan et al., 2014). Jet lag tends to be more debilitating when we fly eastward because our bodies adjust more easily to going to bed later than to going to sleep earlier than normal.

Sleep Deprivation

One of the biggest problems with disrupted circadian rhythms is the corresponding disruptions in the amount and quality of our sleep. Did you know that, compared to 10 years ago, approximately 80% of today's working adults are getting about 38 minutes less sleep each night, or that 50 to 70 million people in the United States suffer some form of a chronic sleep disorder (Ford et al., 2015; Konnikova, 2014; Roenneberg, 2013)? Sleep deprivation is clearly associated with reduced cognitive and motor performance, irritability and other mood alterations, lowered self-control, and increased cortisol levels, which are all signs of stress (Baum et al., 2014; Meldrum et al., 2015; Wright et al., 2015).

Sleep deprivation also increases the risk of cancer, heart diesease, and other illnesses (Mezick et al., 2014; Vitello & Krieger, 2014). It also impairs our immune system, which is one reason adults who get fewer than seven hours of sleep a night are three times as likely to develop a cold as those who sleep at least eight hours a night (Ackermann et al., 2012; Cohen et al., 2009). In addition, sleep-deprived adults are more likely to become obese. This may be because getting inadequate amounts of sleep interferes with the production of hormones that control appetite (Knutson, 2012).

Perhaps the most frightening danger is that chronic, prolonged sleep deprivation may lead to degeneration of key neurons involved in our alertness and proper cognitive functioning, as well as premature aging of our brains (Konnikova, 2014; Veasey & White, 2013). In addition, lapses in attention among sleep-deprived pilots, physicians, truck drivers, and other workers cause serious accidents and cost thousands of lives each year (Bougard et al., 2015; Gonçalves et al., 2015; Pascal, 2014).

The good news is that taking simple measures to increase overall sleep time can offset most of these dangers. For example, researchers have found that "simply" monitoring and reducing grade-school children's screen time, on computers, TVs, and smart phones, led to significant academic, health, and social benefits (Gentile et al., 2014).

Stages of Sleep

The woods are lovely, dark and deep. But I have promises to keep, and miles to go before I sleep.

—ROBERT FROST

Please see entire **FIGURE 5.3, Scientific study of sleep and dreaming,** in your *WileyPLUS* Learning Space course

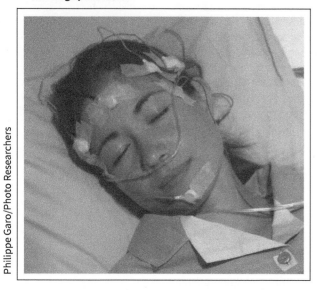

Philippe Garo/Photo Researchers

Having discussed our 24- to 25-hour circadian cycle, and the problems associated with its disruption, we now turn our attention to our patterns of sleep, which consist of four different stages. We begin with an exploration of how scientists study sleep. Surveys and interviews can provide general information, but for more detailed and precise data researchers in sleep laboratories use a number of sophisticated instruments (**Figure 5.3**).

Imagine that you are a participant in a sleep experiment. When you arrive at the sleep lab, you are assigned one of several bedrooms. The researcher hooks you up to various physiological recording devices (**Figure 5.3a**). You will probably need a night or two to adapt to the equipment before the researchers can begin to monitor your typical night's sleep. At first, you enter a relaxed, *presleep* state. As you continue relaxing, your brain's electrical activity slows even further. In the course of about an hour, you progress through several distinct stages of sleep, each progressively deeper (**Figure 5.3b**). Then the sequence begins to reverse. Note that we don't necessarily go through all sleep stages in this exact sequence. But during the course of a night, people usually complete four to five cycles of light to deep sleep and back, each lasting about 90 minutes (**Figure 5.3c**).

NREM and REM Sleep

Note the two important divisions of sleep shown in Figure 5.3b and 5.3c: **non-rapid-eye-movement (NREM) sleep** (*Stages 1, 2,* and *3*) and **rapid-eye-movement (REM) sleep**. During REM sleep, your brain wave patterns are similar to those of a relaxed wakefulness stage, and your eyeballs will move up and down and from left to right. This rapid eye movement is a signal that dreaming is occurring. In addition, during REM sleep your breathing and pulse rates become fast and irregular, and your genitals may show signs of arousal. Yet your musculature is deeply relaxed and unresponsive. Because of these contradictory qualities, REM sleep is sometimes referred to as *paradoxical sleep*.

Dreams very similar to those from REM sleep also occur during NREM sleep, but less frequently (Chellappa et al., 2011; Wamsley et al., 2007). Note how *Stage 1* of NREM sleep is characterized by theta waves and drowsy sleep. During this stage, many people experience sudden muscle movements called *myoclonic jerks* accompanied by a sensation of falling. In *Stage 2 sleep*, muscle activity further decreases and sleep spindles occur, which involve a sudden surge in brain wave frequency. Stages 1 and 2 are relatively light stages of sleep, whereas *Stage 3 sleep* involves the deepest stage of sleep, often referred to as *slow wave sleep (SWS)* or simply *deep sleep*. Sleepers during this deep sleep are very hard to awaken, and if something does wake them, they're generally confused and disoriented at first. This is also a time that sleepwalking, sleep talking, and bedwetting occur. [Note that Stage 3 sleep was previously divided into Stages 3 and 4, but the American Academy of Sleep Medicine (AASM) removed the Stage 4 designation.]

▶ See **TUTORIAL VIDEO, Myths About Sleep, Dreams, Drugs and Hypnosis**

Why Do We Sleep and Dream

There are many myths and misconceptions about why we sleep and dream. Fortunately, scientists have carefully studied what sleep and dreaming do for us and why we spend approximately 25 years of our life in these alternate states of consciousness (ASCs).

Four Sleep Theories

How do scientists explain our shared need for sleep? There are four key theories:

1. Adaptation/protection Sleep evolved because animals need to conserve energy and protect themselves from predators that are more active at night (Drew, 2013; Tsoukalas, 2012). However, as you can see in **Figure 5.4**, animals vary greatly in how much sleep they need each day.

Those with the highest likelihood of being eaten by others, a higher need for food, and the lowest ability to hide tend to sleep the least.

See **FIGURE 5.4, Average daily hours of sleep for different mammals**

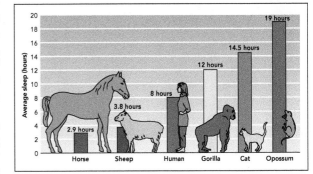

2. Repair/restoration Sleep helps us recuperate from the depleting effects of daily waking activities. Essential chemicals and bodily tissues are repaired or replenished while we sleep. We recover not only from physical fatigue, but also from emotional and intellectual demands (Blumberg, 2015). When deprived of REM sleep, most people "catch up" later by spending more time than usual in this state (the so-called "REM rebound"), which further supports this theory.

Recent research also shows that during sleep our brain repairs and cleans itself of potentially toxic waste products that accumulate while we're awake. Just as our lymph glands clear toxic byproducts from our muscles, the brain's "glymphatic system" (composed of glial cells and cerebrospinal fluid) washes away debris caused by our daily mental processing (Iliff et al., 2012; Konnikova, 2014; Underwood, 2013; Xie et al., 2013).

See **FIGURE 5.5, Aging and the sleep cycle**

3. Growth/development The percentage of deepest sleep (Stage 3) changes over the life span and coincides with changes in the structure and organization of the brain, as well as the release of growth hormones from the pituitary gland—particularly in children. As we age, our brains change less, and we release fewer of these hormones, grow less, and sleep less.

4. Learning/memory Sleep is important for learning and the consolidation, storage, and maintenance of memories (Bennion et al., 2015; Vorster & Born, 2015). This is particularly true for REM sleep, which increases after periods of stress or intense learning. For example, infants and young children, who generally are learning more than

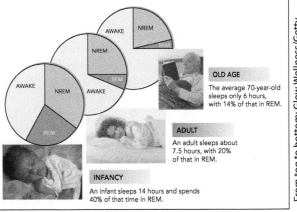

OLD AGE
The average 70-year-old sleeps only 6 hours, with 14% of that in REM.

ADULT
An adult sleeps about 7.5 hours, with 20% of that in REM.

INFANCY
An infant sleeps 14 hours and spends 40% of that time in REM.

From top to bottom: Glow Wellness/Getty Images, Inc.; © Cagri Özgür/iStockphoto; © Jani Bryson/iStockphoto

adults, spend far more of their sleep time in REM sleep (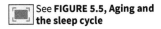 **Figure 5.5**). Interestingly, when we're sleep deprived, we're also more likely to "remember" things that did not actually happen, a phenomenon you'll learn more about in Chapter 7 (Frenda et al., 2014).

Three Dream Theories

Now let's look at three theories of why we dream—and whether dreams carry special meaning or information.

One of the oldest and most scientifically controversial explanations for why we dream is Freud's **wish-fulfillment view**. Freud proposed that unacceptable desires, which are reportedly normally repressed, rise to the surface of consciousness during dreaming. We avoid anxiety, Freud believed, by disguising our forbidden unconscious needs (what Freud called the dream's **latent content**) as symbols (**manifest content**). For example, a journey supposedly symbolizes death; horseback riding and dancing could symbolize sexual intercourse; and a gun might represent a penis.

Most modern scientific research does not support Freud's view (Domhoff, 2003, 2010; Sándor et al., 2014; Siegel, 2010). Critics also say that Freud's theory is highly subjective and that the symbols can be interpreted according to the particular analyst's view or training.

In contrast to Freud, a biological view called the **activation–synthesis hypothesis** suggests that dreams are a by-product of random stimulation of brain cells during REM sleep (Hobson, 1999, 2005; Wamsley & Stickgold, 2010). Alan Hobson and Robert McCarley (1977) proposed that specific neurons in the brain stem fire spontaneously during REM sleep and that the cortex struggles to "synthesize," or make sense of, this random stimulation by manufacturing dreams. This is *not* to say that dreams are totally meaningless. Hobson suggests that even if our dreams begin with essentially random brain activity, our individual personalities, motivations, memories, and life experiences guide how our brains construct the dream.

Have you ever dreamed that you were trying to run away from a frightening situation but found that you could not move? The activation—synthesis hypothesis might explain this dream as random stimulation of the amygdala. As you recall from Chapter 2, the amygdala is a specific brain area linked to strong emotions, especially fear. If your amygdala is randomly stimulated and you feel afraid, you may try to run. But you can't move because your major muscles are temporarily paralyzed during REM sleep. To make sense of this conflict, you might create a dream about a fearful situation in which you were trapped in heavy sand or someone was holding onto your arms and legs.

Finally, other researchers support the **cognitive view of dreams,** which suggests that dreams are simply another type of information processing. That is, our dreams help us organize and interpret our everyday, waking thoughts and experiences. This view of dreaming is supported by research showing strong similarities between dream content and waking thoughts, fears, and concerns (Domhoff, 2003, 2010; Malinowski & Horton, 2014; Sándor et al., 2014).

Like most college students, you've probably experienced what's called "examination-anxiety" dreams. You can't find your classroom, you're running out of time, your pen or pencil won't work, or you've completely forgotten a scheduled exam and show up totally unprepared. (Sound familiar?)

See **FIGURE 5.6, Creative dreaming**

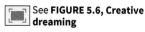

In Sum

The wish fulfillment, activation-synthesis, and cognitive views of dreaming offer three widely divergent perspectives, and numerous questions remain. For example, how would the wish-fulfillment view of dreams explain why human fetuses show REM patterns? On the other hand, how would the activation–synthesis hypothesis explain complicated, story-like dreams or recurrent dreams? Finally, according to the cognitive view, how can we explain dreams that lie outside our everyday experiences? And how is it that the same dream can often be explained by many theories (see 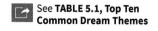 **Figure 5.6**)?

Gender and Dreams

See **TABLE 5.1, Top Ten Common Dream Themes**

Men and women tend to share many of the common dream themes shown in **Table 5.1**. But women are more likely to report dreams of children, family members and other familiar people, household objects, and indoor events. In contrast, men tend to report dreams about strangers, violence, weapons, sexual activity, achievement, and outdoor events (Blume-Marcovici, 2010; Domhoff, 2003, 2010; Mathes et al., 2014; Mazandarani et al., 2013). As a critical thinker, can you see how attitudes toward "proper" male and female gender roles, like caring for children, sex, weapons, and violence, might have affected what the participants were willing to report? Interestingly, a study of WWII prisoners of war (POWs) found that their dreams contained less sexuality and even less aggression than the male norms (Barrett et al., 2014).

Culture and Dreams

Dreams about basic human needs and fears (like sex, aggression, and death) seem to be found in all cultures. Children around the world often dream about large, threatening monsters or wild animals. In addition, dreams around the world typically include more misfortune than good fortune, and the dreamer is more often the victim of aggression than the cause of it (Chang, 2012; Domhoff, 2003, 2007, 2010; Honig & Nealis, 2012; Yu, 2012).

Sleep–Wake Disorders

See **VIRTUAL FIELD TRIP, Diagnosing Sleep Disorders**

In any given year, an estimated 40 million Americans suffer from chronic sleep disorders, and another 30 million experience occasional sleep disorders serious enough to disrupt their daily activities (Morin & Edinger, 2015; Ng et al., 2015; Shin et al., 2014).

As a college student, you'll be particularly interested to know that students who send a higher number of text messages are more likely to suffer from sleep problems, such as taking longer to fall asleep, waking up in the middle of the night, and feeling tired during the day (Murdock, 2013). What causes this link between text messaging and poor sleep quality? Researchers believe that students feel pressure to immediately respond to texts, and may be awakened by alerts from in-coming texts, which can both reduce sleep quality (and quantity).

Based on these statistics and findings, it's clear that almost everyone has difficulty sleeping at some point in his or her lifetime. The most common or serious of these disorders are summarized in ■ **Table 5.2**.

 See **TABLE 5.2, Sleep-Wake Disorders**

Although it's normal to have trouble sleeping before an exciting event, as many as 1 person in 10 may suffer from **insomnia**. Those who suffer from this disorder experience persistent difficulty falling asleep, staying asleep, and/or waking up too early. Nearly everybody has insomnia at some time (Morin et al., 2013; Williamson & Williamson, 2015); a telltale sign is feeling poorly rested the next day. Most people with serious insomnia have other medical or psychological disorders as well (American Psychiatric Association, 2013; Ashworth et al., 2015; Primeau & O'Hara, 2015).

To cope with insomnia, many people turn to nonprescription, over-the-counter sleeping pills, which generally don't work. In contrast, prescription tranquilizers and barbiturates do help people sleep, but they decrease Stage 3 and REM sleep, seriously affecting sleep quality. In the short term, limited use of drugs such as Ambien, Dalmane, Xanax, Halcion, and Lunesta may be helpful in treating sleep problems related to anxiety and acute, stressful situations. However, chronic users run the risk of psychological and physical drug dependence (Mehra & Strohl, 2014; Santamaria & Iranzo, 2014). The hormone *melatonin* may provide a safer alternative. Some research suggests that taking even a relatively small dose (just .3 to .4 milligrams) can help people fall asleep and stay asleep (Hajak et al., 2015; Paul et al., 2015).

Fortunately, there are many effective strategies for alleviating sleep problems without medication. For example, did you know that using an eReader or computer right before you try to fall asleep can make it harder to get to sleep (Chang et al., 2015; van der Lely et al., 2015)? Exposure to light from the electronic screens of a computer or tablet reduces the level of melatonin in the body by about 22%, which makes it more difficult to fall asleep (especially for teenagers).

 See **FIGURE 5.7, Narcolepsy**

Narcolepsy, a sleep disorder characterized by sudden and irresistible onsets of sleep during normal waking hours, afflicts about 1 person in 2,000 and generally runs in families (Kotagal & Kumar, 2013; Williamson & Williamson, 2015). During an attack, REM-like sleep suddenly intrudes into the waking state of consciousness. Victims may experience sudden, incapacitating attacks of muscle weakness or paralysis (known as cataplexy). They may even fall asleep while walking, talking, or driving a car. Although long naps each day and stimulant or antidepressant drugs can help reduce the frequency of attacks, both the causes and cure of narcolepsy are still unknown (■ **Figure 5.7**).

© Juniors/SuperStock

Perhaps the most serious sleep disorder is **sleep apnea**. People with sleep apnea may fail to breathe for a minute or longer and then wake up gasping for breath. When they do breathe during their sleep, they often snore. Sleep apnea seems to result from blocked upper airway passages and/or from the brain's failure to send signals to the diaphragm, thus causing breathing to stop.

Unfortunately, people with sleep apnea are often unaware they have this disorder, and fail to understand how their repeated awakening during the night leave them feeling tired and sleepy during the day. More importantly, they should know that sleep apnea can lead to high blood pressure, strokes, cancer, depression, and heart attacks (Culebras, 2012; Kendzerska et al., 2014; Lavie, 2015).

Treatment for sleep apnea depends partly on its severity. If the problem occurs only when you're sleeping on your back, sewing tennis balls to the back of your pajama top may help remind you to sleep on your side. Because obstruction of the breathing passages is related to obesity and heavy alcohol use (Tan et al., 2015; Yamaguchi et al., 2014), dieting and alcohol restriction are often recommended. For other sleepers, surgery, dental appliances that reposition the tongue, or machines that provide a stream of air to keep the airway open may provide help.

Recent findings suggest that even "simple" snoring (without the breathing stoppage characteristic of sleep apnea) can lead to heart disease and possible death (Deeb et al., 2014; Jones & Benca, 2013). Although occasional mild snoring is fairly normal, chronic snoring is a possible warning sign that should prompt people to seek medical attention.

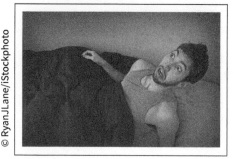

See **FIGURE 5.8, Nightmare or sleep terror?**

© RyanJLane/iStockphoto

Q Answer the **Concept Check** questions.

Two additional sleep disturbances are **nightmares** and **sleep terrors** (■ Figure 5.8). *Sleepwalking*, which sometimes accompanies sleep terrors, usually occurs during NREM sleep. (Recall that large muscles are paralyzed during REM sleep, which explains why sleepwalking normally occurs during NREM sleep.) An estimated 4% of U.S. adults—meaning over 8 million people—have at least one episode of sleepwalking each year (Ohayon et al., 2012). *Sleeptalking* can occur during any stage of sleep, but it appears to arise most commonly during NREM sleep. It can consist of single, indistinct words or long, articulate sentences, and it's even possible to engage some sleep talkers in a limited conversation.

Nightmares, sleep terrors, sleepwalking, and sleeptalking are all more common among young children, but they can also occur in adults, usually during times of stress or major life events (Carter et al., 2014; Plazzi & Nobili, 2014). Patience and soothing reassurance at the time of the sleep disruption are usually the only treatment recommended for both children and adults.

5.3 PSYCHOACTIVE DRUGS

Have you noticed how difficult it is to have a logical, nonemotional discussion about drugs? In our society, where the most popular **psychoactive drugs** are caffeine, tobacco, and ethyl alcohol, people often become defensive when these drugs are grouped with illicit drugs such as marijuana and cocaine. Similarly, marijuana users are disturbed that their drug of choice is grouped with "hard" drugs like heroin.

Understanding Psychoactive Drugs

Most scientists believe that there are good and bad uses of all drugs (Sdrulla et al., 2015). The way drug use differs from drug abuse and how chemical alterations in consciousness affect a person, psychologically and physically, are important topics in psychology.

Alcohol, for example, has a diffuse effect on neural membranes throughout the nervous system. Most psychoactive drugs, however, act in a more specific way: by either enhancing a particular neurotransmitter's effect, as does an **agonist drug,** or inhibiting it, as does an **antagonist drug (Process Diagram 5.1)**.

Is drug abuse the same as drug addiction? The term **drug abuse** generally refers to drug taking that causes emotional or physical harm to oneself or others. Drug consumption among abusers is also typically compulsive, frequent, and intense. **Addiction** is a broad term that refers to a condition in which a person feels compelled to use a specific drug, or engage in almost any type of compulsive activity, from working to surfing the Internet (Sdrulla et al., 2015; Smith, 2015). In fact, the latest version of the *Diagnostic and Statistical Manual (DSM-5)*, which officially classifies mental disorders, now includes *gambling disorders* as part of their substance-related and addictive disorders category. But other disorders, like "sex addiction" or "exercise addiction" were not included due to their controversial nature, and/or insufficient evidence at this time (American Psychiatric Association, 2013; Southern et al., 2015).

Interestingly, some evidence shows that we can become addicted to Facebook and that risky trading in financial markets can create a high indistinguishable from that experienced in drug addiction (Hong & Chiu, 2014; Suissa, 2015; Zaremohzzabieh et al., 2014). Researchers in one study prompted people with a series of statements, such as "You feel an urge to use Facebook more and more," and "You become restless or troubled if you are prohibited from using Facebook" (Andreassen et al., 2012). The degree to which you agree with items like these may indicate that you are addicted, meaning that you feel unreasonably compelled to check and use Facebook throughout the day.

In addition to distinguishing between drug abuse and addiction, many researchers use the term **psychological dependence** to refer to the mental desire or craving to achieve a drug's effects. In contrast, **physical dependence** describes changes in bodily processes that make a drug necessary for minimum daily functioning. Physical dependence appears most clearly when the drug is withheld and the user undergoes **withdrawal** reactions, including physical pain and intense cravings.

Process Diagram 5.1

How Do Agonist and Antagonist Drugs Produce Their Psychoactive Effects?

Most psychoactive drugs produce their mood, energy, and perception-altering effects by changing the body's supply of neurotransmitters. Note how they can alter the synthesis, storage, or release of neurotransmitters (Step 1). They also can change the neurotransmitters' effects on the receiving site of the receptor neuron (Step 2). After neurotransmitters carry their messages across the synapse, the sending neuron normally deactivates the excess, or leftover, neurotransmitter (Step 3). However, when agonist drugs block this process, excess neurotransmitters remain in the synapse, which prolongs the effect of the psychoactive drug.

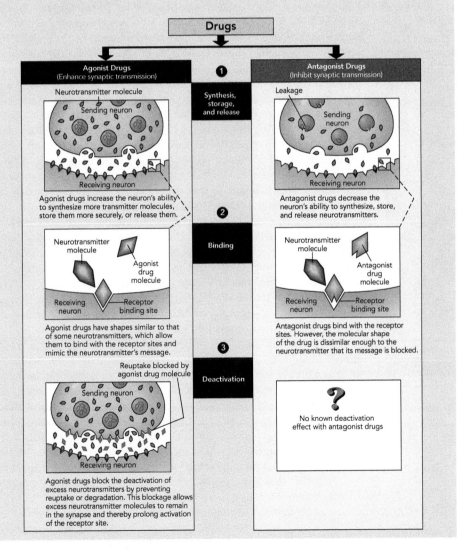

Keep in mind that psychological dependence is no less damaging than physical dependence. The craving in psychological dependence can be strong enough to keep the user in a constant drug-induced state—and to lure an addict back to a drug habit long after he or she has overcome physical dependence. After repeated use of a drug, many of the body's physiological processes adjust to higher and higher levels of the drug, producing a decreased sensitivity called **tolerance**.

Tolerance leads many users to escalate their drug use and experiment with other drugs in an attempt to re-create the original pleasurable altered state. Sometimes using one drug increases tolerance for another, a result known as *cross-tolerance*. Developing tolerance or cross-tolerance does not prevent drugs from seriously damaging the brain, heart, liver, and other organs.

Four Drug Categories

Psychologists divide psychoactive drugs into four broad categories: depressants, stimulants, opiates, and hallucinogens (■ **Table 5.3**).

Depressants, sometimes called "downers," act on the central nervous system to suppress or slow bodily processes and reduce overall responsiveness. Because tolerance and both physical and psychological dependence are rapidly acquired with these drugs, there is strong potential for abuse.

Although alcohol is primarily a depressant, at low doses it has stimulating effects, thus explaining its reputation as a "party drug." As consumption increases, symptoms of drunkenness appear. Alcohol's effects are determined primarily by the amount that reaches the brain (■ **Table 5.4**). Because the liver breaks down alcohol at the rate of about 1 ounce per hour, the number of drinks and the speed of consumption are both very important. People can die after drinking large amounts of alcohol in a short period of time (■ **Figure 5.9**). In addition, men's bodies are more efficient than women's at breaking down alcohol. Even after accounting for differences in size and muscle-to-fat ratio, women have a higher blood-alcohol level than men following equal doses of alcohol.

Alcohol should not be combined with any other drug; combining alcohol and barbiturates—both depressants—is particularly dangerous (Marczinski, 2014). Together, they can relax the diaphragm muscles to such a degree that the person suffocates to death.

Cocaine is a powerful central nervous system stimulant extracted from the leaves of the coca plant. It produces feelings of alertness, euphoria, well-being, power, energy, and pleasure. But it also acts as an *agonist drug* to block the reuptake of our body's natural neurotransmitters that produce these same effects. As you can see in ■ **Figure 5.10**, cocaine's ability to block reuptake allows neurotransmitters to stay in the synapse longer than normal—thereby artificially prolonging the effects and depleting the user's neurotransmitters.

Although cocaine was once considered a relatively harmless "recreational drug," even small initial doses can be fatal because cocaine interferes with the electrical system of the heart, causing irregular heartbeats and, in some cases, heart failure. It also can produce heart attacks, hypertension, and strokes by temporarily constricting blood vessels, as well as cognitive declines and brain atrophy (Kozor et al., 2014; Siniscalchi et al., 2015; Vonmoos et al., 2014). The most dangerous form of cocaine is the smokable, concentrated version known as "crack," or "rock." Its lower price makes it affordable and attractive to a large audience. And its greater potency makes it more highly addictive.

Even legal stimulants can lead to serious problems. For example, cigarette smoking is considered to be the single most preventable cause of death and disease in the United States, and nicotine addiction is the second leading cause of death worldwide (Herbst et al., 2014; Smith, 2015; Toll et al., 2014). Like smoking, chewing tobacco is also extremely dangerous. Sadly, in 2014 fans mourned the loss of Hall of Fame baseball player Tony Gwynn, who died of mouth cancer, which he attributed to his lifelong use of chewing tobacco.

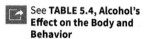
See **TABLE 5.3, Effects of the Major Psychoactive Drugs**

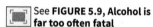
See **TABLE 5.4, Alcohol's Effect on the Body and Behavior**

See **FIGURE 5.9, Alcohol is far too often fatal**

Kyle Bursaw/Daily Chronicle/AP

See **FIGURE 5.10, Cocaine: An agonist drug in action**

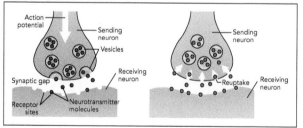

Given these well-known health hazards, why do people ever start using tobacco? Researchers have found that nicotine activates the same brain areas as cocaine (David et al., 2013; Herman et al., 2014; Li et al., 2014). Nicotine's effects—relaxation, increased alertness, and diminished pain and appetite—are so powerfully reinforcing that some people continue to smoke even after having a cancerous lung removed.

See **FIGURE 5.11, How opiates create physical dependence**

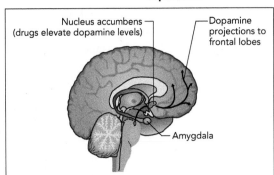

Nucleus accumbens (drugs elevate dopamine levels)

Dopamine projections to frontal lobes

Amygdala

Opiates, or narcotics, which are derived from the opium poppy, are used medically to relieve pain because they mimic the brain's natural endorphins (Chapter 2), which numb pain and elevate mood (Satterly & Anitescu, 2015). This creates a dangerous pathway to drug abuse, however. After repeated flooding with artificial opiates, the brain eventually reduces or stops the production of its own opiates. If the user later attempts to stop, the brain lacks both the artificial and normal level of painkilling chemicals, and withdrawal becomes excruciatingly painful (**Figure 5.11**).

So far, we have discussed three of the four types of psychoactive drugs: depressants, stimulants, and opiates. One of the most intriguing alterations of consciousness comes from **hallucinogens**, drugs that produce sensory or perceptual distortions, including visual, auditory, and kinesthetic hallucinations. Some cultures have used hallucinogens for religious purposes, as a way to experience "other realities" or to communicate with the supernatural. In Western societies, most people use hallucinogens for their reported "mind-expanding" potential.

Hallucinogens are commonly referred to as *psychedelics* (from the Greek for "mind manifesting"). They include mescaline (derived from the peyote cactus), psilocybin (derived from a certain type of mushroom), phencyclidine (chemically derived), and LSD (lysergic acid diethylamide, derived from ergot, a rye mold).

LSD, or "acid," produces dramatic alterations in sensation and perception, including an altered sense of time, synesthesia (blending of the senses), and spiritual experiences. And the amount of LSD the size of an aspirin tablet is enough to produce psychoactive effects in over 3000 people! Perhaps because the LSD experience is so powerful, few people "drop acid" on a regular basis. Nevertheless, LSD can be an extremely dangerous drug. Bad LSD "trips" can be terrifying and may lead to accidents, attempted suicides, and deaths. One 32-year-old man, with no known psychiatric disorder, intentionally removed his own testes after the first and single use of LSD combined with alcohol (Blacha et al., 2013)!

Marijuana, also called cannabis, is classified as a hallucinogen even though it has some properties of a depressant—inducing drowsiness and lethargy—and some of a narcotic—acting as a weak painkiller. In low doses, marijuana produces mild euphoria; moderate doses lead to an intensification of sensory and perceptual experiences, such as the illusion that time is passing slowly. High doses may produce hallucinations, delusions, and distortions of body image (Brewer & Collins, 2014; Hall & Degenhardt, 2014). The active ingredient in marijuana is THC, or tetrahydrocannabinol, which attaches to receptors that are abundant throughout the brain.

Some research has found marijuana to be therapeutic in treating glaucoma (an eye disease), alleviating the nausea and vomiting associated with chemotherapy, and dealing with other health problems (Loflin & Earleywine, 2015; Marczinski, 2014; Robson, 2014).

Marijuana is currently the most commonly used illicit drug in the United States, and it remains relatively controversial for a variety of reasons (Balkin, 2015; Pacek et al., 2015). Despite many states now legalizing marijuana for either medical or recreational use, chronic marijuana use can lead to decreases in IQ and overall cognitive functioning (Meier et al., 2012; Thames et al., 2014). In addition, it contributes to throat and respiratory disorders, impaired lung functioning and immune response, declines in testosterone levels, reduced sperm count, and disruption of the menstrual cycle and ovulation (Copeland et al., 2014; Hall & Degenhardt, 2014; Shakoor et al., 2015; Smith et al., 2015).

Although some research supports the popular belief that marijuana serves as a "gateway" to other illegal drugs, other studies find little or no connection (Kirisci et al., 2013; Mosher & Akins, 2014). Marijuana can be habit forming, but few users experience the intense cravings associated with cocaine or opiates. Withdrawal symptoms are mild because the drug dissolves in the body's fat and leaves the body very slowly, which explains why a marijuana user can test positive for days or weeks after the last use.

Club Drugs

As you may know from television or newspapers, psychoactive drugs like Rohypnol (the "date rape drug," also called "roofies") and MDMA (3,4-methylenedioxy-methylamphetamine, "Molly" or "ecstasy") are among our nation's most popular drugs of abuse. Other "club" drugs, like GHB (gamma-hydroxybutyrate), Ketamine ("special K"), methamphetamine ("ice" or "crystal meth"), "bath salts," and LSD, are also gaining in popularity. Unfortunately, they can have very serious consequences (Dunne et al., 2015; Khey et al., 2014; Stiles et al., 2015; Weaver et al., 2015). For example, recreational use of "ecstasy" is associated with potentially fatal damage to hippocampal cells in the brain, and a reduction in the neurotransmitter serotonin in the brain, which can lead to problems regulating mood, appetite, sleep, and memory (Abadinsky, 2014; Asl et al., 2015).

In addition, club drugs, like all illicit drugs, are particularly dangerous because there are no truth-in-packaging laws to protect buyers from unscrupulous practices. Sellers often substitute unknown cheaper, and possibly even more dangerous, substances for the ones they claim to be selling. Finally, just as "drinking and driving don't mix," club drug use may lead to impaired driving and decision making, as well as risky sexual behaviors and increased risk of sexually transmitted infections. Add in the fact that some drugs, such as Rohypnol, are odorless, colorless, tasteless, and can easily be added to beverages by individuals who want to intoxicate or sedate others, and you can see that the dangers of club drug use go far beyond the drug itself.

Q Answer the **Concept Check** questions.

5.4 UNDERSTANDING MEDITATION AND HYPNOSIS

As we have seen, sleep, dreaming, and psychoactive drugs all lead to altered states of consciousness. Changes in consciousness also occur with meditation and hypnosis.

Meditation

"Suddenly, with a roar like that of a waterfall, I felt a stream of liquid light entering my brain through the spinal cord . . . I experienced a rocking sensation and then felt myself slipping out of my body, entirely enveloped in a halo of light. I felt the point of consciousness that was myself growing wider, surrounded by waves of light" (Krishna, 1999, pp. 4–5).

This is how spiritual leader Gopi Krishna described his experience with **meditation**, a group of techniques designed to focus attention, block out distractions, and produce an altered state of awareness (**Study Organizer 5.2**). Most people in the beginning stages of meditation report a simpler, mellow type of relaxation, followed by a mild euphoria and a sense of timelessness. Some advanced meditators report experiences of profound rapture, joy, and/or strong hallucinations.

See **STUDY ORGANIZER 5.2, Benefits of Meditation**

How can we explain these effects? Brain imaging studies suggest that meditation's requirement to focus attention, block out distractions, and concentrate on a single object, emotion, or word, reduces the number of brain cells that must be devoted to the multiple, competing tasks normally going on within the brain's frontal lobes. This narrowed focus thus explains the feelings of timelessness and mild euphoria.

Research has also verified that meditation can produce dramatic changes in basic physiological processes, including heart rate, oxygen consumption, sweat gland responses, and brain activity. In addition, it's been somewhat successful in reducing pain, anxiety, and stress; lowering blood pressure; and improving cognitive functioning and overall mental health (Ding et al., 2014; Taylor & Abba, 2015; van Vugt, 2015). As you can see in **Study Organizer 5.2b and c**, studies have also found that meditation can change the body's sympathetic and parasympathetic responses, and increase structural support for the sensory, decision-making, emotion regulation, and attention-processing centers of the brain (Esch, 2014; Tang et al., 2014; Xue et al., 2014).

Hypnosis

Relax . . . your eyelids are so very heavy . . . your muscles are becoming more and more relaxed . . . your breathing is becoming deeper and deeper . . . relax . . . your eyes are closing . . . let go . . . relax.

Hypnotists use suggestions like these to begin **hypnosis**, a trance-like state of heightened suggestibility, deep relaxation, and intense focus. Once hypnotized, some people can be convinced that they are standing at the edge of the ocean, listening to the sound of the waves and feeling the ocean mist on their faces. Invited to eat a "delicious apple" that is actually an onion, the hypnotized person may relish the flavor. Told they are watching a very funny or sad movie, hypnotized people may begin to laugh or cry at their self-created visions.

From the 1700s to modern times, entertainers and quacks have used (and abused) hypnosis. Physicians, dentists, and therapists, however, have long employed it as a respected clinical tool. Modern scientific research has removed much of the mystery surrounding hypnosis. A number of features characterize the hypnotic state (Huber et al., 2014; Spiegel, 2015; Yapko, 2015):

- Narrowed, highly focused attention (ability to "tune out" competing sensory stimuli);
- Increased use of imagination and hallucinations;
- A passive and receptive attitude;
- Decreased responsiveness to pain;
- Heightened suggestibility, or a greater willingness to respond to proposed changes in perception ("This onion is an apple").

Today, even with available anesthetics, hypnosis is occasionally used in surgery and for the treatment of cancer, chronic pain, and severe burns (Adachi et al., 2014; Spiegel, 2015; Tan et al., 2015). Hypnosis has found its best use in medical areas such as dentistry and childbirth, where patients have a high degree of anxiety, fear, and misinformation. For example, some studies have found that women who use hypnosis in labor and childbirth experience lower levels of pain and a shorter duration of labor (Beebe, 2014; Madden et al., 2012). Because tension and anxiety strongly affect pain, any technique that helps the patient relax is medically useful.

In psychotherapy, hypnosis can help patients relax, recall painful memories, and reduce anxiety. Despite the many myths about hypnosis, it has been used with modest success in the treatment of phobias and in helping people to lose weight, stop smoking, and improve study habits (Hope & Sugarman, 2015; Iglesias & Iglesias, 2014).

One Final Note

Before going on to the next chapter, we'd like to take an unusual step for authors. We'd like to offer you, our reader, a piece of caring, personal and professional advice about alternate states of consciousness (ASCs). The core problem while you're in any ASC is that you're less aware of external reality, which places you at high risk. This applies to both men and women. Interestingly, we all recognize these dangers while sleeping and dreaming, and we've developed standard ways to protect ourselves. For example, when we're driving on a long trip and start to feel sleepy, we stop for coffee, walk around, and/or rent a hotel room to prevent ourselves from falling asleep behind the wheel.

Our simple advice to you is to follow this same "sleepy driver" logic and standards. If you decide to use drugs, meditate, undergo hypnosis, or engage in any other form of altered consciousness, research the effects and risks of your ASC, and plan ahead for the best options for dealing with it—just like you set up a designated driver before drinking.

Q Answer the **Concept Check** questions.

WP LS Go to your WileyPLUS Learning Space course for video episodes, examples, art, tables, Concept Checks, practice, and other pedagogical resources that will help you succeed in this course.

Reading for

LEARNING

WP LS Go to your WileyPLUS Learning Space course for video episodes, examples, art, tables, Concept Checks, practice, and other pedagogical resources that will help you succeed in this course.

6.1 CLASSICAL CONDITIONING

See **VIRTUAL FIELD TRIP, The Search Dog Foundation**

Although most people think of learning as something formal, like what occurs in the classroom, psychologists see the term as much broader. We define **learning** as *a relatively permanent change in behavior or mental processes caused by experience*. This relative permanence applies to bad habits, like texting while driving or procrastinating about studying, as well as to useful behaviors and emotions, such as training guide dogs for the blind or falling in love.

Learning also involves change and experience. From previous experiences, you may have learned that you "can't do well on essay exams," or that you "can't give up talking on the phone or texting while driving." The good news is that since learning is only "relatively" permanent, with new experiences, like practicing the study skills sprinkled throughout this text, and actively practicing turning off your phone in the car, mistaken beliefs and bad habits can be replaced with new, more adaptive ones.

We begin this chapter with a study of one of the earliest forms of learning, *classical conditioning*, made famous by Pavlov's salivating dogs.

Beginnings of Classical Conditioning

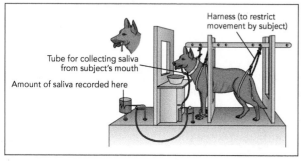
See **FIGURE 6.1, Pavlov's experimental setup**

Harness (to restrict movement by subject)

Tube for collecting saliva from subject's mouth

Amount of saliva recorded here

Why does your mouth water when you stare at a large slice of delicious cake or a juicy steak? The answer was accidentally discovered in the early 1900s by Russian physiologist Ivan Pavlov. His initial plan was to study the role of saliva in digestion through the use of a tube attached to dogs' salivary glands (■ **Figure 6.1**).

During these experiments, one of Pavlov's students noticed that even before receiving the actual food, many dogs began salivating at the mere sight of the food, the food dish, the smell of the food, or even just the sight of the person who usually delivered the food! Pavlov's genius was in recognizing the importance of this "unscheduled" salivation. He realized that the dogs were not only responding on the basis of hunger (a biological need), but also as a result of experience or learning.

Excited by this accidental discovery, Pavlov and his students conducted several follow up experiments, including sounding a tone on a tuning fork just before food was placed in the dogs' mouths. After several pairings of the tone and food, dogs in the laboratory began to salivate after just hearing the tone alone.

Pavlov and later researchers found that many things can become conditioned stimuli for salivation if they are paired with food—a bell, a buzzer, a light, and even the sight of a circle or triangle drawn on a card. This type of learning, called **classical conditioning**, develops through involuntary, passive, paired associations. More specifically, a neutral stimulus (such as the tone on a tuning fork) comes to elicit a response after repeated pairings with a naturally occurring stimulus (like food).

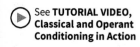
See **TUTORIAL VIDEO, Classical and Operant Conditioning in Action**

To fully understand classical conditioning, and how it applies to our everyday life, the first step is to recognize that **conditioning** is simply another word for *learning*. Next, we need to explain that classical conditioning is a three-step process—*before*, *during*, and *after conditioning*. This process is explained in detail below, and visually summarized in **Process Diagram 6.1**.

Process Diagram 6.1

The Three Basic Steps of Classical Conditioning

Although Pavlov's initial experiment used a metronome, a ticking instrument designed to mark exact time, his best-known method (depicted here) involved a tone from a tuning fork. As you can see in this diagram, the basic process of classical conditioning is simple. Just as you've been classically conditioned to respond to your cell phone's tones, or possibly to just the sight of a pizza box, Pavlov's dogs learned to respond to the tuning fork's tone. Unfortunately, many students get confused by these technical terms. So here's a tip that might help: The actual stimuli (tone and food) remain the same—only their names change from *neutral* to *conditioned* or from *unconditioned* to *conditioned*. A similar name change happens for the response (salivation)—from *unconditioned* to *conditioned*.

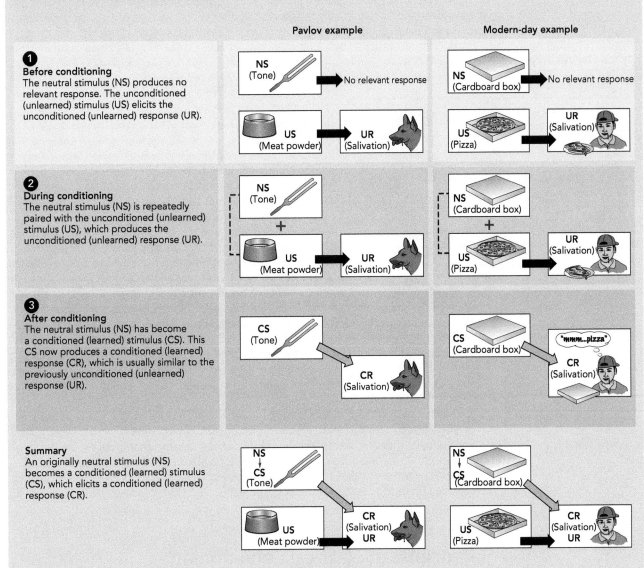

① Before conditioning
The neutral stimulus (NS) produces no relevant response. The unconditioned (unlearned) stimulus (US) elicits the unconditioned (unlearned) response (UR).

② During conditioning
The neutral stimulus (NS) is repeatedly paired with the unconditioned (unlearned) stimulus (US), which produces the unconditioned (unlearned) response (UR).

③ After conditioning
The neutral stimulus (NS) has become a conditioned (learned) stimulus (CS). This CS now produces a conditioned (learned) response (CR), which is usually similar to the previously unconditioned (unlearned) response (UR).

Summary
An originally neutral stimulus (NS) becomes a conditioned (learned) stimulus (CS), which elicits a conditioned (learned) response (CR).

STEP 1 Before conditioning, the sound of the tone does NOT lead to salivation, which makes the tone a **neutral stimulus (NS)**. Conversely, food naturally brings about salivation, which makes food an *unlearned*, **unconditioned stimulus (US)**. The initial reflex of salivation also is *unlearned*, so it is called an **unconditioned response (UR)**.

STEP 2 During conditioning, the tuning fork is repeatedly sounded right before the presentation of the food (US).

STEP 3 After conditioning, the tone alone will bring about salivation. At this point, we can say that the dog is *classically conditioned*. The previously neutral stimulus (NS) (the tone) has now become a *learned*, **conditioned stimulus (CS)**, that produced a *learned*, **conditioned response (CR)** (the dog's salivation). [Note that the "R" in (UR) in Step 1, and the (CR) in this Step 3, refers to both "reflex" and "response."]

In sum, the overall goal of Pavlov's classical conditioning was for the dog to learn to associate the tone with the unconditioned stimulus (food), and then to show the same response (salivation) to the tone as to the food.

So what does a salivating dog have to do with your everyday life? Classical conditioning is a fundamental way that all animals, including humans, learn (Chance, 2014; Martin & Pear, 2014). The human feelings of excitement and/or compulsion to gamble, love for your parents (or significant other), drooling at the sight of chocolate cake or pizza, and the almost universal fear of public speaking all largely result from classical conditioning.

How do we learn to be afraid of public speaking, or of typically harmless things like mice and elevators? In a now-famous experiment, John Watson and Rosalie Rayner (1920) demonstrated how a fear of rats could be classically conditioned.

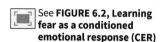
See **FIGURE 6.2, Learning fear as a conditioned emotional response (CER)**

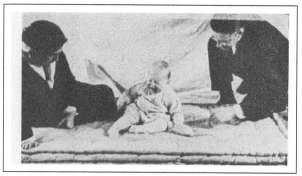
Benjamin Harris

In this study, a healthy 11-month-old child, later known as "Little Albert," was first allowed to play with a white laboratory rat (▣ **Figure 6.2**). Like most other infants, Albert was initially curious and reached for the rat, showing no fear. Knowing that infants are naturally frightened by loud noises, Watson stood behind Albert and when he reached for the rat, Watson banged a steel bar with a hammer. The loud noise obviously frightened the child, and made him cry. The rat was paired with the loud noise only seven times before Albert became classically conditioned, and demonstrated fear of the rat even without the noise. The rat had become a CS that brought about the CR (fear).

This deliberate experimental creation of what's now called a **conditioned emotional response (CER)** remains a classic in psychology. Note, however, that it has been heavily criticized, and would never be allowed under today's experimental standards (Bausell, 2015; Ethical Principles of Psychologists, 2010; Ollendick et al., 2012). The research procedures used by Watson and Rayner violated several ethical guidelines for scientific research (Chapter 1). They not only deliberately created a serious fear in a child, but they also ended their experiment without *extinguishing* (removing) it. In addition, the researchers have been criticized because they did not measure Albert's fear objectively. Their subjective evaluation raises doubt about the actual degree of fear conditioned.

Despite such criticisms, this study of Little Albert showed us that many of our likes, dislikes, prejudices, and fears are examples of *conditioned emotional responses (CERs)*. For example, if your romantic partner always uses the same shampoo, simply the smell of that particular shampoo may soon elicit a positive response. In Chapter 13, you'll discover how Watson's research later led to powerful clinical tools for eliminating exaggerated and irrational fears of a specific object or situation, known as *phobias* (Donovan et al., 2015; Kilic et al., 2014; Martin & Pear, 2014).

Principles of Classical Conditioning

Fear of dentists and dental work is very common. How did it develop? Imagine being seated in a dental chair, and, even though the drill is nowhere near you, its sound immediately makes you feel anxious. Your anxiety is obviously not innate. Little babies don't cringe at the sound of a dental drill, unless it's very loud. Your fear of the drill, and maybe dentistry in general, involves one or more of the six principles summarized in **Study Organizer 6.1**, and discussed in detail below. As you'll discover in Chapters 12 and 13, the good news is that even people with serious dental fears can learn to overcome them.

1. Acquisition After Pavlov's original (accidental) discovery of classical conditioning, he conducted numerous experiments beyond the basic **acquisition** phase, a general term for the initial learning (acquisition) of the stimulus-response (S-R) relationship.

2. Generalization One of Pavlov's most interesting findings was that stimuli similar to the original conditioned stimulus (CS) also can elicit the conditioned response (CR). For example, after first conditioning dogs to salivate to the sound of low-pitched tones, Pavlov later demonstrated that the dogs would also salivate in response to higher-pitched tones. Similarly, after Watson and Rayner's conditioning experiment, "Little Albert" learned to fear not only rats, but also a rabbit, dog, and a bearded Santa Claus mask. This spreading (or *generalizing*) of the CR to similar stimuli, even though they have never been paired with the US, is called stimulus **generalization** (Chance, 2014; McSweeney & Murphy, 2014).

See **FIGURE 6.3, Six key principles of classical conditioning**

3. Discrimination In contrast, just as Pavlov's dogs learned to generalize and respond to similar stimuli in a similar way, they also learned how to *discriminate* between similar stimuli. For example, when he gave the dogs food following a high-pitched tone, but not when he used a low-pitched tone, he found that they learned the difference between the two tones, and only salivated to the high-pitched one. Likewise, "Little Albert" learned to recognize differences between rats and other stimuli, and presumably overcame his fear of them. This learned ability to distinguish (discriminate), and NOT respond to a new stimulus, as if it were the previously conditioned stimulus (CS), is known as stimulus **discrimination**.

4. Extinction What do you think happened when Pavlov repeatedly sounded the tone without presenting food? The answer is that the dogs' salivation gradually declined, a process Pavlov called **extinction (in classical conditioning)**, which refers to the gradual diminishing of a CR when the US is withheld or removed. Without continued association with the US, the CS loses its power to elicit the CR.

5. Spontaneous recovery Once extinguished, is a conditioned response (CR) gone for good? It's important to note that extinction is not complete unlearning. It does not fully "erase" the learned connection between the stimulus and the response (Bouton & Todd, 2014; Willcocks & McNally, 2014). Pavlov found that sometimes, after a conditioned response had apparently been extinguished, if he sounded the tone once again, the dogs would salivate. This reappearance of a previously extinguished conditioned response (CR) is called **spontaneous recovery** (see the **Applying Psychology**).

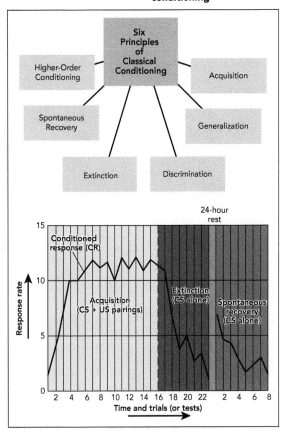

6. Higher-order conditioning The phenomenon of **higher-order conditioning** takes basic classical conditioning one step higher. Also known as "second order conditioning," this process refers to a situation in which a previously neutral stimulus (like a tone) was first made into a conditioned stimulus by pairing it with an unconditioned stimulus (like food). The next step, or second order conditioning, then uses that previously conditioned stimulus as a basis for creating a NEW conditioned stimulus (like a flashing light) that produces it's own conditioned response. In short, a new CS is created by pairing it with a previously created CS (**Process Diagram 6.2**).

See **STUDY ORGANIZER 6.1, Six Principles and Applications of Classical Conditional to a CER**

Q Answer the **Concept Check** questions.

Process Diagram 6.2

Higher-Order Conditioning and Everyday Life

Children are not born salivating to the sight of McDonald's golden arches. So why do they beg adults to take them to "Mickey D's" after simply seeing an ad showing the golden arches? It's because of *higher-order conditioning*, which occurs when a new conditioned stimulus (CS) is created by pairing it with a previously conditioned stimulus (CS).

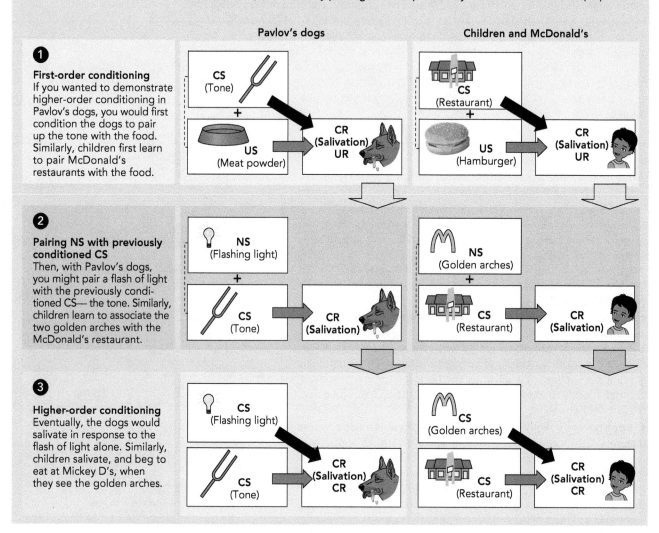

❶ First-order conditioning
If you wanted to demonstrate higher-order conditioning in Pavlov's dogs, you would first condition the dogs to pair up the tone with the food. Similarly, children first learn to pair McDonald's restaurants with the food.

❷ Pairing NS with previously conditioned CS
Then, with Pavlov's dogs, you might pair a flash of light with the previously conditioned CS— the tone. Similarly, children learn to associate the two golden arches with the McDonald's restaurant.

❸ Higher-order conditioning
Eventually, the dogs would salivate in response to the flash of light alone. Similarly, children salivate, and beg to eat at Mickey D's, when they see the golden arches.

(Pavlov's dogs)
- CS (Tone) + US (Meat powder) → CR (Salivation) UR
- NS (Flashing light) + CS (Tone) → CR (Salivation)
- CS (Flashing light) / CS (Tone) → CR (Salivation) CR

(Children and McDonald's)
- CS (Restaurant) + US (Hamburger) → CR (Salivation) UR
- NS (Golden arches) + CS (Restaurant) → CR (Salivation)
- CS (Golden arches) / CS (Restaurant) → CR (Salivation) CR

6.2 OPERANT CONDITIONING

As we've just seen, classical conditioning is based on what happens *before* we *involuntarily* respond: Something happens to us, and we learn a new response. In contrast, **operant conditioning** is based on what happens *after* we *voluntarily* perform a behavior (Chance, 2014; Olson & Hergenhahn, 2013). In other words, we do something, and learn from the consequences (▣ **Figure 6.4**).

Classical and operant conditioning are also known as **associative learning**. As the name implies, they both occur when an organism makes a connection, or *association*, between two events. During classical conditioning, an association is made between two stimuli, whereas in operant conditioning the association is made between a response and its consequences.

See **FIGURE 6.4, Classical versus operant conditioning**

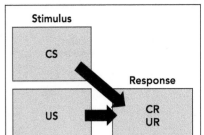

The key point to remember is that *consequences* are the heart of operant conditioning. In classical conditioning, consequences are irrelevant—Pavlov's dogs still got to eat whether they salivated or not. But in operant conditioning, the organism voluntarily performs a behavior (an *operant*) that produces a consequence—either reinforcement or punishment—and the behavior either increases or decreases. It's also very important to note that **reinforcement** is the adding or taking away of a stimulus following a response, which *increases* the likelihood of that response being repeated. **Punishment** is the adding or taking away of a stimulus following a response, which *decreases* the likelihood of that response being repeated. (We'll return to these terms in the next section.)

Beginnings of Operant Conditioning

In the early 1900s, Edward Thorndike, a pioneer of operant conditioning, was the first to identify that the frequency of a behavior is controlled by its consequences (Thorndike, 1911). Today this is known as Thorndike's **law of effect**, which further clarifies that any behavior followed

See **FIGURE 6.5, Thorndike's law of effect**

by pleasant consequences is likely to be repeated, whereas any behavior followed by unpleasant consequences is likely to be stopped. Thorndike's findings were based on his study of cats in puzzle boxes (**Figure 6.5**).

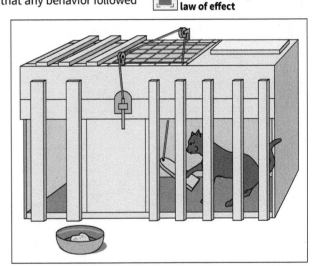

B. F. Skinner later extended Thorndike's law of effect to more complex behaviors. However, he carefully avoided Thorndike's use of terms like *pleasant* and *unpleasant* because they are subjective, and not directly observable. Furthermore, Skinner argued that such words make unfounded assumptions about what an organism feels or wants, and imply that behavior is due to conscious choice or intention. Skinner believed that to understand behavior, we should consider only external, observable stimuli and responses. We must look outside the learner, not inside.

Skinner also talked about reinforcement and punishment in terms of *increasing* or *decreasing* the likelihood of the response being repeated. If a toddler whines for candy, and the parent gives in, the child's whining will likely increase. But what if the parent yelled at the child for whining, yet still gave him or her the lollipop? The child might feel both happy to get the candy, and sad because the parent is upset. Because we can't know the full extent of the child's internal, mixed feelings, it's cleaner (and more scientific) to limit our focus to observable behaviors and consequences. If the child's whining for lollipops increases, we can say that whining was reinforced. If not, then it was punished.

In keeping with his focus on external, observable stimuli and responses, Skinner emphasized that reinforcement and punishment should always be presented *after* the behavior of interest has occurred. This was because Skinner believed that the only way to know how we have influenced an organism's behavior is to check whether it increases or decreases. As he pointed out, we too often think we're reinforcing or punishing behavior, when we're actually doing the opposite.

For example, imagine yourself as an elementary school teacher with a quiet student in your class. You want to encourage her to interact more in class, so you praise the student the few times she does speak up. But what if the student is so painfully shy that she's embarrassed by this extra attention? If so, you may actually decrease the number of times the child talks in your class. As Skinner said, we need to see whether the target's actual responses increase or decrease—NOT what we *think* the other person should like or will do. Can you also see why it's important to remember that what is reinforcing or punishing for one person may not be so for another?

Reinforcement versus Punishment

Reinforcers, which increase a response, can be a powerful tool in all parts of our lives. Psychologists group them into two types, primary and secondary. **Primary reinforcers** are any *unlearned*, innate stimuli that reinforce, and increase the probability of a response (like

See **TUTORIAL VIDEO, Understanding Reinforcement vs. Punishment** food, water, and sex). **Secondary reinforcers** are any *learned* stimuli that reinforce, and increase the probability of a response (like money, praise, and attention). The key point is that "primary" is another word for unlearned, whereas "secondary" means learned.

In addition, both primary and secondary reinforcers can produce **positive reinforcement** or **negative reinforcement**, depending on whether certain stimuli are added or taken away ( **Table 6.1**). We admit that this terminology is very confusing because positive normally means something "good" and negative generally means something "bad." But recall that Skinner cautioned us to avoid subjective terms like good and bad, or pleasant and unpleasant because they are not external and directly observable. Instead, he used *positive* and *negative* in line with other scientific terminology. You'll find this section much easier if you always remember that "positive" is simply adding something [+], and "negative" is taking something away [−].

See **TABLE 6.1**, How Reinforcement Increases (or Strengthens) Behavior

As with reinforcement, there are two kinds of punishers—primary and secondary. **Primary punishers** are any unlearned, innate stimuli that punish, and decrease the probability of a response (e.g., hunger, thirst, discomfort). **Secondary punishers** are any learned stimuli that punish, and decrease the probability of a response (e.g., poor grades, a traffic ticket, loss of an important promotion). Also, like in reinforcement, there are two kinds of punishment—positive and negative. **Positive punishment** is the addition (+) of a stimulus that decreases (or weakens) the likelihood of the response occurring again. **Negative punishment** is the taking away (−) of a stimulus, which decreases (or weakens) the likelihood of the response occurring again (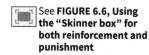 **Table 6.2**).

See **TABLE 6.2**, How Punishment Decreases (or Weakens) Behavior

At this point, it's also very important to emphasize that negative reinforcement is NOT punishment. In fact, the two concepts are actually the complete opposite of one another. Reinforcement (both positive and negative) *increases* a behavior, whereas punishment (both positive and negative) *decreases* a behavior. (To check your understanding of the principles of both reinforcement and punishment, see **Figure 6.6**.)

See **FIGURE 6.6**, Using the "Skinner box" for both reinforcement and punishment

Problems with Punishment

Keep in mind that punishment is a tricky concept that's difficult to use appropriately and effectively. We often think we're punishing, yet the behaviors continue. Similarly, we too often mistakenly think we're reinforcing when we're actually punishing. The key thing to remember is that punishment, by definition, is a process that adds or takes away something, which causes a behavior to decrease. If the behavior does not decrease, it's NOT punishment! So it's always important to watch the actual behavior, and not what we think should happen. Thus, if parents keep taking away their child's privileges, yet the child's misbehaviors increase, the parents may think they're punishing, while the child is enjoying the extra attention. Similarly, dog owners who call their dogs several times, and then yell at them for finally coming to them are actually punishing the desired behavior—coming when called.

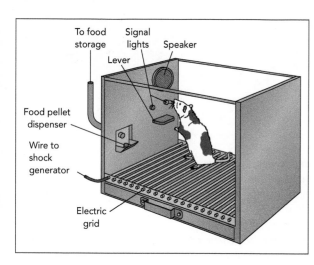

To food storage · Signal lights · Speaker · Lever · Food pellet dispenser · Wire to shock generator · Electric grid

In addition to these problems with punishment, to be effective it should always be *clear*, *direct*, *immediate*, and *consistent*. However, this is extremely hard to do. Police officers cannot stop every driver each and every time they speed. And parents can't scold a child each time he or she curses.

See **TABLE 6.3**, Potential Side Effects of Punishment

We do recognize that there are situations when punishment is necessary, such as when a child takes something that doesn't belong to him or her. However, even in limited circumstances like this, it can still have at least seven important drawbacks (see 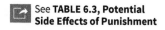 **Table 6.3**).

After considering all these potential problems with punishment, you may be feeling a bit overwhelmed and wondering what to do instead. The most important reminder is that punishment teaches us *what not to do*, whereas reinforcement teaches us *what to do*.

Principles of Operant Conditioning

Earlier, we discussed the six principles of classical conditioning. In this section, we explore six principles of operant conditioning: *generalization, discrimination, extinction, shaping, schedules of reinforcement*, and *superstition*. (Note that some of these principles are very similar to those in classical conditioning.)

1. Generalization   Like *generalization (in classical conditioning),* which occurs when the CR is elicited not only by the CS, but also by stimuli similar to the CS, **generalization (in operant conditioning)** refers to responding to a new stimulus, as if it is the original, previously conditioned stimulus (CS). A pigeon that's been trained to peck at a green light, might also peck at a red light. And, a young child who is rewarded for calling her father "Daddy," might generalize and call all men "Daddy." [Study tip: Note that in classical conditioning the CR is *involuntarily elicited*, whereas in operant conditioning the CR is a *voluntary response*.]

2. Discrimination   *Discrimination (in classical conditioning)* refers to the learned ability to distinguish (discriminate) between stimuli that differ from the CS. Likewise, **discrimination (in operant conditioning)** refers to the learned ability to distinguish (discriminate) between stimuli that increase or decrease a behavior, and then to respond accordingly. A pigeon might be punished after pecking at a green light, and not for pecking at a red light. As a result, it would quickly learn to only peck at red, and to stop pecking at green. Similarly, a child who is only reinforced for calling her father "Daddy," will quickly learn to stop calling all men "Daddy."

3. Extinction   Recall that *extinction (in classical conditioning)* involves a gradual diminishing of the conditioned response (CR) when the unconditioned stimulus (US) is withheld or removed. In contrast, **extinction (in operant conditioning)** refers to a gradual diminishing of a response when it is no longer reinforced. Skinner quickly taught pigeons to peck at a certain stimulus using food as a reward (Bouton & Todd, 2014; Cohen-Hatton & Honey, 2013). However, once the reinforcement stopped, the pigeons quickly stopped pecking. How does this apply to human behavior? If a local restaurant stops serving our favorite dishes, we'll soon stop going to that restaurant. Similarly, if we routinely ignore compliments or kisses from a long-term partner, he or she may soon stop giving them.

4. Shaping   It's easy to see how reinforcement is better than punishment for teaching and learning most things. But how would you teach someone complex behaviors, like how to play the piano or to speak a foreign language? How do seals in zoos and amusement parks learn how to balance beach balls on their noses, or how to clap their flippers together on command from the trainers?

For new and complex behaviors such as these, which aren't likely to occur naturally, **shaping** is the key. Skinner believed that shaping, or *rewarding successive approximations,*

See **FIGURE 6.7, Can shaping teach a monkey to water ski?**

explains a variety of abilities that each of us possesses, from eating with a fork, to playing a musical instrument. Parents, athletic coaches, teachers, therapists, and animal trainers all use shaping techniques (Armstrong et al., 2014; McDougall, 2013; Meredith et al., 2014).

For example, Momoko, a female monkey, is famous in Japan for her water skiing, deep sea diving, and other amazing abilities (■ **Figure 6.7**). Her animal trainers used the successive steps of shaping to teach her these skills. First, they reinforced Momoko (with a small food treat) for standing or sitting on the water ski. Then they reinforced her each time she accomplished a successive step in learning to water ski.

Kaku Kurita/Getty Images, Inc.

5. Schedules of Reinforcement   Now that we've discussed how we learn complex behaviors through shaping, you may want to know how to maintain it. When Skinner was training his animals, he found that learning was most rapid if the correct response was reinforced every time it occurred—a pattern called **continuous reinforcement**.

Although most effective during the initial training/learning phase, continuous reinforcement unfortunately also leads to rapid *extinction*—the gradual diminishing of a response when it is no longer reinforced. Furthermore, in the real world, continuous reinforcement is generally not practical or economical. When teaching our children, we can't say, "Good job! You brushed your teeth!" every morning for the rest of their lives. As an employer, we can't give a bonus for every task our employees accomplish. For pigeons in the wild, and people in the real world, behaviors are almost always reinforced only occasionally and unpredictably—a pattern called **partial (or intermittent) reinforcement**.

Given the impracticality, and near impossibility, of continuous reinforcement, let's focus on the good news. Partially reinforced behaviors are far more resistant to extinction. Skinner found that pigeons that were reinforced on a continuous schedule would continue pecking approximately 100 times after food was removed completely—indicating extinction. In contrast, pigeons reinforced on a partial schedule continued to peck thousands of times (Skinner, 1956). Moving from pigeons to people, have you noticed that small children often whine and/or throw temper tantrums for that tempting treat in the grocery store even though their parents only occasionally give in? Can you see how the whining or tantrums have been intermittently reinforced AND why the child persists? If you'd like an adult example, consider the problem of persistent gambling (■ **Figure 6.8**).

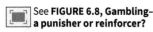
See **FIGURE 6.8, Gambling—a punisher or reinforcer?**

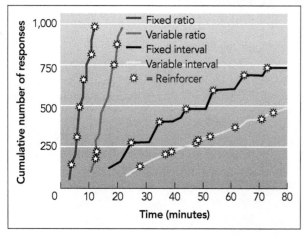

Air Images/Shutterstock

How can you personally benefit from this information? Try using continuous reinforcement for initial mastery of a task. Once the task is well learned, you'll want to move on to a partial schedule—thanks to its strong resistance to extinction. For example, if you want to improve your study habits, start by reinforcing yourself immediately and consistently for every desired response (e.g., letting yourself get a snack, or check your text messages, after you've studied for 20 minutes). Later, when you feel that you're at your desired level of studying behaviors, move yourself to a partial schedule (e.g., allow yourself a larger treat intermittently). Note that you should only use this approach for *desired* behaviors. Allowing yourself to watch TV or go out with friends, rather than studying, puts your undesirable, procrastination behavior on a partial schedule of reinforcement, and makes procrastinating difficult to extinguish.

See **FIGURE 6.9, Which schedule is best?**

When using partial reinforcement, it's also important to note that some partial **schedules of reinforcement** are better suited for maintaining or changing behavior than others (Craig et al., 2014; Jessel & Borrero, 2014; Thraikill & Bouton, 2015). There are four schedules—**fixed ratio (FR)**, **variable ratio (VR)**, **fixed interval (FI)**, and **variable interval (VI)**. **Table 6.4** defines these terms, compares their respective response rates, and provides examples.

How do we know which schedule to choose? The type of partial schedule selected depends on the type of behavior being studied, and on the speed of learning desired (Dack et al., 2010; Lubar, 2015). For example, suppose you want to teach your dog to sit. First, you could reinforce your dog with a cookie every time he sits (continuous reinforcement). To make his training more resistant to extinction, you then could switch to a partial reinforcement schedule. Using the *fixed ratio* schedule, you would offer a cookie only after your dog sits a certain number of times. As you can see in ■ **Figure 6.9**, a fixed ratio leads to the highest overall response rate. But each of the four types of partial schedules has different advantages and disadvantages (see again ■ **Table 6.4**).

See **TABLE 6.4, Four Schedules of Partial (Intermittent) Reinforcement**

6. Superstition In addition to shaping and schedules of reinforcement, conditioning has other important applications to our everyday life. For example, when B. F. Skinner (1948, 1992) was conducting his experiments on the various schedules of reinforcement, he discovered something surprising—*accidental reinforcement could lead to superstitious behaviors!* He set the feeding mechanism in the cages of eight pigeons to release food once every 15 seconds. No matter what the birds did, they were reinforced at 15-second intervals. Interestingly, six of the pigeons acquired behaviors that they repeated over and over, even though the behaviors were not necessary to receive the food. For example, one pigeon kept turning in counterclockwise circles, and another kept making jerking movements with its head.

Why did the pigeons engage in such repetitive and unnecessary behavior? Recall that a *reinforcer* increases the probability that a response just performed will be repeated. Skinner was not using the food to reinforce any particular behavior. However, the pigeons associated the food with whatever behavior they were engaged in when the food was randomly dropped into the cage. Thus, if the bird was circling counterclockwise when the food was presented, it would repeat that motion to receive more food.

See **FIGURE 6.10,**
Superstition

David E. Klutho/Sports Illustrated/Getty Images

Like Skinner's pigeons, we humans also believe in many superstitions that may have developed from accidental or coincidental reinforcement (■ **Figure 6.10**). As a critical thinker, can you see how superstitions can also be explained by *illusory correlations* and the *confirmation bias*, first discussed in Chapter 1? Imagine you're a student who believes that sitting in a certain position in the classroom causes you to get better exam scores. This superstition may have begun when you happened to be sitting in a certain spot when you first got an unexpectedly high exam score. You then undoubtedly took note of this random coincidence (illusory correlation), and decided to always choose a similar seat in the future. Your superstition is further maintained by the fact that you later focused on all the times you gained higher scores by sitting in that particular spot, and ignored the disconfirming evidence from times when it did not lead to higher scores (the confirmatory bias).

Comparing Operant and Classical Conditioning

Are you feeling overwhelmed with all the important (and seemingly overlapping) terms and concepts for both operant and classical conditioning? This is a good time to stop and carefully review ■ **Table 6.5**, which summarizes all the key terms and compares these two major types of conditioning. Keep in mind that although it's convenient to divide classical and operant conditioning into separate categories, almost all behaviors result from a combination of both forms of conditioning (Watson & Tharp, 2007).

See **TABLE 6.5, Comparing**
Classical and Operant
Conditioning

Answer the
Concept Check questions.

6.3 COGNITIVE-SOCIAL LEARNING

So far, we have examined learning processes that involve associations between a stimulus and an observable behavior. However, many psychologists feel that all of learning can not be explained solely by operant and classical conditioning. **Cognitive-social learning theory** (also called observational learning or cognitive-behavioral theory) emphasizes the roles of thinking and social learning. It also incorporates the general concepts of conditioning, but rather than relying on a simple S–R (stimulus and response) model, this theory emphasizes the interpretation or thinking that occurs within the organism: S–O–R (stimulus–organism–response).

According to this view, humans have attitudes, beliefs, expectations, motivations, and emotions that affect learning. Furthermore, humans and many nonhuman animals are social creatures that are capable of learning new behaviors through the observation and imitation of others. For example, children of parents with anxiety disorders tend to observe and model this behavior, and become more anxious themselves (Aktar et al., 2014). On a more positive note, children with at least one physically active parent are much more likely to be physically active themselves (Voss & Sandercock, 2013).

Insight and Latent Learning

As you recall, early behaviorists likened the mind to a "black box" whose workings could not be observed directly. In the early 1900s, German psychologist Wolfgang Köhler wanted to look inside the box. He believed that there was more to learning—especially learning to solve a complex problem—than responding to stimuli in a trial-and-error fashion.

In one of a series of experiments, Köhler placed a banana just outside the reach of a caged chimpanzee. To reach the banana, the chimp had to use a stick placed near the cage

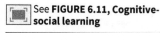
See **FIGURE 6.11, Cognitive-social learning**

to extend its reach. Köhler noticed that the chimp did not solve the problem in a random trial-and-error fashion. Instead, he seemed to sit and think about the situation for a while. Then, in a flash of **insight**, the chimp picked up the stick, and maneuvered the banana to within its grasp (Köhler, 1925). Köhler called this *insight learning* because some internal mental event, which he could only describe as "insight," went on between the presentation of the banana and the use of the stick to retrieve it. (See ▣ **Figure 6.11** for another example of how Köhler's chimps solved a similar, "out-of-reach banana" problem.)

Like Köhler, Edward C. Tolman believed that previous researchers underestimated human and nonhuman animals' cognitive processes and cognitive learning. He noted that, when allowed to roam aimlessly in an experimental maze, with no food reward at the end, rats seemed to develop a **cognitive map**, or mental image of the maze.

To experimentally test his hypothesis of cognitive maps, Tolman allowed one group of rats to aimlessly explore a maze, with no reinforcement. A second group was reinforced with food whenever they reached the end of the maze. The third group was not rewarded during the first 10 days of the trial, but starting on day 11, they found food at the end of the maze.

As expected from simple operant conditioning, the first and third groups were slow to learn the maze, whereas the second group, which had reinforcement, showed fast, steady improvement. However, when the third group started receiving reinforcement (on the 11th day), their learning quickly caught up to the group that had been reinforced every time (Tolman & Honzik, 1930). This showed that the nonreinforced rats had been thinking, and building cognitive maps of the area, during their aimless wandering. Their hidden, **latent learning** only showed up when there was a reason to display it (the food reward).

Cognitive maps and latent learning are not limited to rats. For example, a chipmunk will pay little attention to a new log in its territory (after initially checking it for food). When a predator comes along, however, the chipmunk will head directly for the log, and hide beneath it. Recent experiments provide additional clear evidence of latent learning, and the existence of internal, cognitive maps, in both human and nonhuman animals (Han & Becker, 2014; Nadel, 2013; Redish & Ekstrom, 2013; Wang et al., 2014). Do you remember your first visit to your college campus? You probably just wandered around checking out the various buildings, without realizing you were "latent learning" and building your own "cognitive maps." This exploration undoubtedly came in handy when you later needed to find your classes and the cafeteria!

Observational Learning

In addition to classical and operant conditioning and cognitive processes (such as insight and latent learning), we also learn many things through **observational learning**, which is also called imitation or modeling. From birth to death, observational learning is very important to our biological, psychological, and social survival (the *biopsychosocial model*). Watching others helps us avoid dangerous stimuli in our environment, teaches us how to think and feel, and shows us how to act and interact socially (Hora & Klassen, 2012; Martin & Pear, 2014; Williamson & Brand, 2014).

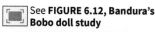
See **FIGURE 6.12, Bandura's Bobo doll study**

Courtesy Albert Bandura

Much of our knowledge about the power of observational learning initially came from the work of Albert Bandura and his colleagues (Bandura et al., 1961; Bandura & Walters, 1963). Wanting to know whether children learn to be aggressive by imitation and modeling, Bandura and his colleagues set up several experiments. They allowed children in the experimental group to watch a live or televised adult model punch, throw, and hit a large inflated Bobo doll (▣ **Figure 6.12 left**). Children in the control group watched as the model played with Tinkertoys and ignored the Bobo doll.

Later, the children were allowed to play in the same room with the same Bobo doll. As Bandura hypothesized, children who had seen the live or televised aggressive model were much more aggressive with the Bobo doll than children who had not seen the modeled aggression (▣ **Figure 6.12 right**). In other words, "Monkey see, monkey do."

See **STUDY ORGANIZER 6.2, Bandura and Observational Learning**

Thanks to the Bobo doll studies and his other experiments, Bandura established that observational learning requires at least four separate processes: *attention, retention, reproduction*, and *motivation*.

Cognitive-Social Learning and Everyday Life

We use cognitive-social learning in many ways in our everyday lives, yet one of the most powerful examples is frequently overlooked—*media influences*. Experimental and correlational research clearly show that when we watch television, go to movies, and read books, magazines, or websites that portray people of color, women, and other groups in demeaning and stereotypical roles, we often learn to expect these behaviors and to accept them as "natural." Exposure of this kind initiates and reinforces the learning of prejudice (Dill & Thill, 2007; Jackson, 2011; Scharrer & Ramasubramanian, 2015).

In addition to prejudice and stereotypes, watching popular media also teaches us what to eat, what toys to buy, what homes and clothes are most fashionable, and what constitutes "the good life." When a TV commercial shows children enjoying a particular cereal, and beaming at their mom in gratitude (and mom is smiling back), both children and parents in the audience are participating in a form of observational learning. They learn that they, too, will be rewarded for buying the advertised brand (with happy children). Unfortunately, the popular media also may negatively affect our ideal body image, starting at a very young age.

Q Answer the **Concept Check** questions.

See **FIGURE 6.13, How our brains respond to reinforcement versus punishment**

6.4 BIOLOGY OF LEARNING

Now that we've discussed how we learn through classical conditioning, operant conditioning, and cognitive-social learning, we need to explore the key biological factors in all forms of learning. We know that for changes in behavior to persist over time, lasting biological changes must occur within the organism. In this section, we will examine neurological and evolutionary influences on learning.

Neuroscience and Learning

Each time we learn something, either consciously or unconsciously, that experience creates new synaptic connections and alterations in a wide network of our brain structures, including the cortex, cerebellum, hippocampus, hypothalamus, thalamus, and amygdala. Interestingly, it appears that somewhat different areas of our brains respond to reinforcement and punishment (Jean-Richard-Dit-Bressel & McNally, 2015; Matsumoto & Hikosaka, 2009; Ollmann et al., 2015). (See ▣ **Figure 6.13**).

Evidence that learning changes brain structure first emerged in the 1960s, from studies of animals raised in enriched versus deprived environments (▣ **Figure 6.14**). Compared with rats raised in a stimulus-poor environment, those raised in a colorful, stimulating "rat Disneyland" had a thicker cortex, increased nerve growth factor (NGF), more fully developed synapses, more dendritic branching, and improved performance on many tests of learning and memory (Cheng et al., 2014; Lima et al., 2014; Speisman et al., 2013).

Admittedly, it is a big leap from rats to humans, but research suggests that the human brain also responds to environmental conditions. For example, both children and adults exposed to stimulating environments generally perform better on intellectual and perceptual tasks than those who are in restricted environments (Petrosini et al., 2013; Schaeffer et al., 2014; Schaie, 1994, 2008).

Unfortunately, babies who spend their early weeks and months of life in an orphanage, and receive little or no one-on-one care or attention, show deficits in the cortex of the brain, indicating that early environmental conditions may have a lasting impact on brain and cognitive development (Moutsiana et al., 2015; Nelson et al., 2014; Schoenmaker et al., 2014). However, research also shows that children who are initially placed in an orphanage, but are later adopted or placed into foster care—where they receive more individual attention—show some improvements in brain development.

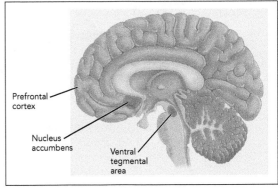

Prefrontal cortex
Nucleus accumbens
Ventral tegmental area

See **FIGURE 6.14, Environmental enrichment and the brain**

PhotoDisc Green/Getty Images

Top-Pet-Pics/Alamy

Mirror Neurons and Imitation

See **FIGURE 6.15, Infant imitation–evidence of mirror neurons?**

Interesting research has identified another neurological influence on learning—a previously unknown type of neuron responsible for imitation and observational learning. For example, when an adult models a facial expression, even newborn infants will respond with a similar expression (**Figure 6.15**). How can newborn infants so easily imitate the facial expressions of others? Using fMRIs and other brain-imaging techniques, researchers have identified specific **mirror neurons**, which are a type of neuron that fires (or activates) when an action is performed, as well as when observing the actions or emotions of others. They're believed to be responsible for human empathy and imitation (Ahlsén, 2008; Braadbaart et al., 2014; Brooks & Meltzoff, 2014; Haker et al., 2013). These neurons are found in several key areas of the brain, and are believed to be responsible for how we identify with what others are feeling, and then imitate their actions. When we see another person in pain, one reason we empathize and "share their pain" may be that our mirror neurons are firing. Similarly, if we watch others who are smiling, our mirror neurons make it harder for us to frown.

Mirror neurons were first discovered by neuroscientists who implanted wires in the brains of monkeys to monitor areas involved in planning and carrying out movement (Ferrari et al., 2005; Rizzolatti, 2014; Rizzolatti et al., 1996, 2008). When these monkeys moved and grasped an object, specific neurons fired. But they also fired when the monkeys simply observed another monkey performing the same or similar tasks.

Have you noticed how spectators at an athletic event sometimes slightly move their arms or legs in synchrony with the athletes? Mirror neurons may be the underlying biological mechanism for this imitation. Deficiencies in these neurons also might help explain the emotional deficits of children and adults with autism or schizophrenia, who often misunderstand the verbal and nonverbal cues of others (Enticott et al., 2012; Guo et al., 2014; Lee et al., 2014; Rizzolatti, 2012).

Although scientists are excited about the promising links between mirror neurons and human and nonhuman animal thoughts, feelings and actions, we do not yet know the full extent of their influence, or how they develop (Caramazza et al., 2014; Hickok, 2014; Thomas, 2012). However, we do know that thanks to our mirror neurons we're born prepared to imitate, and imitation is essential to survival in our complex, highly developed social world.

Evolution and Learning

Because animals can be classically conditioned to salivate to tones and operantly conditioned to perform a variety of novel behaviors, such as a seal balancing a ball on its nose, learning theorists initially believed that the fundamental laws of conditioning would apply equally to almost all species and all behaviors. However, researchers have discovered that each species' biological predispositions enable it to more quickly learn associations that enhance its survival.

See **FIGURE 6.16, Taste aversion**

For example, when a food or drink is associated with nausea or vomiting, that particular food or drink more readily becomes a conditioned stimulus (CS) that triggers a conditioned **taste aversion**, a dislike or avoidance of certain foods or drinks. Like other classically conditioned responses, taste aversions develop involuntarily, and undoubtedly provide an important evolutionary advantage (**Figure 6.16**). Being biologically prepared to quickly associate nausea with certain foods or drinks is obviously adaptive because it helps us avoid that specific food or drink, and similar ones, in the future (Buss, 2015; Goldfinch, 2015; Swami, 2011).

One of the earliest research teams in this area, John Garcia and his colleague Robert Koelling (1966), produced a taste aversion in lab rats by pairing sweetened water (NS) and a shock (US), which in turn produced nausea (UR). After being conditioned, and then recovering from the illness, the rats refused to drink the sweetened water (CS) because of the conditioned taste aversion. Remarkably, however, Garcia and Koelling also discovered that only certain neutral stimuli could produce the nausea. Pairings of a noise (NS) with the shock (US) produced no taste aversion.

Similarly, perhaps because of the more "primitive" evolutionary threat posed by snakes, darkness, spiders, and heights, people tend to more easily develop phobias of these stimuli, compared to guns, knives, or electrical outlets. Furthermore, both adults and very young children appear to have an innate ability to very quickly identify the presence of a snake. However, they're less able to quickly identify other (non-life-threatening) objects, including a caterpillar, flower, or toad (LoBue & DeLoache, 2008; Young et al., 2012). We apparently inherit a built-in (innate) readiness to form associations between certain stimuli and responses—but not others. This is known as **biological preparedness**.

Just as Garcia and Koelling could only produce some classically conditioned noise–nausea associations, other researchers have found that an animal's natural behavior patterns can interfere with operant conditioning. For example, early researchers tried to teach a chicken to play baseball (Breland & Breland, 1961). Through shaping and reinforcement, the chicken first learned to pull a loop that activated a swinging bat, and then learned to time its response to actually hit the ball. But instead of running to first base, it would chase the ball as if it were food. Regardless of the lack of reinforcement for chasing the ball, the chicken's natural, predatory behavior took precedence. This tendency for a conditioned behavior to revert (drift back) to innate response patterns is known as **instinctive drift**.

In this chapter, we've discussed three general types of learning, classical, operant, and cognitive-social. We've also examined the biological effects on learning. What are the most important "take home messages" from this chapter? First, as humans, we have the ability to learn and change! Using what you've discovered in this chapter, we hope you'll remember to avoid punishment whenever possible, and "simply" reinforce desired behaviors. This basic principle can also be applied on a national and global scale. If we all work together to remove inappropriate reinforcers, we can truly change the world. Admittedly, this sounds grandiose and simplistic. But we sincerely believe in the power of the material in this chapter. Your life, and the world around you, can be significantly improved with perseverance and a thoughtful application of learning principles.

It's not that I'm so smart, it's just that I stay with problems longer.

—ALBERT EINSTEIN

Q Answer the **Concept Check** questions.

WP LS Go to your WileyPLUS Learning Space course for video episodes, examples, art, tables, Concept Checks, practice, and other pedagogical resources that will help you succeed in this course.

Reading for

MEMORY

7.1 THE NATURE OF MEMORY

See **VIRTUAL FIELD TRIP:**
USA Memory Championships

Memory is learning that persists over time. It allows us to learn from our experiences and to adapt to ever-changing environments. Without it, we would have no past or future. Yet our memories are also highly fallible. Although some people think of memory as a gigantic library, or an automatic video recorder, our memories are not exact recordings of events. Instead, memory is highly *selective* (Baddeley et al., 2015; Chen & Wyble, 2015). As discussed in Chapter 5, we only pay attention to, and remember, a small fraction of the information we're exposed to each day. Memory is also a **constructive process** through which we actively organize and shape information as it is being encoded, stored, and retrieved (Herriot, 2014; Karanian & Slotnick, 2015). As expected, this construction often leads to serious errors and biases, which we'll discuss throughout the chapter. Would you like personal proof of the constructive nature of your memory?

See **TUTORIAL VIDEO:**
Constructing Memories

Memory Models

To understand memory (and its constructive nature), you need a model of how it operates. Over the years, psychologists have developed numerous models for memory, and we'll focus on the two most common ones.

Encoding, Storage, and Retrieval (ESR) Model

According to the **encoding, storage,** and **retrieval (ESR) model**, the barrage of information that we encounter every day goes through three basic operations: *encoding, storage*, and *retrieval*. Each of these processes represents a different function that is closely analogous to the parts and functions of a computer (**Process Diagram 7.1**).

To input data into a computer, you begin by typing letters and numbers on the keyboard. The computer then translates these keystrokes into its own electronic language. In a roughly similar fashion, your brain **encodes** sensory information (sound, visual images, and other senses) into a neural code (language) it can understand and use. Like the computer hard drive, your brain then stores the information for later retrieval.

This Process Diagram contains essential information NOT found elsewhere in the text, which is likely to appear on quizzes and exams. Be sure to study it CAREFULLY!

Process Diagram 7.1

Encoding, Storage, and Retrieval

The encoding, storage, and retrieval (ESR) model of memory is often compared to a computer's information processing system.

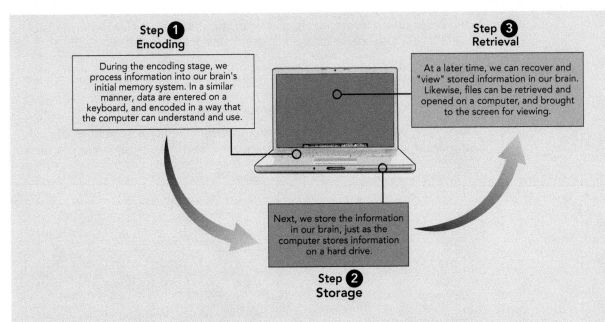

Step 1 **Encoding**

During the encoding stage, we process information into our brain's initial memory system. In a similar manner, data are entered on a keyboard, and encoded in a way that the computer can understand and use.

Step 3 **Retrieval**

At a later time, we can recover and "view" stored information in our brain. Likewise, files can be retrieved and opened on a computer, and brought to the screen for viewing.

Next, we store the information in our brain, just as the computer stores information on a hard drive.

Step 2 **Storage**

Test Your Critical Thinking

1. Human memory is often compared to the workings of a computer. Based on your own experiences, what might be the advantages and disadvantages of this comparison?

2. Imagine a pill that could turn your brain into a perfect computer with absolute recall and unlimited storage of all information. Would you take the pill? Why or why not?

To successfully encode, the first step is to pay attention. And, using *selective attention* (Chapters 4 and 5), we can deliberately focus our attention on information we want to remember. The *depth of processing* (or the effort made to understand the material) is the next important step to better encoding (Craik & Lockhart, 1972; Craik & Tulving, 1975). Given that deeper processing produces the greatest encoding and overall memory, try thinking hard about the material, relate it to other previously learned material, put the information in your own words, and/or talk about it with others. The term "**levels of processing**" refers to a continuum ranging from shallow to intermediate to deep, with deeper processing leading to improved encoding, storage, and retrieval.

Once information is *encoded*, it must be **stored**. Computer information is normally stored on a flash drive or hard drive, whereas human information is stored in the brain.

Finally, information must be **retrieved**, or taken out of storage. We retrieve stored information by going to files on our computer, or to "files" in our brain. Keep this model in mind. To do well in college, or almost any other pursuit, you must successfully encode, store and retrieve a large amount of facts and concepts. Throughout this chapter, we'll discuss ways to improve your memory during each of these processes.

Three-Stage Memory Model

Perhaps the most popular and long-lasting memory model is the one derived from the *Atkinson-Shiffrin theory*, known as the **three-stage memory model** (Atkinson & Shiffrin, 1968; Eichenbaum, 2013; Li, 2016). Like the previous ESR model, it also has been compared to a computer, with an input, process, and output. However, according to this model, three different storage "boxes," or memory stages, hold and process information. Each stage has a different purpose, duration, and capacity (**Process Diagram 7.2**). Let's discuss each stage in more detail.

Process Diagram 7.2

The Traditional Three-Stage Memory Model

Each "box" represents a separate memory system that differs in purpose, duration, and capacity. When information is not transferred from sensory memory or short-term memory, it is assumed to be lost. Information stored in long-term memory can be retrieved, and sent back to short-term memory for use.

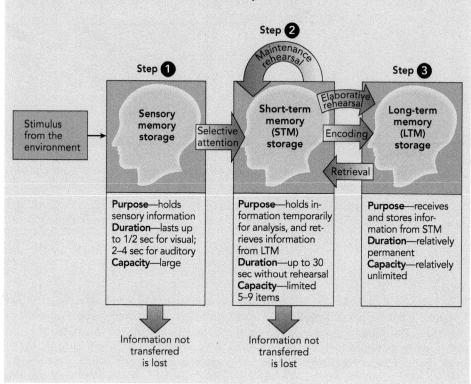

Step **1**

Stimulus from the environment

Sensory memory storage

Selective attention

Step **2**

Maintenance rehearsal

Short-term memory (STM) storage

Elaborative rehearsal

Encoding

Retrieval

Step **3**

Long-term memory (LTM) storage

Purpose—holds sensory information
Duration—lasts up to 1/2 sec for visual; 2–4 sec for auditory
Capacity—large

Purpose—holds information temporarily for analysis, and retrieves information from LTM
Duration—up to 30 sec without rehearsal
Capacity—limited 5–9 items

Purpose—receives and stores information from STM
Duration—relatively permanent
Capacity—relatively unlimited

Information not transferred is lost

Information not transferred is lost

Sensory Memory

Everything we see, hear, touch, taste, and smell must first enter our **sensory memory**. Once it's entered, the information remains in sensory memory just long enough for our brain to locate relevant bits of data, and transfer them to the next stage of memory. For visual information, known as *iconic memory*, the visual image (icon) stays in sensory memory only about one-half a second before it rapidly fades away.

In an early study of iconic sensory memory, George Sperling (1960) flashed an arrangement of 12 letters like the ones in 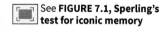 **Figure 7.1** for 1/20 of a second. Most people, he found, could recall only 4 or 5 of the letters. But when instructed to report just the top, middle, or bottom row, depending on whether they heard a high, medium, or low tone, they reported almost all the letters correctly. Apparently, all 12 letters are held in sensory memory right after they are viewed, but only those that are immediately attended to are processed.

Like the fleeting visual images in iconic memory, auditory stimuli (what we hear) is also temporary, but a weaker "echo," or *echoic memory*, of this auditory input lingers for up to four seconds (Erviti et al., 2015; Kojima et al., 2014; Neisser, 1967). Why are iconic and auditory memories so fleeting? We cannot process all

See **FIGURE 7.1, Sperling's test for iconic memory**

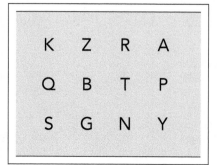

K Z R A

Q B T P

S G N Y

incoming stimuli, so lower brain centers take only a few seconds to "decide" if the information is important enough to promote to conscious awareness.

Early researchers believed that sensory memory had an unlimited capacity. However, later research suggests that sensory memory does have limits and that stored images are fuzzier than once thought (Cohen, 2014; Franconeri et al., 2013; Howes & O'Shea, 2014).

Short-Term Memory (STM)

The second stage of memory processing, **short-term memory (STM)**, temporarily stores and processes sensory stimuli. If the information is important, STM organizes and sends this information along to long-term memory (LTM). Otherwise, information decays and is lost. STM also retrieves stored memories from LTM.

The capacity and duration of STM are limited to 5 to 9 bits of information, and less than 30 seconds. To extend the *capacity* of STM, you can use a technique called **chunking**, which involves grouping separate pieces of information into larger, more manageable units (Gilbert et al., 2015; Miller, 1956). Have you noticed that your credit card, Social Security, and telephone numbers are almost always grouped into three or four units separated by hyphens? The reason is that it's easier to remember numbers in chunks rather than as a string of single digits.

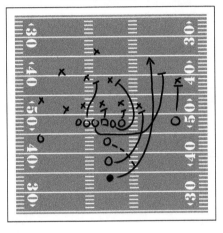

See **FIGURE 7.2, Chunking in football**

Chunking also helps athletes. What do you see when you observe this arrangement of players from a page of a football playbook (**Figure 7.2**)? To the novice eye, it looks like a random assembly of lines and arrows, and a naive person generally would have little or no understanding or appreciation of the skills required in football. But experienced players and seasoned fans generally recognize many or all of the standard plays. To them, the scattered lines form meaningful patterns—classic arrangements that recur often. Just as you group the letters of this sentence into meaningful words and remember them long enough to understand the meaning of the sentence, expert football players group the different football plays into easily recalled patterns (or chunks).

You can also extend the *duration* of your STM almost indefinitely by consciously "juggling" the information—a process called **maintenance rehearsal**. You are using maintenance rehearsal when you look up a phone number and repeat it over and over until you key in the number.

People who are good at remembering names also know how to take advantage of maintenance rehearsal. They repeat the name of each person they meet, aloud or silently, to keep it active in STM until they can move it on to LTM. They also make sure that other thoughts (such as their plans for what to say next) don't intrude.

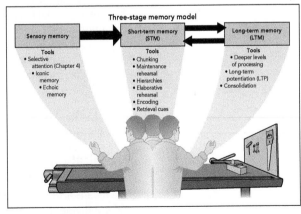

See **FIGURE 7.3, How working memory might work**

Working Memory

As you can see in **Figure 7.3**, short-term memory is more than just a passive, temporary "holding area." It's an active "workbench" where our brains are constantly processing information. Therefore, many researchers now prefer the term **working memory** (Baddeley, 1992, 2007; Baddeley et al., 2015; Barak & Tsodyks, 2014). All our conscious thinking occurs in this "working memory," and the manipulation of information that occurs here helps explain some individual differences in memory skills, as well as certain memory errors and false constructions described in this chapter. For example, researchers have found that people who play action video games show higher levels of visual, working memory capacity than those who play a control (non-action) video game (Blacker et al., 2014). This suggests that certain types of video games may provide mental training exercises that boost particular types of memory.

Long-Term Memory (LTM)

Once information has been transferred from STM, it is organized and integrated with other information in **long-term memory (LTM)**. LTM serves as a storehouse for information that

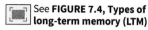
See **FIGURE 7.4, Types of long-term memory (LTM)**

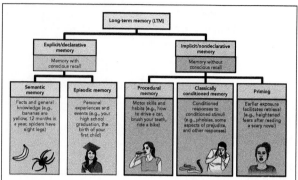

See **TUTORIAL VIDEO: Organizing Long-Term Memories**

must be kept for long periods. When we need the information, it is sent back to STM for our conscious use. Compared with sensory memory and STM, LTM has relatively unlimited *capacity* and *duration* (Eichenbaum, 2013). But, just as with any other possession, the better we label and arrange our memories, the more readily we'll be able to retrieve them.

How do we store the vast amount of information we collect over a lifetime? Several types of LTM exist (**Figure 7.4**).

Explicit/declarative memory refers to intentional learning or conscious knowledge. If asked to remember your phone number or your mother's name, you can easily state (*declare*) the answer directly (*explicitly*).

Explicit/declarative memory can be further subdivided into two parts. **Semantic memory** is memory for general knowledge, rules, events, facts, and specific information. It is our mental encyclopedia. In contrast, **episodic memory** is like a mental diary. It records the major events (*episodes*) in our lives. Some of our episodic memories are short-lived, whereas others can last a lifetime.

Have you ever wondered why you, like most adults, can recall almost nothing of the years before you reached age 3? Research suggests that a concept of self, sufficient language development, and growth of the frontal lobes of the cortex (along with other structures) may be necessary for us to encode and retrieve early events many years later (Bauer & Larkina, 2014; Feldman, 2014; Pathman & Bauer, 2013). Interestingly, research has shown that older adults describe their most important memories as occurring between the ages of 17 and 24, in part because many major life transitions—such as getting married, attending college, starting a first job, and having children—happen during this period of time (Steiner et al., 2014).

Implicit/nondeclarative memory refers to unintentional learning or memories that are independent of conscious recall. Try telling someone else how you tie your shoelaces without demonstrating the actual behavior. Because your memory of this skill is beyond conscious recall (*implicit*), and hard to describe (*declare*) in words, this type of memory is sometimes referred to as *implicit/nondeclarative*.

Implicit/nondeclarative memory consists of *procedural* motor skills, like tying your shoes or riding a bike, as well as *classically conditioned memory* responses, such as fears and prejudices (Chapter 6). It also includes **priming**, in which prior exposure to a stimulus (*prime*) facilitates recall of information already in storage (Cesario, 2014; Clark et al., 2014). Priming often occurs even when we do not consciously remember being exposed to the prime. Have you ever felt nervous being home alone while reading a Stephen King novel, experienced sadness after hearing about a tragic event in the news, or developed amorous feelings while watching a romantic movie? These are all examples of how the situation we are in may influence our mood, in conscious or less conscious ways. Given this new insight into how priming can "set you up" for certain emotions, can you see how those who haven't studied psychology might be more likely to mislabel or overreact to their emotions?

Improving Long-Term Memory (LTM)

Here are three of the very best ways to improve LTM: *organization, rehearsal,* and *retrieval*.

Organization

To successfully encode information for LTM, we need to *organize* material into hierarchies. This means arranging a number of related items into broad categories that we further divide and subdivide. (This organizational strategy for LTM is similar to the strategy of chunking material in STM.) For instance, grouping small subsets of ideas together (creating subheadings under larger, main headings), makes the material in this book more understandable and *memorable*.

Admittedly, organization takes time and work. But you'll be happy to know that some memory organization and filing is done automatically while you sleep (Bell et al., 2014; Cona et al., 2014; Nielsen et al., 2015). Unfortunately, despite claims to the contrary, research shows that we can't recruit our sleeping hours to memorize new material, such as a foreign language. (See again the *Myth Busters* in stages of sleep, Chapter 5.)

Rehearsal

Like organization, *rehearsal* also improves encoding for both STM and LTM. If you need to hold information in STM for longer than 30 seconds, you can simply keep repeating it (maintenance rehearsal). But storage in LTM requires **elaborative rehearsal**.

This type of rehearsal involves forming a number of different connections of new material, and linking them to previously stored information. It's obviously easier to remember something if we associate it with something we already know. This is why we use so many analogies in this text. For example, earlier we compared the encoding, storage, and retrieval (ECR) model of memory to the workings of a computer, knowing that most of our readers are relatively familiar with computers.

Elaborative rehearsal was first proposed by Craik and Lockhart (1972) in their *levels of processing* model of memory. As discussed earlier, the term refers to the fact that we process our memories on a continuum from shallow, to intermediate, to deep, and that deep levels of processing result in improved encoding, storage, and retrieval. How does this apply to your everyday life? A recent study found that students who took notes on laptops performed worse on conceptual questions than students who took notes on paper (Mueller & Oppenheimer, 2014). Why? The researchers suggested that students who take notes using a laptop tend to just transcribe lectures verbatim (*shallow processing*), rather than reframing lecture material in their own words (*deeper processing*).

Retrieval

Finally, effective *retrieval* is critical to improving LTM. There are two types of **retrieval cues** (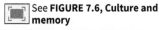 **Figure 7.5**). *Specific* cues require you only to *recognize* the correct response. *General* cues require you to *recall* previously learned material by searching through all possible matches in LTM—a much more difficult task.

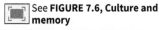
See **FIGURE 7.5, Recall versus recognition**

Whether cues require recall or only recognition is not all that matters. Imagine that while house hunting, you walk into a stranger's kitchen and are greeted with the unmistakable smell of freshly baked bread. Instantly, the aroma transports you to your grandmother's kitchen, where you spent many childhood afternoons doing your homework. You find yourself suddenly thinking of the mental shortcuts your grandmother taught you to help you learn your multiplication tables. You hadn't thought about these little tricks for years, but somehow a whiff of baking bread brought them back to you. Why?

Antonio M. Rosario/Photographer's Choice/Getty Images

In this imagined, baking bread episode, you have stumbled upon the **encoding-specificity principle** (Tulving & Thompson, 1973). In most cases, we're able to remember better when we attempt to recall information in the *same* context in which we learned it (Grzybowski et al., 2014; Lin et al., 2013; Unsworth et al., 2012). Have you noticed that you tend to do better on exams when you take them in the same seat and classroom in which you originally studied the material? This happens because the matching location acts as a retrieval cue for the information.

See **FIGURE 7.6, Culture and memory**

People also remember information better if their moods during learning and retrieval match (Forgas & Eich, 2013; Rokke & Lystad, 2014). This phenomenon, called *mood congruence*, occurs because a given mood tends to evoke memories that are consistent with that mood. When you're sad (or happy or angry), you're more likely to remember events and circumstances from other times when you were sad (or happy or angry).

In addition, memory retrieval is most effective when we are in the same state of consciousness as we were in when the memory was formed. For example, people who are intoxicated will better remember events that happened in a previous drunken state, compared to when they are sober. This is called *state-dependent retrieval* or *state-dependent memory* (Jafari-Sabet et al., 2014; Sanday et al., 2013; Zarrindast et al., 2014).

© Ferdinando Scianna/Magnum Photos, Inc.

Q | Answer the **Concept Check** questions.

Finally, as illustrated in 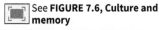 **Figure 7.6**, cultural factors can play a role in how well we remember and retrieve certain memories.

7.2 FORGETTING

 See **TUTORIAL VIDEO: How Could I Forget?**

We've all had numerous experiences with *forgetting*—the inability to remember information that was previously available. We misplace our keys, forget the name of a familiar person, and even miss exams! Although forgetting can be annoying and even catastrophic, it's also adaptive. If we remembered everything we ever saw, heard, or read, our minds would be overwhelmed with useless information.

Forgetting Curve

Psychologists have long been interested in how and why we forget. Hermann Ebbinghaus first introduced the experimental study of learning and forgetting in 1885. As you can see in ▣ **Figure 7.7**, his research found that forgetting occurs soon after we learn something, and then gradually tapers off (Ebbinghaus, 1885).

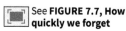 See **FIGURE 7.7, How quickly we forget**

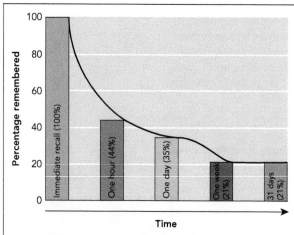

If this dramatic "curve of forgetting" discourages you from studying, keep in mind that meaningful material is much more memorable than nonsense syllables. Furthermore, after some time had passed, and Ebbinghaus thought he had completely forgotten the list, he discovered that *relearning* it took less time than the initial learning took. Similarly, if your college requires you to repeat some of the math or foreign language courses you took in high school, you'll be happily surprised by how much you recall, and how much easier it is to relearn the information, the second time around.

In addition to the good news about relearning, research shows that *visual imagery* (the images we've added to this text, and the ones you create while listening to lectures or reading this text), and the *self-reference effect* (applying information to your personal life), greatly improve LTM and decrease forgetting (Collins et al., 2014; Cunningham et al., 2014; Paivio, 1995).

Before going on, it's important to discuss the large number of software programs and websites, such as Lumosity, promoted on the Internet and television ads, which promise to dramatically improve our memory and decrease forgetting. The developers and promoters of these programs claim they have the potential to optimize and revolutionize the way we learn (Schroers, 2014; Weir, 2014). But will they?

Interestingly, many of these "new" programs are based on the older, well-established principle of **distributed practice**, in which studying or practice is broken up into a number of short sessions over a longer period of time. As you first discovered in Chapter 1, and will hear more about in this and other chapters, distributed practice involves repeated reviews, and "drill and practice" of new material. Many students think they can use *massed practice* or "cramming," and wait until the the last minute to study a large body of material right before an exam. However, distributed practice is widely recognized as one of the very best tools for learning and grade improvement (Dunlosky et al., 2013; Kornmeier et al., 2014; Küpper-Tetzel, 2014). In response to these research findings, we've built in numerous opportunities for distributed practice and *Self-Tests* throughout this text.

Theories of Forgetting

As mentioned earlier, the ability to forget is essential to the proper functioning of memory, and psychologists have developed several theories to explain why forgetting occurs: *decay, interference, motivated forgetting, encoding failure,* and *retrieval failure* (**Process Diagram 7.3**). Each theory focuses on a different stage of the memory process or a particular type of problem in processing information.

In *decay theory*, memory is processed and stored in a physical form—for example, in a network of neurons. Connections between neurons probably deteriorate over time, leading to forgetting. This theory explains why skills and memory degrade if they go unused ("use it or lose it").

Process Diagram 7.3

Why We Forget: Five Key Theories

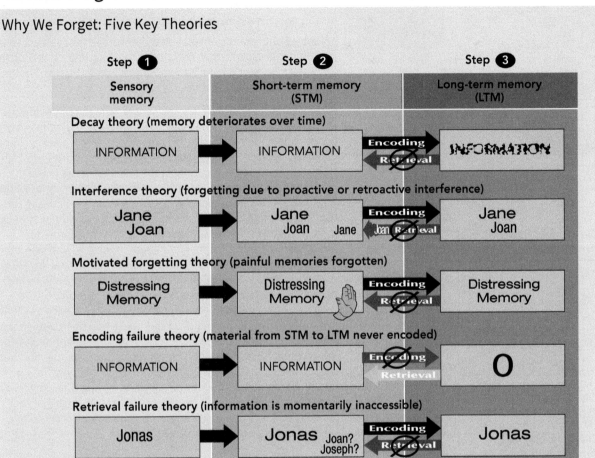

Note: If you want to remember the five theories, think of how forgetting involves memories that grow "dimmer." Note that the first letter of each theory has almost the same spelling—D-I-M-E-R.

In *interference theory*, forgetting is caused by two competing memories, particularly memories with similar qualities. At least two types of interference exist: *retroactive* and *proactive* (🖥 **Figure 7.8**). When new information disrupts (*interferes* with) the recall of old, "retro" material, it is called **retroactive interference** (acting backward in time). Learning your new phone number may cause you to forget your old phone number (prior information is forgotten). Conversely, when old information disrupts (*interferes with*) the recall of new information, it is called **proactive interference** (acting forward in time). Old information (like the Spanish you learned in high school) may interfere with your ability to learn and remember material from your new college course in French.

Motivated forgetting theory is based on the idea that we forget some information for a reason. According to Freudian theory, people forget unpleasant or anxiety-producing information, either consciously or unconsciously, such as the box of cookies you ate last night.

In *encoding failure theory*, our sensory memory receives the information and passes it to STM. But during STM, we may overlook precise details, and may not fully encode it, which would result in a failure to pass along a complete memory for proper storage in LTM.

See **FIGURE 7.8, Retroactive interference and proactive interference**

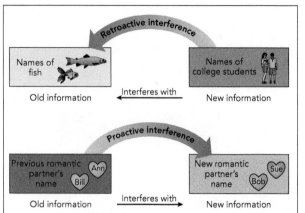

According to *retrieval failure theory*, memories stored in LTM aren't forgotten. They're just momentarily inaccessible. For example, the **tip-of-the-tongue (TOT) phenomenon**—a strong, confident feeling that you know something, while not being able to retrieve it at the moment—is known to result from interference, faulty cues, and high emotional arousal.

Factors in Forgetting

Since Ebbinghaus's original research, scientists have discovered numerous factors that contribute to forgetting. Three of the most important are the *misinformation effect*, the *serial-position effect*, and *source amnesia.*

1. Misinformation Effect Many people (who haven't studied this chapter or taken a psychology course) believe that when they're recalling an event, they're remembering it as if it were a kind of instant replay. However, as you know, our memories are highly fallible, and filled with personal constructions that we create during encoding, storage, and retrieval. Research on the **misinformation effect** shows that misleading information that occurs *after an event* may further alter and revise those constructions. Can you see how this is another example of *retroactive interference*? Our original memories are forgotten or altered because of misleading, post-event information.

See **FIGURE 7.9, The serial-position effect**

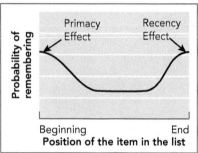

For instance, in one study, participants completed an interview in one room and then answered questions about it in another room (Morgan et al., 2013). Participants who received neutral questions like "Was there a telephone in the room?" answered accurately for the most part, making errors on only 10% of the questions. However, other participants were asked questions such as "What color was the telephone?" which falsely implied that there had been a telephone in the room. Of these respondents, 98% "remembered" a telephone. Other experiments have created false memories by showing participants doctored photos of themselves taking a completely fictitious hot-air balloon ride, or by asking participants to simply imagine an event, such as having a nurse remove a skin sample from their finger. In these and similar cases, a large number of participants later believed that the misleading information was correct, and that the fictitious or imagined events actually occurred (Hinze et al., 2014; Kirk et al., 2015; Wilford et al., 2014).

See **FIGURE 7.10, The power of negative political campaigns**

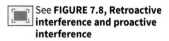

2. Serial-Position Effect Stop for a moment, and write down the names of all the U.S. presidents that you can immediately recall. How did you do? Research shows that most people recall presidents from the beginning of history (e.g., Washington, Adams, Jefferson), and the more recent (e.g., Clinton, Bush, Obama). This is known as the **serial-position effect** (**Figure 7.9**). We tend to recall items at the beginning (*primacy effect*) and the end (*recency effect*) better than those in the middle of the list. And when we do remember those in the middle, it's normally because they are associated with significant events, such as Lincoln and the Civil War (Bonk & Healy, 2010; Hurlstone et al., 2014; Morrison et al., 2014; Roediger & DeSoto, 2014).

3. Source Amnesia Each day we read, hear, and process an enormous amount of information, and it's easy to get confused about how we learned who said what to whom, and in what context. Forgetting the origin of a previously stored memory is known as **source amnesia** (Ferrie, 2015; Leichtman, 2006; Paterson et al., 2011). (See **Figure 7.10**.)

Q Answer the **Concept Check** questions.

7.3 BIOLOGICAL BASES OF MEMORY

How Biology Affects Memory

A number of biological changes occur when we learn and remember. Among them are neuronal and synaptic changes and hormonal changes, which are the first topics in this section.

Then we discuss where memories are activated and located in our brain. We end with an exploration of the biology of memory loss.

Neuronal and Synaptic Changes

In Chapter 6, we discussed how learning and memory both modify our brain's neural networks. For instance, when learning to play a sport like tennis repeated practice builds specific neural pathways that make it easier and easier for us to get the ball over the net. These same pathways later enable us to save and remember how to play the game the next time we go out onto the tennis court.

How do these changes, called **long-term potentiation (LTP)**, occur? They happen in at least two ways. First, as early research with rats raised in enriched environments showed (Rosenzweig et al., 1972), repeated stimulation of a synapse can strengthen the synapse by causing the dendrites to grow more spines (Chapter 6). This results in more synapses, more receptor sites, and more sensitivity.

See **FIGURE 7.11, How does a sea slug learn and remember?**

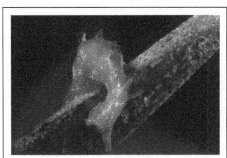

© Wolfgang Pölzer/Alamy Inc.

Second, when learning and memory occur, there is a measurable change in the amount of neurotransmitter released, which thereby increases the neuron's efficiency in message transmission. Research with *Aplysia* (sea slugs) clearly demonstrates this effect (■ **Figure 7.11**).

Further evidence comes from research with genetically engineered "smart mice," which have extra receptors for a neurotransmitter named NMDA (N-methyl-d-aspartate). These mice perform significantly better on memory tasks than do normal mice (Lin et al., 2014; Plattner et al., 2014; Tsien, 2000).

Although it is difficult to generalize from sea slugs and mice, research also supports the idea that LTP is one of the major biological mechanisms underlying learning and memory in humans (Baddeley et al., 2015; Panja & Bramham, 2014; Shin et al., 2010).

Hormonal Changes and Emotional Arousal

As discussed in Chapter 2, our sympathetic nervous system and endocrine system naturally release neurotransmitters and hormones when we're excited or stressed–the "fight or flight" response. Strong emotions also release these same hormones, and they can have both positive and negative effects on how we encode, store, and retrieve our memories (Conway, 2015; Emilien & Durlach, 2015; Trammel & Clore, 2014).

The enhancing effect of hormones on memory is particularly apparent in what are known as **flashbulb memories (FBMs)** (Brown & Kulik, 1977). During important, unexpected, and/or threatening emotional events, like the 9/11 terrorist attack on New York City, our brains appear to form automatic, vivid, detailed, and seemingly permanent memories of an emotionally significant moment or event. In fact, researchers have found that people have retained their FBMs of the 9/11 attack for as long as ten years, and that their confidence in these memories have remained high (Hirst et al., 2015). We also sometimes create uniquely personal (and happy) FBMs during highly emotional, important events, like getting married or graduating from college (■ **Figure 7.12**).

See **FIGURE 7.12, A common flashbulb memory**

Ariel Skelley//Getty Images

How does this happen? It's as if our brains command us to take "flash pictures" of this moment in time in order to capture it for long-term storage. As we've just seen, a flood of neurotransmitters and hormones helps create strong, immediate memories. In addition, we also replay these memories in our minds again and again, which further encourages stronger and more lasting memories. Can you see how this makes adaptive sense? The surge of hormones apparently serves the protective function of automatically setting us up to "pay attention, learn, and remember!"

Research shows that our FBMs for specific details, particularly the time and place the emotional event occurred, are fairly accurate (Rimmele et al., 2012). However, they also suffer the same alterations and decay as all other forms of memory. They're NOT perfect recordings of events (Hirst et al., 2015; Lanciano et al., 2010; Schmidt, 2012). For example, shortly after the death of Michael Jackson, researchers asked participants to report on their

FBMs and other reactions to the news of his death. When these same people were interviewed again 18 months later, researchers found that despite several discrepancies in their memories, confidence in their personal accuracy remained high (Day & Ross, 2014).

In sum, what separates FBMs from ordinary, everyday memories is their vividness and our subjective confidence in their accuracy. But confidence is not the same as accuracy—an important point we'll return to in the last part of this chapter.

Emotional arousal can sometimes intensify our memories, as in the case of FBMs. However, it also can interfere with their encoding, storage, and retrieval (Schwabe & Wolf, 2014; Trammell & Clore, 2014; Zoladz et al., 2014). For instance, have you ever "blanked out" during a final exam or while giving a speech? If so, you now understand how emotional arousal affects your own memory.

What's the key take-home message about memory and emotional arousal? We all need to remember that our memory processes are impaired during high emotional arousal. We sometimes say and do things during these times that we later regret. News reports are filled with stories of people becoming dangerously confused during fires or other emergencies because they panic and forget important survival tips, such as the closest exit routes. Can you see why airlines and fire departments routinely provide safety and evacuation drills, or how it's dangerous to drive when we're arguing with a loved one, or to discipline our children when we're very angry? Recognizing that we're "not in our right minds" during times of high emotional arousal may save our lives—and our relationships!

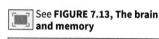

See **FIGURE 7.13, The brain and memory**

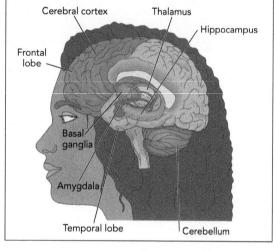

Cerebral cortex

Thalamus

Hippocampus

Frontal lobe

Basal ganglia

Amygdala

Temporal lobe

Cerebellum

Memory and Our Brain

Now that we've looked at neuronal and synaptic changes that happen during memory formation, and hormonal changes that occur during emotional arousal, we need to closely examine what's happening in our brains. For example, the hormones that accompany strong emotions make more glucose energy available for overall brain activity. They also initiate reactions in key areas of our brain, particularly the *amygdala* (a part of the limbic system associated with emotions). Although early memory researchers believed that memory was *localized*, or stored and activated, in a few specific areas of the brain, modern research shows us that it occurs throughout our brains. The most prominent areas are highlighted in ▦ **Figure 7.13**.

Biology and Memory Loss

The leading cause of neurological disorders—including memory loss—among young U.S. men and women between the ages of 15 and 25 is *traumatic brain injury (TBI)*. These injuries most commonly result from car accidents, falls, blows, or gunshot wounds.

See **FIGURE 7.14, Professional sports and brain damage**

TBI happens when the skull suddenly collides with another object. Compression, twisting, and distortion of the brain inside the skull all cause serious and sometimes permanent damage to the brain. The frontal and temporal lobes often take the heaviest hit because they directly collide with the bony ridges inside the skull (Chapter 2).

One of the most troubling, and controversial, causes of TBIs is blows to the head during sports participation (Pearce et al., 2015; Solomon & Zuckerman, 2015). Both professional and nonprofessional athletes frequently experience *concussions*, a mild form of TBI, and multiple concussions can lead to *chronic traumatic encephalopathy (CTE)*. Sadly, the frequency of sports related brain injuries may have been grossly underestimated (Baugh et al., 2015), and a growing body of research connects these multiple brain injuries to diseases and disorders like Alzheimer's, depression, and even suicide (▦ **Figure 7.14**).

Until recently, there was no reliable way to quickly identify and diagnose initial brain injuries early after trauma. However, scientists using MRI brain scans (Chapter 2) are now able to identify these injuries by looking at significant damage to the blood-brain barrier (BBB). (The BBB is a semipermeable membrane that normally blocks dangerous substances from penetrating through to our brain.) Using this new technique, researchers have identified an association between playing football and an increased risk of BBB pathology (Weissberg et al., 2014). And experts are now working to find ways to target and treat these "leaky" BBBs, thereby allowing earlier detection and more informed decisions by athletes and others who suffer from brain injuries.

Two Forms of Amnesia

Now that we know a little more about brain injuries, let's examine their direct effects on memory loss—specifically *amnesia*. Consider this true account of amnesia. When H.M. was 27, he underwent brain surgery to correct his severe epileptic seizures. Although the surgery improved his medical problem, now something was clearly wrong with H.M.'s long-term memory. When his uncle died, he grieved in a normal way. But soon after, he began to ask why his uncle never visited him. H.M. had to be repeatedly reminded of his uncle's death, and each reminder would begin a new mourning process. H.M. lived another 55 years not recognizing the people who cared for him daily. Each time he met his caregivers, read a book, or ate a meal, it was as if for the first time (Barbeau et al., 2010; Carey, 2008; Corkin, 2002). Sadly, H.M. died in 2008—never having regained his long-term memory.

What would it be like to be H.M.—existing only in the present moment, unable to learn and form new memories? It's important to note that being completely amnesic about your

See **FIGURE 7.15, Two types of amnesia**

past and not knowing who you are is a common plot in movies and television. However, real-life amnesia generally doesn't cause a loss of self-identity. Instead, the individual typically has trouble retrieving old memories or forming new ones. These two forms of amnesia are called *retrograde* and *anterograde* (▣ **Figure 7.15**). In **retrograde amnesia** (acting backward in time), the individual is unable to retrieve information from the past. For example, after an accident, someone may not remember (is amnesic) for events that occurred *before* the brain injury. However, the same person has no trouble remembering things that happened after the injury. As the name implies, only the old, "retro," memories are lost.

Retrograde Amnesia

Old memories formed *before* an event (e.g., a car accident) are LOST.

Old information

New memories formed *after* an event (e.g., a car accident) are OKAY.

New information

What causes retrograde amnesia? In cases where the individual is only amnesic for the events right before the brain injury, the cause may be a failure of *consolidation*. We learned earlier that during long-term potentiation (LTP), our neurons change to accommodate new learning and memory formation. We also know that new memories are initially "fragile" and sensitive to disruption. They require a series of processes in order for these neural changes to become fixed and stable in long-term memory (LTM), a process known as **consolidation**. Like heavy rain on wet cement, the brain injury "wipes away" unstable memories because the cement has not yet hardened (*retrograde amnesia*).

In contrast to retrograde amnesia, in which people cannot recall the past, victims of **anterograde amnesia** (acting forward in time) cannot form new memories. These people are unable to transfer new information from STM to LTM. For example, after H.M.'s surgery, he was unable to retain (or was *amnesic* for) new information. Anterograde amnesia generally results from a surgical injury or from diseases, such as chronic alcoholism and senile dementia, which is a form of severe mental deterioration in old age. Continuing our analogy with cement, anterograde amnesia would be like having permanently hardened cement, which prevents the laying down of new long-term memories.

Keep in mind that retrograde amnesia is normally temporary and somewhat common, such as what happens to football players after a head injury. In contrast, anterograde amnesia is relatively rare and most often permanent. However, patients often show surprising abilities to learn and remember implicit/nondeclarative tasks (such as procedural motor skills like mowing a lawn). Also note that some individuals have both forms of amnesia. For example, H.M. (the man introduced earlier) suffered primarily from anterograde amnesia. However, he also had a mild memory loss for events in his life that happened the year or two *before* the operation (*retrograde amnesia*) (Annese et al., 2014; Barbeau et al., 2010; Reisberg, 2014).

See **VIRTUAL FIELD TRIP:**
Alzheimer's Treatment
Center

See **FIGURE 7.16, The effect**
of Alzheimer's disease on
the brain

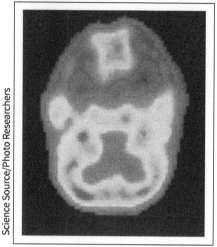

Science Source/Photo Researchers

Q | Answer the
Concept Check questions.

Like traumatic brain injuries, disease can alter the physiology of the brain and nervous system, affecting memory processes. For instance, *Alzheimer's disease (AD)* is a progressive mental deterioration that occurs most commonly in old age (■ **Figure 7.16**). The most noticeable early symptoms are disturbances in memory, which become progressively worse until, in the final stages, the person fails to recognize loved ones, needs total nursing care, and ultimately dies from the disease.

Alzheimer's does not attack all types of memory equally. A hallmark of the disease is an extreme decrease in *explicit/declarative memory* (Howes & O'Shea, 2014; Irish et al., 2011; Müller et al., 2014). Individuals with Alzheimer's fail to recall facts, information, and personal life experiences. However, they still retain some *implicit/nondeclarative* memories, such as simple, classically conditioned responses, and procedural tasks, like brushing their teeth.

Brain autopsies of people with Alzheimer's show unusual *tangles* (structures formed from degenerating cell bodies) and *plaques* (structures formed from degenerating axons and dendrites). Hereditary Alzheimer's generally strikes its victims between the ages of 45 and 55. Some experts believe the cause of Alzheimer's is primarily genetic and age related. However, like many other diseases, it undoubtedly results from a mixture of multiple factors (Arshavsky, 2014; Korczyn, 2014; Liu et al., 2014).

Unfortunately, there is currently no effective way to diagnose early Alzheimer's. However, new research based on tell-tale changes in the human retina is promising (Tsai et al., 2014). Individuals with Alzheimer's disease may also benefit from beginning a healthy diet and exercise program. One encouraging study found that 9 out of 10 patients with Alzheimer's who adopted such a program showed substantial improvement in memory and cognitive function (Bredesen, 2014).

7.4 MEMORY DISTORTIONS AND IMPROVEMENT

"Remembrance of things past is not necessarily the remembrance of things as they were."

—MARCEL PROUST

At this point in your life, you've undoubtedly experienced a painful breakup of a serious love relationship and/or witnessed similar breakups among your close friends. During these breakups, did you wonder how the reported experiences of two people in the same partnership could be so different? Why would each partner reconstruct his or her own personal memory of the relationship?

Explaining Memory Distortions

There are several reasons why we shape, rearrange, and distort our memories. One of the most common is our need for *logic* and *consistency*. When we're initially forming new memories or sorting through old ones, we fill in missing pieces, make "corrections," and rearrange information to make it logical and consistent with our previous experiences or personal desires. If you left a relationship because you found a new partner, you might rearrange your memories to suit your belief that you are just a weak-willed person who can't stay faithful, or that you were mismatched from the beginning. However, if you were the one left behind, you might reconstruct your memories, and now remember that your previous partner was a manipulative "player" from the beginning. Of course, not all memory reconstructions are so dramatic. However, it's important that we recognize and remember the fallibility of our memory processes during initial encoding, current storage, and future retrieval.

We also shape and construct our memories for the sake of *efficiency*. We edit, summarize, augment, and tie new information in with related memories in LTM. For instance, when taking notes during lectures, you can't record every word. Instead, you edit and summarize the information, and (hopefully) tie it into other related material. Unfortunately, your notes may miss important details that later trip you up during exams. Similarly, when we need to retrieve the stored information, we often leave out seemingly unimportant elements or misremember the details.

Despite all their problems and biases, our memories are normally fairly accurate and serve us well in most situations. They have evolved to encode, store, and retrieve general and/or vital information, such as the location of various buildings on our college campus, or the importance of looking both ways when we cross the street. However, when faced with tasks that require encoding, storing, and retrieving precise details like those in a scholarly text, remembering names and faces of potential clients, or recalling where we left our keys, our brains are not as well-equipped.

Memory and the Criminal Justice System

When our memory errors come into play within the criminal justice system, they may lead to wrongful judgments of guilt or innocence—with possible life or death consequences!

In the past, one of the best forms of trial evidence a lawyer could have was an *eyewitness*—"I was there; I saw it with my own eyes." Unfortunately, research has identified several problems with eyewitness testimony (Benton et al., 2014; Loftus, 1993, 2011; National Resource Council, 2014). For example, if multiple eyewitnesses talk to one another after a crime, they may "remember" and corroborate erroneous details that someone else reported, which is why police officers try to separate eyewitnesses while taking their reports.

See **TUTORIAL VIDEO: Eyewitness Testimony**

See **FIGURE 7.17, The dangers of eyewitness testimony**

As a critical thinker, can you see how the details and problems we discussed earlier about flashbulb memories (FBMs) also might apply to eyewitness testimony? Traumatic events, like watching a crime, often become FBMs for eyewitnesses. Despite high confidence in their personally vivid memories, they can make serious errors, such as identifying an innocent person as the perpetrator (■ **Figure 7.17**).

Problems with eyewitness recollections are so well established that judges now allow expert testimony on the unreliability of eyewitness testimony, and routinely instruct jurors on its limits (Benton et al., 2014; Cutler & Kovera, 2013; Pezdek, 2012). If you serve as a member of a jury or listen to accounts of crimes in the news, remind yourself of these problems. Also, keep in mind that research participants in eyewitness studies generally report their inaccurate memories with great self-assurance and strong conviction (DeSoto & Roediger, 2014; Goodwin et al., 2013; Morgan & Southwick, 2014).

© Michael DeMocker/The Times-Picayune/Landov

See **FIGURE 7.18, Understanding and improving eyewitness testimony**

Interestingly, research now suggests that the accuracy of eyewitness testimony can be improved if people are asked to make very fast judgments (Brewer et al., 2012). In fact, giving people only a few seconds to identify the culprit in a lineup increases the accuracy of such identifications by 20 to 30%, compared to allowing people to take as long as they want to make a decision. In addition, simply asking people to close their eyes when they're tying to remember leads to greater accuracy in both audio and visual details (Nash et al., 2015). ■ **Figure 7.18** offers further insights on eyewitness testimony.

False Versus Repressed Memories

"We invent memories. Without thinking. If we tell ourselves something happened often enough, we start to believe it, and then we can actually remember it".

—S.J. WATSON

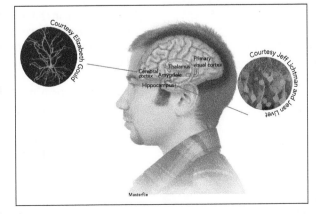

Like eyewitness testimony, false memories can have serious legal, personal, and social implications. Have you heard the famous and true story of psychologist Elizabeth Loftus? When she was 14, her mother drowned in the family's pool. Decades later, a relative told Elizabeth that she, Elizabeth, had been the one to find her mother's body. Despite her initial shock, Elizabeth's memories slowly started coming back. Her recovery of these gruesome childhood memories, although painful, initially brought great relief. It also seemed to explain why she had always been fascinated by the topic of memory.

Then her brother called to say there had been a mistake! The relative who told Elizabeth that she had been the one to discover her mother's body later remembered—and other relatives confirmed—that it had actually been Aunt Pearl, not Elizabeth. Loftus, a world renowned expert on memory distortions, had unknowingly created her own *false memory*.

Like this personal experience from Elizabeth Loftus, extensive research has shown that it's relatively easy to create false memories (Lindner & Henkel, 2015; Loftus & Cahill, 2007; Lynn et al., 2015). For example, a recent study found that 70% of participants had false memories of committing a serious crime (such as theft or assault), simply after being asked to generate such a memory in the lab (Shaw & Porter, 2015).

Even more troubling is the finding that once false memories have been formed, they can multiply over time—and last for years. Researchers in one study showed participants pictures of an event, such as a girl's wallet being stolen (Zhu et al., 2012). Participants then read a series of statements about the event, which included both accurate information (for instance, the person who took the girl's wallet was a man) and false information (the person who took the girl's wallet put it in his pants pocket). (In reality, the picture showed him hiding the wallet in his jacket). Initially, after reading these statements, participants identified only 31% of the false events as having occurred. However, when participants were asked 1½ years later which events had occurred, they identified 39% of the false statements as true, indicating that false memories not only last, but can even multiply over time. Can you see how this is an example of one of the factors in forgetting, *source amnesia,* and why this research has dangerous legal implications for eyewitness testimony?

Do you also recall our earlier discussion of the *misinformation effect,* and how experimenters created a false memory of seeing a telephone in a room? Participants who were asked neutral questions, such as "Was there a telephone in the room?" made errors on only 10% of the queries (Morgan et al., 2013). In contrast, when participants were asked "What color was the telephone?" falsely implying that a telephone had been in the room, 98% "remembered" it being there. Interestingly, research using brain scans have shown that different areas of the brain are activated during true versus false memories (■ **Figure 7.19**).

Are you wondering how all this research applies to our everyday life? In addition to serious legal problems with eyewitness testimony, false memories also affect our political memories. On a personal level, it's important to remember that everyone is vulnerable to creating and believing false memories. Just because something feels true, doesn't mean that it is.

Repressed Memories—True or False?

Creating false memories may be somewhat common, but can we recover true memories that are buried in childhood? *Repression* is the supposed unconscious coping mechanism by which we prevent anxiety-provoking thoughts from reaching consciousness (Chapter 11). According to some research, repressed memories are *actively* and *consciously* "forgotten" in an effort to avoid the pain of their retrieval (Anderson et al., 2004; Boag, 2012). Others suggest that some memories are so painful that they exist only in an *unconscious* corner of the mind, making them inaccessible to the individual (Haaken, 2010; Mancia & Baggott, 2008). In both cases, therapy supposedly would be necessary to unlock the hidden memories.

Repression is a complex and controversial topic in psychology. No one doubts that some memories are forgotten and later recovered. What some question is the idea that *repressed memories* of painful experiences (especially childhood sexual abuse) are stored in the unconscious mind, and their role in judicial processes (Howe & Knott, 2015; Lambert et al., 2010; Loftus & Cahill, 2007).

Critics suggest that most people who have witnessed or experienced a violent crime, or are adult survivors of childhood sexual abuse, have intense, persistent and highly-conscious memories. They have trouble *forgetting*, not remembering. Some also wonder whether

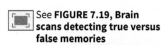
See **FIGURE 7.19, Brain scans detecting true versus false memories**

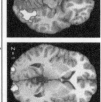

therapists may inadvertently create false memories in their clients during therapy. They propose that if a clinician even suggests the possibility of abuse, the client's own *constructive processes* may lead him or her to create a false memory. For example, the client might start to incorporate portrayals of abuse from movies and books into his or her own memory, forgetting their original sources and eventually coming to see them as reliable.

This is not to say that all psychotherapy clients who recover memories of sexual abuse (or other painful incidents) have invented those memories. In fact, some research suggests that children may remember experiencing sexual abuse, but not understand or recognize those behaviors as abuse until adulthood (McNally, 2012).

Interestingly, a recent research report suggests that the so-called, "memory wars" may be getting less heated (Patihis et al., 2014). Comparing attitudes in the 1990s to today, these researchers found less belief in repressed memories among mainstream psychologists, and undergraduates with greater critical-thinking abilities. Keep in mind, while researchers continue exploring the mechanisms underlying delayed remembering, we must be careful not to ridicule or condemn people who recover true memories of abuse, or other tragic events. In the same spirit, we must protect innocent people from wrongful accusations that come from false memories. Hopefully, with continued research (and perhaps new technology) we may someday better protect the interests of both the victim and the accused.

Memory Tools for Student Success

One of the many beauties of our human brain is that we can recognize the limits and problems of memory, and then develop appropriate coping mechanisms. Just as our ancestors domesticated wild horses and cattle to overcome the physical limits of the human body, we can develop similar approaches to improve our mental limits—especially those responsible for fine detail. The following tools for memory improvement (**Study Organizer 7.1**), some of which were mentioned earlier, are particularly helpful for college success.

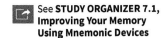
See **TUTORIAL VIDEO: Enhancing Your Memory**

See **STUDY ORGANIZER 7.1, Improving Your Memory Using Mnemonic Devices**

A Final Word

As we have seen throughout this chapter, our memories are remarkable—yet highly fickle. Recognizing the frailties of memory will make us better jurors in the courtroom, more informed consumers, and more thoughtful, open-minded parents, teachers, students, and friends.

Unfortunately, sometimes our memories are better than we would like. Traumatic, and extremely emotional, memories can persist even when we would very much like to forget. Though painful, these memories can sometimes provide important insights. As Elizabeth Loftus suggests in a letter to her deceased mother:

> I thought then [as a 14-year-old] that eventually I would get over your death. I know today that I won't. But I've decided to accept that truth. What does it matter if I don't get over you? Who says I have to? David and Robert still tease me: "Don't say the M word or Beth will cry." So what if the word mother affects me this way? Who says I have to fix this? Besides, I'm too busy (Loftus, 2002, p. 70).

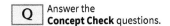
Answer the **Concept Check** questions.

Go to your WileyPLUS Learning Space course for video episodes, examples, art, tables, Concept Checks, practice, and other pedagogical resources that will help you succeed in this course.

Reading for

THINKING, LANGUAGE, AND INTELLIGENCE

8.1 THINKING

If you go on to major in psychology, you'll discover that researchers often group thinking, language, and intelligence under the larger umbrella of **cognition**, the mental activities of acquiring, storing, retrieving, and using knowledge (Groome et al., 2014; Matlin, 2016). Cognition is also a core theme throughout this text—especially Chapters 4, 5, 6, 7 and 14. In this section of Chapter 8, we'll begin our discussion of cognition with *thinking*—what it is and where it's located.

Every time we take in information and mentally act on it, we're thinking. These thought processes are both localized and distributed throughout our brains in networks of neurons. For instance, during problem solving or decision making, our brains are most active in the *prefrontal cortex*. This region associates complex ideas; makes plans; forms, initiates, and allocates attention; and supports multitasking. In addition to the localization of thinking processes, the prefrontal cortex also links to other areas of the brain, such as the limbic system (Chapter 2), to synthesize information from several senses (Garrett, 2015; Harding et al., 2015). Studies also note that successful psychotherapy leads to changes in limbic system activity–apparently reflecting positive changes in thought processes (Haas et al., 2015; Viviani et al., 2015). Now that we know where thinking occurs, we need to discuss its basic components.

Cognitive Building Blocks

Imagine yourself lying, relaxed, in the warm, gritty sand on an ocean beach. Do you see palms swaying in the wind? Can you smell the salty sea, and taste the dried salt on your lips? Can you hear children playing in the surf? What you've just created is a *mental image*, a mental representation of a previously stored sensory experience, which includes visual, auditory, olfactory, tactile, motor, and gustatory imagery (McKellar, 1972). We all have a mental space where we visualize and manipulate our sensory images (Laeng et al., 2014; Naselaris et al., 2015).

In addition to mental images, our thinking also requires forming **concepts**, or mental representations of a group or category (Lagarde et al., 2015; Pape et al., 2015). Concepts can be concrete (like car and concert) or abstract (like intelligence and beauty). They are essential to thinking and communication because they simplify and organize information. Normally, when you see a new object or encounter a new situation, you relate it to your existing conceptual structure and categorize it according to where it fits. For instance, if you see a metal box with four wheels driving on the highway, you know it is a car, even if you've never seen that particular model before. How do we learn concepts? As you can see in ▣ **Figure 8.1**, they develop through the environmental interactions of three major building blocks—*prototypes, artificial concepts,* and *hierarchies* (Ferguson & Casasola, 2015; McDaniel et al., 2014):

Problem Solving

Many years ago in Los Angeles, a 12-foot-high tractor-trailer reportedly got stuck under a bridge that was 6 inches too low. After hours of towing, tugging, and pushing, the police and transportation workers were stumped. Then a young boy happened by and asked, "Why don't you let some air out of the tires?" It was a simple, creative suggestion—and it worked.

See **FIGURE 8.1, The three key building blocks of concepts**

Stephen St. John/NG Image Collection

Our lives are filled with problems—some simple, some difficult. In all cases, problem solving requires moving from a given state (the problem) to a goal state (the solution), a process that usually has three steps: *preparation*, *production*, and *evaluation* (Bourne et al., 1979). Note in **Process Diagram 8.1** that during the preparation stage, we identify and separate relevant from irrelevant facts, and define the ultimate goal.

Next, during the production stage, we generate possible solutions, called hypotheses, by using *algorithms* and *heuristics*. **Algorithms**, as problem-solving strategies, are logical, step-by-step procedures, that if followed correctly, will always lead to an eventual solution. But they are not practical in many situations. **Heuristics** are simplified "rules of thumb" based on experience that are often used as a shortcut for problem solving. They're much faster and more adaptive because they free up our thinking, so that we're better able to direct our cognitive resources elsewhere. Unfortunately, they do not guarantee a solution. See the **Applying Psychology** for sample problem-solving heuristics, and how they might apply to your current or future career.

See **TABLE 8.1: Three Problem-Solving Heuristics and Your Career**

This Process Diagram contains essential information NOT found elsewhere in the text, which is likely to appear on quizzes and exams. Be sure to study it CAREFULLY!

Process Diagram 8.1

Three Steps to the Goal

There are three stages of problem solving that help you attain a goal, such as moving to a new home.

❶ Preparation
Begin by clarifying the problem using these three steps in preparation.

- Define the ultimate goal.

- Outline your limits and/or desires.

- Separate the negotiable from the nonnegotiable.

Move to a new home close to work.

✓ Must allow pets.
✓ Must be close enough to walk.
✓ I prefer a house to an apartment building.
✓ Fireplaces are nice.

✓ Must allow pets.
✓ Must be close enough to walk.
* I prefer a house to an apartment building.
* Fireplaces are nice.

❷ Production
Next, test your possible paths and solutions with one or both of these methods.

- Use an **algorithm**, a logical step-by-step procedure that, if followed correctly, will eventually solve the problem. But algorithms may take a long time—especially for complex problems.

- Use a **heuristic**, a simple rule for problem solving that does not guarantee a solution, but offers a likely shortcut to it.

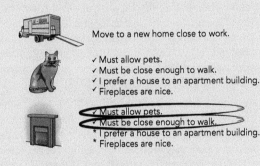

Look at every ad in the paper and call all of those that allow pets.

Work backwards from the solution—start by drawing a 1-mile radius around work to narrow the search.

❸ Evaluation
Did your possible solutions solve the problem?

- If no, then you must return to the production and/or preparation stages.

- If yes, then take action to achieve your goal.

Finally, during the evaluation stage we judge the hypotheses generated during the production stage against the criteria established in the preparation stage.

In addition to the three basic steps of preparation, production, and evaluation, we also sometimes solve problems with a sudden flash of *insight*, like Köhler's chimps that stacked boxes to reach the bananas (Chapter 6). On other occasions, just relaxing, and setting our problem aside for a while, an *incubation period*, often leads to a solution without further conscious thought.

Five Potential Barriers to Problem Solving

See **TUTORIAL VIDEO: Barriers to Problem-Solving**

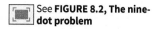
See **FIGURE 8.2, The nine-dot problem**

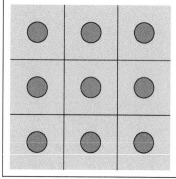

As we've just seen, algorithms, heuristics, insight, and an incubation period all help us solve problems in our daily life. In this section, we'll discuss five potential barriers to effective problem solving. Why "potential"? It's because most of these factors have both positive and negative influences:

1. *Mental sets* Why are some problems so difficult to solve? The reason may be that we often stick to problem-solving strategies that have worked in the past, called **mental sets**, rather than trying new, possibly more effective ones (◻ **Figure 8.2**).

2. *Functional fixedness* We also sometimes fail to see solutions to our problems because we tend to view objects as functioning only in the usual or customary way—a phenomenon known as **functional fixedness** (Ness, 2015; Wright et al., 2015). When a child uses sofa cushions to build a fort, or you use a table knife instead of a screwdriver to tighten a screw, you both have successfully avoided functional fixedness. Similarly, the individual who discovered a way to retrofit diesel engines to allow them to use discarded restaurant oil as fuel has overcome functional fixedness—and may become very wealthy! For practice overcoming functional fixedness, see ◻ **Figure 8.3**.

See **FIGURE 8.3, Overcoming functional fixedness**

3. *Confirmation bias* Have you wondered why members of the U.S. Congress can't seem to solve serious national problems, like our deteriorating bridges and highways? Or why we can't resolve ongoing disputes with our roommates or spouses? It may be that we all fall victim to the tendency to search for and use information that supports our personal beliefs and ideas, while refuting or discounting the beliefs and ideas of our opponents. As first discussed in Chapters 1 and 4, this inclination to seek confirmation for our preexisting beliefs, and to overlook contradictory evidence, is known as **confirmation bias** (Knobloch-Westerwick et al., 2015; Nickerson, 1998; Spratt et al., 2015).

4. *Availability heuristic* In the fall of 2014, many Americans became highly concerned about contracting the Ebola virus, given the media's intense coverage of its occurrence in Africa—and its relatively high mortality rate. One nationwide poll revealed that nearly two-thirds of Americans said they were concerned about an Ebola epidemic in the United States. Yet in reality, only two people in the United States died of Ebola that year, and both of these people had actively cared for patients with this virus in Africa. Can you imagine how intense media coverage of Africa's Ebola crisis (in which thousands of people died from this virus) led people to overestimate their individual risk of acquiring this disease? Media coverage also helps explain why people are overly concerned about shark attacks, and overly optimistic about winning the lottery. These are just three of many examples of the **availability heuristic**, in which we take a mental shortcut, and make estimates of the frequency or likelihood of an event based on information that is most readily *available* in our memories. In other words, we give greater credence to information and examples that readily spring to mind (Mase et al., 2015; Tversky & Kahneman, 1974, 1993).

5. *Representativeness heuristic* Have you ever been walking in the woods, and immediately froze or jumped away because you thought you saw a dangerous snake, when in fact it was just a twisted stick on the ground? If so, this would be an example of the **representativeness heuristic**, in which we estimate the probability of an event based on how well something matches (or *represents*) an existing prototype or stereotype in our minds (Lien & Yuan, 2015; Peteros & Maleyeff, 2015). We all have a prototype of a snake in our mind and the twisted stick matches this prototype. Can you see how this is also an example of the availability heuristic? While walking in the woods, you're primed to look out for snakes, and the sight of the twisted stick brought immediate images of a snake to your mind.

Pros and Cons of These Five Potential Barriers

To understand how each these five factors can both help and hinder problem solving, let's try a few applications to your personal life. First, how did you decide to go to college, or to choose a particular major? If you once dreamed of being a physician, did you change your mind because you got a low score on your college entrance exams? Can you see how your previous strategy and *mental set* of not studying very hard may have allowed you to slide by in high school, yet negatively affected you on college entrance exams?

In addition, note how fixating on your low entrance exam scores (possible *functional fixedness*) may have blocked you from seeing other alternatives to becoming a physician. Similarly, the *availability* and *representative heuristics* may have provided you with ready images and prototypes of future failures in the college courses required to be a physician. Finally, once you decided NOT to pursue your career dream, did you fall victim to the *confirmation bias,* and only look for information that confirmed that decision (e.g., the hard work and high cost of all those years of medical training), while simultaneously ignoring or discounting the contradictory evidence (e.g., the high salary, job security, and job satisfaction of that occupation)? In short, if you're doubting your ability to be a physician, or any other high-achieving professional, because you fear you're not smart enough, be sure to talk to successful people in your desired career. You'll undoubtedly discover that personal traits, and character strengths, like self-control, motivation, and perseverance are generally better predictors of achievement than college entrance exams or IQ scores (Chua & Rubenfeld, 2014; Dweck, 2007; Mischel, 2014).

Just as each of these biases and heuristics may affect your career path, they also affect numerous areas of your life both inside and outside of college. For instance, researchers in one study examined the reported treatment outcomes in existing medical research records (Hrobjartsson et al., 2013). In some of these studies, the person who evaluated the effectiveness of specific medical treatments did not know whether the patient received a particular treatment or a placebo (they were "blind" to the condition). (See Chapter 1.) In other cases, however, the researcher who evaluated the treatment was aware of which treatment each patient received. Can you guess their findings? "Aware" evaluators judged the treatments to be more effective than the "non-aware" evaluators. Do you understand why? Medical doctors, therapists, and others who treat our ailments all have preexisting beliefs based on their professional training and clinical experience. Unfortunately, once they develop a preferred treatment, they may only seek confirmatory evidence of their preexisting beliefs. As a result of this type of observer bias, certain treatments may be judged as being more effective than they actually are, and possibly better options will be ignored.

The problem isn't just that participants in research settings often fall victim to the confirmation bias, but how it occurs in everyday life. Like gamblers who keep putting coins into slot machines, we all have preexisting beliefs and biases that may lead us to focus only on our "hits" and ignore our "misses," with consequences ranging from small to catastrophic.

On the other hand, cognitive strategies, such as the availability and representativeness heuristics, provide mental shortcuts that are generally far more likely to help than to hurt us (Pohl et al., 2013). They allow immediate "inferences that are fast, frugal, and accurate" (Todd & Gigerenzer, 2000, p. 736). If you note that several houses on your street have safety bars on their windows, you might be motivated to add your own safety bars, and thereby decrease your chances of being burglarized. Likewise, if you're hiking in an area with dangerous snakes, and you see a twisted stick on the ground, it's smart to initially freeze or jump away. When faced with an immediate decision, we often don't have time to investigate all the options. We need to make quick decisions based on the currently available information.

The important take home message is to be aware of, and try to avoid, these potential barriers to problem solving. Also, keep in mind that an open mind, critical thinking, and a willingness to think through complex issues are the foundations of good problem solving and clear thinking.

Creativity

Everyone exhibits a certain amount of creativity in some aspects of life. Even when doing ordinary tasks, like planning an afternoon of errands, you are being somewhat creative. Similarly, if you've ever used a plastic garbage bag as a temporary rain jacket, or placed a thick, college textbook on a chair as a booster seat for a child, you've found creative solutions to problems.

Conceptions of **creativity** are obviously personal and depend on our culture, but most agree that a creative solution or performance generally produces original, appropriate, and valued outcomes in a novel way. Three characteristics are also associated with creativity: *originality*, *fluency*, and *flexibility*. Thomas Edison and his colleagues' development of the first "practical" light bulb offers a prime example of each of these characteristics (■ **Table 8.2**).

How do we measure creativity? Most tests focus on **divergent thinking**, a type of thinking that produces many solutions to the same problem (Baer, 2013; van de Kamp et al., 2015). In contrast to **convergent thinking**, a type of thinking that seeks the single best solution to a problem, divergent thinking is open-ended and focused on generating unusual, novel solutions. For example, when assigned a research paper, many students have trouble coming up with something on their own. They're thinking "inside the box" when they focus on the specific problem of completing the assignment in a way that will please the professor. This is convergent thinking. Instead, most professors want to develop creative, divergent thinking, and encourage (force?) their students to dig deeper to come up with their own unique ideas.

Psychologists have developed several methods to test for divergent thinking. The *Unusual Uses Test* requires you to think of as many uses as possible for an object, such as a brick. In the Anagrams Test, you're asked to reorder the letters in a word to make as many new words as possible.

How can we increase general creativity? Growing evidence suggests numerous skills are important, including divergent thinking, problem solving, perspective taking, cognitive flexibility, intelligence, openness, conscientiousness, and numerous affective processes (Li et al., 2015; Plucker et al., 2015; Sowden et al., 2015). For children, unstructured free play time is also very important because it helps them express and regulate emotions, which in turn helps develop their own capacities to feel, express, regulate, and think about emotions (Russ, 2014; Russ & Wallace, 2013).

Can you see why this research on the value of play is particularly important given the increasing pressure on parents and schools to emphasize science, math, and structured activities? Play appears to build the skills essential to success in the arts, entrepreneurship, and even fields like science and engineering. Researchers also suggest that because it allows safe practice for skills necessary for adult activities, play provides an evolutionary advantage to both human and nonhuman animals (Bateson & Martin, 2013; Tsai, 2015).

In addition to these traits, and free play time, did you know that even taking a simple walk will increase creativity? (See ■ **Figure 8.4**.) Research also suggests that creativity requires the coming together of at least seven interrelated resources, as shown in ■ **Table 8.3**. Can you think of other ways to increase it in your own life?

See **TABLE 8.2: Three Elements of Creative Thinking**

See **FIGURE 8.4, Increasing creativity?**

See **TABLE 8.3: Resources of Creative People**

Answer the **Concept Check** questions.

ThinkStock/Getty Images

8.2 LANGUAGE

Using **language** enables us to mentally manipulate symbols, thereby expanding our thinking. Whether it's spoken, written, or signed, language also allows us to communicate our thoughts, ideas, and feelings (Harley, 2014; Jandt, 2016).

Language Characteristics

To produce language, we first build words using **phonemes** [FO-neems] and **morphemes** [MOR-feems]. Then we string words into sentences using rules of **grammar**, such as *syntax* and *semantics* (**Process Diagram 8.2**).

Process Diagram 8.2

Building Blocks of Language

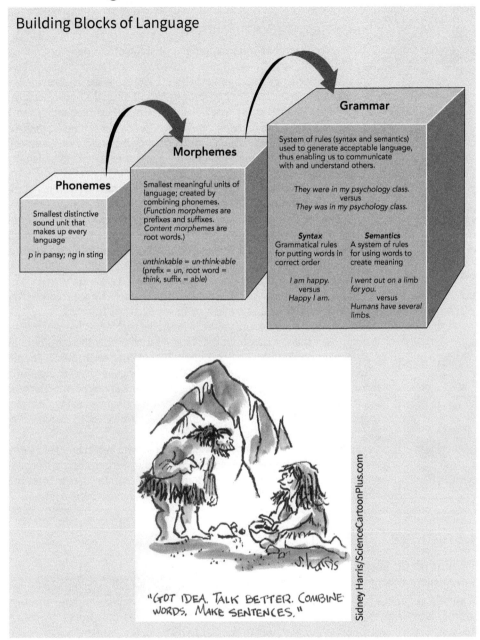

Grammar

System of rules (syntax and semantics) used to generate acceptable language, thus enabling us to communicate with and understand others.

They were in my psychology class.
versus
They was in my psychology class.

Syntax
Grammatical rules for putting words in correct order

I am happy.
versus
Happy I am.

Semantics
A system of rules for using words to create meaning

I went out on a limb for you.
versus
Humans have several limbs.

Morphemes

Smallest meaningful units of language; created by combining phonemes. (*Function morphemes* are prefixes and suffixes. *Content morphemes* are root words.)

unthinkable = *un·think·able* (prefix = *un*, root word = *think*, suffix = *able*)

Phonemes

Smallest distinctive sound unit that makes up every language

p in pansy; *ng* in sting

"GOT IDEA. TALK BETTER. COMBINE WORDS. MAKE SENTENCES."

Sidney Harris/ScienceCartoonPlus.com

What happens in our brains when we produce and comprehend language? Language, just like our thought processes, is both localized and distributed throughout our brain (▣ **Figure 8.5**). For example, the amygdala is active when we engage in a special type of language—cursing or swearing. Why? Recall from Chapter 2 that the amygdala is linked to emotions, especially fear and rage. So it's logical that the brain regions activated by swearing or hearing swear words would be the same as those for fear and aggression.

In Figure 8.5, also note *Broca's area* (which is responsible for speech generation) and *Wernicke's area* (which controls language comprehension). Keep in mind, however, that other parts of our brain are also activated during different types of language generation and listening.

See **FIGURE 8.5, Language and our brain**

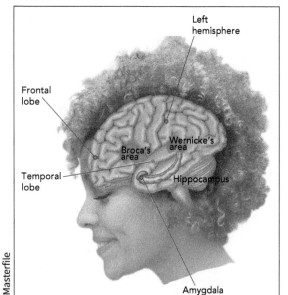

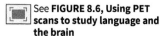

Masterfile

How do we know which parts of the brain are involved with language? Scientists can track brain activity through a colored *positron emission tomography (PET) scan*. Injection of the radioactive isotope oxygen-15 into the bloodstream of the participant makes areas of the brain with high metabolic activity "light up" in red and orange on the scan (**Figure 8.6**).

Language and Thought

Does the fact that you speak English instead of German—or Chinese instead of Swahili—determine how you reason, think, and perceive the world? Linguist Benjamin Whorf (1956) believed so. As evidence for his *linguistic relativity hypothesis*, Whorf offered a now classic example: Because Inuits (previously known as Eskimos) supposedly have many words for snow (*apikak* for "first snow falling," *pukak* for "snow for drinking water," and so on), they reportedly perceive and think about snow differently from English speakers, who have only one word—*snow*.

Though intriguing, Whorf's hypothesis has not fared well. He apparently exaggerated the number of Inuit words for snow (Pullum, 1991) and ignored the fact that English speakers have a number of terms to describe various forms of snow, such as *slush*, *sleet*, *hard pack*, and *powder*. Other research has directly contradicted Whorf's hypothesis. For instance, Eleanor Rosch (1973) found that although people of the Dani tribe in New Guinea possess only two color names—one indicating cool, dark colors, and the other describing warm, bright colors—they discriminate among multiple hues as well as English speakers do.

See **FIGURE 8.6, Using PET scans to study language and the brain**

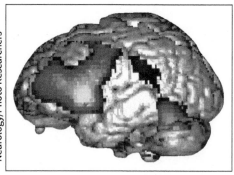

Wellcome Department of Cognitive Neurology/Photo Researchers

Whorf apparently was mistaken in his belief that language *determines* thought. But there is no doubt that language *influences* thought (Bylund & Athanasopoulos, 2015; Ettlinger et al., 2014; Zhong et al., 2015). People who speak multiple languages report that the language they're currently using affects their sense of self, and how they think about events (Berry et al., 2011; Lai & Narasimhan, 2015). For example, when speaking Chinese, they tend to conform to Chinese cultural norms. However, when speaking English, they tend to adopt Western norms.

Our words also influence the thinking of those who hear them. That's why companies avoid *firing* employees. Instead, employees are *outplaced* or *non-renewed*. Similarly, the military uses terms like *preemptive strike* to cover the fact that they attacked first and *tactical redeployment* to refer to a retreat.

Language Development

Prelinguistic Stage

See **TABLE 8.4: Language Acquisition**

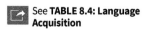

Although each child's language development varies in timing, virtually all children follow a similar sequence (see **Table 8.4**). Beginning at birth, a child communicates through crying, facial expressions, eye contact, and body gestures. Babies only hours old begin to "teach" their caregivers when and how they want to be held, fed, and played with. Babies even start to learn language before they are born. In one study, researchers played sounds from two different languages—English and Swedish—to newborn babies in both the United States and Sweden (Moon et al., 2013). These babies were hooked up to a computer that played various vowel sounds each time the baby sucked on a specially designed pacifier. Half the babies heard sounds from the language they'd been exposed to in utero, whereas the others heard vowels from a different language. In both countries, the babies who heard the foreign vowels sucked more frequently than those who heard sounds that matched their previous in utero experience. The researchers concluded that before birth, these newborns, like all newborns, become familiar—through listening to their mother's voice—with the sounds in their native language, and are now more interested in hearing novel sounds!

Linguistic Stage

After the prelinguistic stage, infants quickly move toward full language acquisition (see again Table 8.4). The various stages within this table are believed to be universal, meaning that all children progress through similar stages regardless of the culture they're born into, or what language(s) they ultimately learn to speak. By age 5, most children have mastered basic grammar and typically use about 2,000 words (a level of mastery considered adequate for getting by in any given culture). Past this point, vocabulary and grammar gradually improve throughout life (Levey, 2014; Oller et al., 2014).

See **VIRTUAL FIELD TRIP:** Baby Sign Language

Theories of Language Development

Some theorists believe that language capability is innate, primarily a matter of maturation. Noam Chomsky (1968, 1980) suggests that children are "prewired" with a neurological ability known as a *language acquisition device (LAD)* that enables them to analyze language and to extract the basic rules of grammar. This mechanism needs only minimal exposure to adult speech to unlock its potential. As evidence for this *nativist position*, Chomsky observes that children everywhere progress through the same stages of language development at about the same ages. He also notes that babbling is the same in all languages and that deaf babies babble just like hearing babies.

Nurturists argue that the nativist position doesn't fully explain individual differences in language development. They hold that children learn language through a complex system

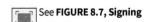 See **FIGURE 8.7, Signing**

of rewards, punishments, and imitation. For example, parents smile and encourage any vocalizations from a very young infant. Later, they respond even more enthusiastically when the infant babbles "mama" or "dada." In this way, caregivers unknowingly use *shaping* (Chapter 6) to help babies learn language.

Animals and Language?

Without question, nonhuman animals communicate. They regularly send warnings, signal sexual interest, share locations of food sources, and so on. But can nonhuman animals master the complexity of human language? Since the 1930s, many language studies have attempted to answer this question by probing the language abilities of chimpanzees, gorillas, and other animals (Beran et al., 2014; Scott-Phillips, 2015; Zuberbühler, 2015).

One of the most successful early studies was conducted by Beatrice and Allen Gardner (1969), who recognized chimpanzees' manual dexterity and ability to imitate gestures. The Gardners used American Sign Language (ASL) with a chimp named Washoe. By the time Washoe was 4 years old, she had learned 132 signs and was able to combine them into simple sentences such as "Hurry, gimme toothbrush" and "Please tickle more." The famous gorilla Koko also uses ASL to communicate; she reportedly uses more than 1,000 signs (■ **Figure 8.7**).

Ron Cohn/The Gorilla Foundation/koko.org

 See **FIGURE 8.8, Computer-aided communication**

In another well-known study, a chimp named Lana learned to use symbols on a computer to get things she wanted, such as food, a drink, and a tickle from her trainers, and to have her curtains opened (Rumbaugh et al., 1974) (■ **Figure 8.8**).

Dolphins also are often the subject of interesting language research (Güntürkün, 2014; Kuczaj et al., 2015). Communication with dolphins is done by means of hand signals or audible commands transmitted through an underwater speaker system. In one typical study, trainers gave dolphins commands made up of two- to five-word sentences, such as "Big ball—square—return," which meant that they should go get the big ball, put it in the floating square, and return to the trainer (Herman et al., 1984). By varying the syntax (the order of the words) and specific content of the commands, the researchers showed that dolphins are sensitive to these aspects of language.

Michael Nichols/NG Image Collection

Scientists disagree about how to interpret the findings on chimps, apes, and dolphins. Most believe that nonhuman animals definitely communicate, but they're not using true language because they don't convey subtle meanings, use language creatively, or communicate at an abstract level. Other critics propose that these animals do not truly understand language, but are simply operantly conditioned (Chapter 6) to imitate symbols to receive rewards. Opponents also contend that data regarding animal language has not always been well documented (Beran et al., 2014; Savage-Rumbaugh, 1990; Terrace, 1979).

Proponents of animal language respond that apes can use language creatively and have even coined some words of their own. For example, Koko signed "finger bracelet" to describe a ring and "eye hat" to describe a mask (Patterson & Linden, 1981). Proponents also argue that, as demonstrated by the dolphin studies, animals can be taught to understand basic rules of sentence structure. As you can see, the jury is still out on whether or not nonhuman animals use "true" language or not. Stay tuned!

⬛ Q Answer the **Concept Check** questions.

8.3 INTELLIGENCE

Many people equate intelligence with "book smarts." For others, the definition of intelligence depends on the characteristics and skills that are valued in a particular social group or culture (Goldstein et al., 2015; Plucker & Esping, 2014; Suzuki et al., 2014). For example, the Mandarin word that corresponds most closely to the word *intelligence* is a character meaning "good brain and talented" (Matsumoto & Juang, 2013). The word is also associated with traits like imitation, effort, and social responsibility (Keats, 1982). An experiment carried out in seven countries even found that smiling versus non-smiling affected judgments of intelligence. Interestingly. German respondents perceived smiling individuals as being more intelligent, whereas Chinese participants judged smilers as less intellgent (Krys et al., 2014).

Even among Western psychologists there is considerable debate over the definition of intelligence. In this discussion, we rely on a formal definition of **intelligence**—*the global capacity to think rationally, act purposefully, profit from experience, and deal effectively with the environment* (Wechsler, 1944, 1977).

The Nature of Intelligence

In the 1920s, British psychologist Charles Spearman first observed that high scores on separate tests of mental abilities tend to correlate with each other. Spearman (1923) thus proposed that intelligence is a common skill set, which he termed **general intelligence (*g*)**. He believed that *g* underlies all intellectual behavior, including reasoning, solving problems, and performing well in all areas of cognition. Spearman's work laid the foundations for today's standardized intelligence tests (Bouchard, 2014; Cooper, 2015; Woodley of Menie & Madison, 2015).

Multiple Intelligences

About a decade later, L. L. Thurstone (1938) proposed 7 primary mental abilities: verbal comprehension, word fluency, numerical fluency, spatial visualization, associative memory, perceptual speed, and reasoning. J. P. Guilford (1967) later expanded this number, proposing that as many as 120 factors are involved in the structure of intelligence.

Around the same time, Raymond Cattell (1963, 1971) reanalyzed Thurstone's data and suggested that two subtypes of *g* exist:

- **Fluid intelligence (*gf*)** refers to the ability to think quickly and abstractly, and to solve novel problems. Fluid intelligence is relatively independent of education or experience. And, like most other biological capacities, it declines with age (Gerstorf et al., 2015; Klein et al., 2015).

- **Crystallized intelligence (*gc*)** refers to the store of knowledge and skills gained through experience and education (Blanch, 2015; Sternberg, 2014, 2015). Crystallized intelligence tends to increase over the life span.

Today there is considerable support for the concept of *g* as a measure of academic smarts. However, many contemporary cognitive theorists believe that intelligence is a collection of many, separate, specific abilities.

The fact that brain-damaged patients often lose some intellectual abilities, while retaining others, suggests to psychologist Howard Gardner that different intelligences are located in discrete areas throughout the brain. According to *Gardner's theory of multiple intelligences* (1983, 2011), people have different profiles of intelligence because they are stronger in some areas than others (■ **Table 8.5**). They also use their intelligences differently to learn new material, perform tasks, and solve problems.

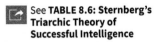

See **TABLE 8.5: Gardner's Multiple Intelligences and Possible Careers**

Have you heard from others that you're naturally good at something like writing, math, or spatial skills? Gardner's research shows that most people possess one or more natural intelligences important to success in various occupations. Carefully consider each of the multiple intelligences in Table 8.5 and how they might help guide you toward a satisfying career.

Robert Sternberg's *triarchic theory of successful intelligence* also assumes multiple abilities. As shown in ■ **Table 8.6**, Sternberg believes there are three separate, learned aspects of intelligence: (1) *analytic*, (2) *creative*, and (3) *practical* (Sternberg, 1985, 2015).

See **TABLE 8.6: Sternberg's Triarchic Theory of Successful Intelligence**

Sternberg emphasizes the process underlying thinking, rather than just the product. He also stresses the importance of applying mental abilities to real-world situations, rather than testing mental abilities in isolation. In short, Sternberg avoids the traditional idea of intelligence as an innate form of "book smarts." Instead, he emphasizes successful intelligence as the learned ability to adapt to, shape, and select environments in order to accomplish personal and societal goals.

Emotional Intelligence

Finally, in addition to the numerous forms of multiple intelligences we just discussed, Daniel Goleman's research (1995, 2000, 2008) and best-selling books have popularized the concept of **emotional intelligence (EI)**, based on original work by Peter Salovey and John Mayer (1990). Emotional intelligence (EI) is generally defined as *the ability to perceive, understand, manage, and utilize emotions accurately and appropriately.*

See **TUTORIAL VIDEO: Emotional Intelligence**

Have you ever wondered why some people who are very intelligent, in terms of "book smarts," still experience frequent conflicts and repeated failures in their friendships and work situations? Proponents of EI have suggested that traditional measures of human intelligence ignore a crucial range of abilities that characterize people who are high in EI, and tend to excel in real life: self-awareness, impulse control, persistence, zeal and self-motivation, empathy, and social deftness (Higgs & Dulewicz, 2014; Ruiz-Arranda et al., 2014; Stein & Deonarine, 2015).

Although the idea of emotional intelligence is very appealing, critics fear that a handy term like EI invites misuse, but their strongest reaction is to Goleman's proposals for widespread teaching of EI. For example, Paul McHugh, director of psychiatry at Johns Hopkins University, suggests that Goleman is "presuming that someone has the key to the right emotions to be taught to children. We don't even know the right emotions to be taught to adults" (cited in Gibbs, 1995, p. 68).

Measuring Intelligence

Different IQ tests approach the measurement of intelligence from different perspectives. However, most are designed to predict grades in school. Let's look at the most commonly used IQ tests.

See **FIGURE 8.9, The normal curve**

The *Stanford-Binet Intelligence Scale* is loosely based on the first IQ tests developed in France around the turn of the twentieth century by Alfred Binet. In the United States, Lewis Terman (1916) developed the Stanford-Binet test (at Stanford University) to assess the intellectual ability of U.S.-born children ages 3 to 16. The test is revised periodically—most recently in 2003. The test is administered individually and consists of such tasks as copying geometric designs, identifying similarities, and repeating number sequences.

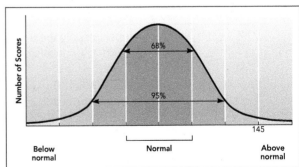

After administering the individual test to a large number of people, researchers discovered that their scores typically are distributed in a **normal distribution** and a symmetrical, bell-shaped curve (■ **Figure 8.9**). This means that a majority of the scores fall in the middle of the curve and a few scores fall on the extremes. Interestingly, height and weight also follow this same normal distribution.

See **TUTORIAL VIDEO: Understanding IQ**

See **FIGURE 8.10, Items similar to those on the Wechsler Adult Intelligence Scale (WAIS)**

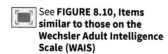

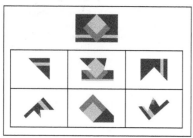

See **TUTORIAL VIDEO: Standards for Psychological Tests**

In the original version of the Stanford-Binet, results were expressed in terms of a **mental age (MA)**, which is a person's level of mental development relative to that of others. For example, if a 7-year-old's score equaled that of an average 8-year-old, the child was considered to have a MA of 8. To determine the child's **intelligence quotient (IQ)**, MA was divided by the child's chronological age (CA) (actual age in years) and multiplied by 100.

The most widely used intelligence test today, the *Wechsler Adult Intelligence Scale (WAIS)*, was developed by David Wechsler in the early 1900s (Wechsler, 1939). He later created a similar test for school-aged children. Like the Stanford-Binet, Wechsler's tests yield an overall intelligence score. However, Wechsler's tests also include separate index scores related to four specific areas: verbal comprehension, perceptual reasoning, working memory, and processing speed. See ▣ **Figure 8.10** for samples of the perceptual reasoning test items.

Today, most intelligence test scores are expressed as a comparison of a single person's score to a national sample of similar-aged people. Even though the actual IQ is no longer calculated using the original formula comparing mental and chronological ages, the term *IQ* remains as a shorthand expression for intelligence test scores.

What makes a good test? How are the tests developed by Binet and Wechsler any better than those published in popular magazines and presented on television programs? To be scientifically acceptable, all psychological tests must fulfill three basic requirements (Dombrowski, 2015; Jackson, 2016; Suzuki et al., 2014):

- **Standardization** in intelligence tests (as well as personality, aptitude, and most other tests) involves following a certain set of uniform procedures. First, every test must have *norms*, or average scores, developed by giving the test to a representative sample of people (a diverse group of people who resemble those for whom the test is intended). Second, testing procedures must be standardized. All test takers must be given the same instructions, questions, and time limits, and all test administrators must follow the same objective score standards.

- **Reliability** is usually determined by retesting participants to see whether their test scores change significantly. Retesting can be done via the *test–retest method*, in which participants' scores on two separate administrations of the same test are compared, or via the *split-half method*, which splits a test into two equivalent parts (such as odd and even questions) and determines the degree of similarity between the two halves.

- **Validity** is the ability of a test to measure what it is designed to measure. The most important type of validity is *criterion-related validity*, or the accuracy with which test scores can be used to predict another variable of interest (known as the criterion). Criterion-related validity is expressed as the *correlation* (Chapter 1) between the test score and the criterion. If two variables are highly correlated, then one can be used to predict the other. Thus, if a test is valid, its scores will be useful in predicting an individual's behavior in some other specified situation. One example is using intelligence test scores to predict grades in college.

Can you see why a test that is standardized and reliable but not valid is worthless? For instance, a test for skin sensitivity may be easy to standardize (the instructions specify exactly how to apply the test agent), and it may be reliable (similar results are obtained on each retest). But it certainly would not be valid for predicting college grades.

Extremes in Intelligence

One of the best methods for judging the validity of a test is to compare people who score at the extremes. Despite the uncertainties discussed in the previous section, intelligence tests provide one of the major criteria for assessing mental ability at the extremes— specifically, for diagnosing *intellectual disability* and *mental giftedness*.

The clinical label *intellectually disabled* (previously referred to as *mentally retarded*) is applied when someone has significant deficits in general mental abilities, such as reasoning, problem solving, or academic learning. And these deficits typically result in impairments of adaptive functioning, including communication, social participation, and personal independence (American Psychiatric Association, 2013; Kumin, 2015).

Fewer than 3% of people are classified as having an intellectual disability (■ **Table 8.7**). Of this group, 85% have only mild intellectual disability, and many become self-supporting, integrated members of society. Furthermore, people can score low on some measures of intelligence and still be average or even gifted in others (Treffert, 2014; Werner & Roth, 2014). The most dramatic examples are people with *savant syndrome*. People with savant syndrome generally score very low on IQ tests (usually between 40 and 70), yet they demonstrate exceptional skills or brilliance in specific areas, such as rapid calculation, art, memory, or musical ability (■ **Figure 8.11**).

See **TABLE 8.7**: Degrees of Intellectual Disability

See **FIGURE 8.11, Savant syndrome—an unusual form of intelligence**

© Justin Sutcliffe/Redux Pictures

Some forms of intellectual disability stem from genetic abnormalities, such as Down syndrome, fragile-X syndrome, and phenylketonuria (PKU). Other causes are environmental, including prenatal exposure to alcohol and other drugs, extreme deprivation or neglect in early life, and brain damage from physical trauma, such as auto accidents or sports injuries. However, in many cases, there is no known cause of the intellectual disability.

At the other end of the intelligence spectrum are people with especially high IQs (typically defined as being in the top 1 or 2%). In the early 1900s, Lewis Terman identified 1,500 gifted children—affectionately nicknamed the "Termites"—with IQs of 140 or higher (Terman, 1925). He and his colleagues then tracked their progress through adulthood. The number who became highly successful professionals was many times the number a random group would have provided (Kreger Silverman, 2013; Plucker & Esping, 2014; Terman, 1954). Those who were most successful tended to also have extraordinary motivation, and someone at home or school who was especially encouraging (Goleman, 1980). Unfortunately, like members of all groups, these "Termites" also became alcoholics, got divorced, and committed suicide at close to the national rate (Campbell & Feng, 2011; Leslie, 2000; Terman, 1954). In sum, a high IQ is no guarantee of success in every endeavor. As mentioned earlier, research shows that personal traits, and character strengths, like self-control, motivation, and perseverance, may be the strongest predictors of overall achievement and well-being (Chua & Rubenfeld, 2014; Dweck, 2007; Mischel, 2014).

See **VIRTUAL FIELD TRIP: Down Syndrome Connection**

See **VIRTUAL FIELD TRIP: High IQ Society**

Nature, Nurture, and IQ

Psychologists have long debated several important questions related to intelligence: How is brain functioning related to intelligence? What factors—environmental or hereditary—influence an individual's intelligence? These questions, and the controversies surrounding them, are discussed in this section.

See **FIGURE 8.12, Do intelligent brains work more efficiently?**

The Brain's Influence on Intelligence

A basic tenet of neuroscience is that all mental activity (including intelligence) results from neural activity in the brain. Most recent research on the biology of intelligence has focused on brain functioning. For example, neuroscientists have found that people who score highest on intelligence tests also respond more quickly on tasks requiring perceptual judgments (Hofman, 2015; Sternberg, 2014, 2015; Wagner et al., 2014).

In addition to a faster response time, research using positron emission tomography (PET) scans to measure brain activity (Chapter 2) suggests that intelligent brains work smarter, or more efficiently, than less-intelligent brains (Jung & Haier, 2007; Neubauer et al., 2004) (■ **Figure 8.12**).

Does size matter? It makes logical sense that bigger brains would be smarter; after all, humans have larger brains than less intelligent species, such as dogs. (Some animals, such as whales and dolphins, have larger brains than humans, but our brains are larger relative to our body size.) In fact, brain-imaging studies have found a significant

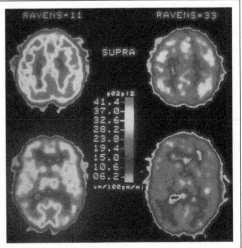

Courtesy Richard J. Haier, University of California-Irvine

correlation between brain size (adjusted for body size) and intelligence (Bouchard, 2014; Møller & Erritzøe, 2014). On the other hand, Albert Einstein's brain was no larger than normal (Witelson et al., 1999). In fact, some of Einstein's brain areas were actually smaller than average, but the area responsible for processing mathematical and spatial information was 15% larger than average.

See **FIGURE 8.13, Genetic and environmental influences on IQ**

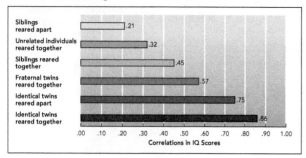

See **FIGURE 8.14, Genetics versus environment**

Differences *within* groups are due almost entirely to genetics (the seed).

Poor soil

Fertile soil

Differences *between* groups are due almost *entirely* to environment (the soil).

See **FIGURE 8.15, Brain sex differences**

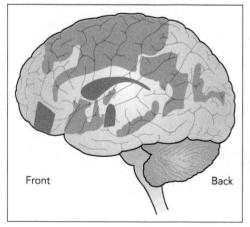

Front

Back

See **TABLE 8.8: Problem-Solving Tasks Favoring Women and Men**

Genetic and Environmental Influences on Intelligence

Similarities in intelligence between family members are due to a combination of hereditary (shared genetic material) and environmental factors (similar living arrangements and experiences). Researchers interested in separating out the role of heredity in intelligence often focus on identical (monozygotic) twins because they share 100% of their genetic material, as shown in ▣ **Figure 8.13**. For example, the long-running Minnesota Study of Twins, an investigation of identical twins raised in different homes, and reunited only as adults, found that genetic factors appear to play a surprisingly large role in the IQ scores of identical twins (Bouchard, 2014; Johnson & Bouchard, 2007).

However, such results are not conclusive. Adoption agencies tend to look for similar criteria in their choice of adoptive parents. Therefore, the homes of these "reared apart" twins were actually quite similar. In addition, these twins shared the same nine-month prenatal environment, which also might have influenced their brain development and intelligence (Felson, 2014; White et al., 2002).

Studies have also shown that early malnutrition can retard a child's intellectual development, curiosity, and motivation for learning (Schoenmaker et al., 2015; Waber et al., 2014). In addition, children who lack reliable care and stable attachment, or who experience abuse in the first few years of life, show not only lower intelligence, but also less empathy for others and a greater vulnerability to later substance abuse and addiction (Luby et al., 2012). In contrast, compared to the average child, children who are enrolled in high-quality preschool programs, and are regularly read to by their parents, show increases in IQ.

In short, intelligence is like a rubber band. Heredity equips each of us with innate intellectual capabilities (our personal rubber band). But our environment helps deteriorate or stretch this band, which significantly influences whether or not we reach our full intellectual potential (Protzko et al., 2013).

Between Versus Within Group Differences

As we've just seen, intelligence in general does show a high degree of heritability. However, it's VERY important to recognize that heritability cannot explain *between*-group differences! Note the overall difference between the average height of plants on the left and those on the right in ▣ **Figure 8.14**. Just as we cannot say that the difference *between* these two groups of plants is due to heredity, we similarly cannot say that differences in IQ *between* any two groups of people are due to heredity. Note also the considerable variation in height *within* the group of plants on the left and those *within* the group on the right. Just as some plants are taller than others, there are individuals who score high on IQ tests and individuals who score low. Always remember that the greatest differences in IQ scores occur when we compare individuals *within* groups—not *between* groups.

What about the widely publicized differences between the sexes, such as those in verbal and math skills? Research using brain scans, autopsies, and volumetric measurements has found several sex differences in the brains of men and women (▣ **Figure 8.15**). For example, two areas in the frontal and temporal lobes, associated with language skills, are generally larger in women than in men. In contrast, a region in the parietal lobes, correlated with manipulating spatial relationships and mathematical abilities, is typically larger in men than in women (Garrett, 2015; Ingalhalikar et al., 2014). ▣ **Table 8.8** illustrates the tasks researchers have used to demonstrate these and other sex differences.

How do we explain this? Evolutionary psychologies often suggest that sex differences like these may be the product of gradual genetic adaptations (Buss, 2015; Ingalhalikar et al., 2014). In ancient societies, men were most often the "hunters," while women were almost always the "gatherers." Therefore, the male's superiority on many spatial tasks and target-directed motor skills (see again Table 8.8) may have evolved from the adaptive demands of hunting, whereas activities such as food gathering, childrearing, and domestic tool construction and manipulation may have contributed to the female's language superiority and fine motor coordination.

Some critics, however, suggest that evolution progresses much too slowly to account for this type of behavioral adaptation. Furthermore, there is wide cross-cultural variability in gender differences, and explanations of these differences are difficult to test scientifically (Halpern, 2014; Miller & Halpern, 2014; Newcombe, 2010). Other research has found that simply having both women and men play action-packed video games for a few short weeks almost completely removes the previously reported gender role differences in some spatial tasks (Feng et al., 2007).

In addition to possible gender differences in verbal and math skills, there is an ongoing debate in our country over the reported differences in IQ scores between various ethnic groups. Unfortunately, some of the strongest proponents of the "heritability of intelligence" argument seem to ignore the "fertile soil" background of the groups who score highest on IQ tests. As an open-minded, critical thinker, carefully consider these important research findings:

- *Environmental and cultural factors may override genetic potential and later affect IQ test scores.* Like plants that come from similar seeds, but are placed in poor versus enriched soil, children of color are more likely to grow up in lower socioeconomic conditions, which may hamper their true intellectual potential. Also, in some ethnic groups and economic classes, a child who excels in school may be ridiculed for trying to be different from his or her classmates. Moreover, if children's own language and dialect do not match their education system, or the IQ tests they take, they are obviously at a disadvantage (Davies et al., 2014; Suzuki et al., 2014; von Stumm & Plomin, 2015). Furthermore, environmental stress can also lead to short-term decreases in test scores.

- *Traditional IQ tests may be culturally biased.* If standardized IQ tests contain questions that reflect American middle-class culture, they will discriminate against test takers whose language, knowledge, and experience differ from those of the majority (Chapman et al., 2014; Stanovich, 2015). Researchers have attempted to create a *culture-fair* or *culture-free* IQ test, but they have found it virtually impossible to do. Past experiences, motivation, test-taking abilities, and previous experiences with tests are powerful influences on IQ scores.

- *Intelligence (as measured by IQ tests) is not a fixed trait.* Around the world, IQ scores have increased over the past half century. This well-established phenomenon, known as the *Flynn effect*, may be due to improved nutrition, better public education, more proficient test-taking skills, and rising levels of education for a greater percentage of the world's population (Baker et al., 2015; Flynn, 1987, 2010; Flynn et al., 2014). Fortunately, research also shows that simply believing that intelligence is NOT a fixed trait is correlated with higher academic grades, and fewer feelings of helplessness (Castella & Byrne, 2015; Romero et al., 2014).

- *Selectively highlighting IQ scores from certain race or ethnic groups is deceptive because race and ethnicity, like intelligence itself, are almost impossible to define.* Depending on the definition we use, there are between 3 and 300 races, and no race is pure in a biological sense (Humes & Hogan, 2015; Kite, 2013). Furthermore, like President Barack Obama, Tiger Woods, and Mariah Carey, many people today self-identify as multiracial.

- *Negative stereotypes about women and people of color can cause some group members to doubt their abilities.* This phenomenon, called **stereotype threat**, may, in turn, reduce their intelligence test scores (Boucher et al., 2015; Franceschini et al., 2014; Steele & Aronson, 1995).

In the first study of stereotype threat, Claude Steele and Joshua Aronson (1995) recruited Black and White college students (with similar ability levels) to complete a difficult verbal exam. In one group of students, who were told that the exam was diagnostic of their intellectual abilities, Blacks underperformed in relation to Whites. However, in a second group, who were NOT told that the exam was diagnostic, there were no differences between the two group scores.

See **FIGURE 8.16, The dynamics of stereotype threat**

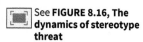

This and other research suggests that stereotype threat occurs because some members of stereotyped groups become aware of the stereotype, and begin to doubt their abilities. This creates anxiety, which in turn leads to a self-fulfilling prophecy and decreased performance. Other members of stereotyped groups respond to stereotype threat by *disidentifying*, telling themselves they don't care about the test scores (Major et al., 1998; Rothgerber & Wolsiefer, 2014). Unfortunately, this attitude reduces motivation and also leads to decreased performance (■ **Figure 8.16**).

See **FIGURE 8.17, "Obama effect" versus stereotype threat**

REUTERS/Larry Downing/Landov

Stereotype threat affects many social groups, including Blacks, women, Native Americans, Latinos, low-income people, elderly people, and white male athletes (Franceschini et al., 2014; Hively & El-Alayhi, 2014; Steele & Aronson, 1995). This research helps explain some group differences in intelligence and achievement tests. As such, it underscores why relying solely on such tests to make critical decisions affecting individual lives—as in college admissions, hiring, or promotions—is unwarranted and possibly even unethical.

What's the good news? First, some early research suggested that having Barack Obama as a positive role model improved academic performance by Blacks—thus offsetting the stereotype threat (Marx et al., 2009) (■ **Figure 8.17**). Second, people who have the opportunity to self-affirm, or validate, their identities in some type of meaningful way do not show the negative effects of stereotype threat. For example, Black first-year college students who receive information about how to feel more connected to their college or university show higher GPAs. They also experience better health three years later, as compared to those who did not receive the "how to connect" information (Walton & Cohen, 2011). Similarly, social-belonging and affirmation-training interventions have improved the "chilly climate" for women in male-dominated fields, like engineering (Walton et al., 2015). The interventions also had a positive effect on their social experiences, relationships, and overall achievement in the field.

So what's the most important take-away message? In this chapter, we've explored three cognitive processes (*thinking, language,* and *intelligence*), each of which is greatly affected by numerous interacting factors. To solve problems, be creative, communicate with others, and adapt to our environment requires various mental abilities, most of which are not genetic, and can be developed through authentichappiness.sas.upenn.edu.

Q Answer the **Concept Check** questions.

WP LS Go to your WileyPLUS Learning Space course for video episodes, examples, art, tables, Concept Checks, practice, and other pedagogical resources that will help you succeed in this course.

Reading for

LIFE SPAN DEVELOPMENT

9

WP LS Go to your WileyPLUS Learning Space course for video episodes, examples, art, tables, Concept Checks, practice, and other pedagogical resources that will help you succeed in this course.

9.1 STUDYING DEVELOPMENT

Just as some parents carefully watch and study their child's progress throughout his or her life span, the field of **developmental psychology** studies age-related physical and psychological processes from conception to death, or "womb to tomb" (■ **Table 9.1**). Their studies have led to three key theoretical issues.

See **TABLE 9.1**, Life Span Development

See **TUTORIAL VIDEO:** Understanding Development in Context

Theoretical Issues

Almost every area of research in human development frames questions around three major issues (Bergin & Bergin, 2015; Bornstein et al., 2014; Cavanaugh & Blanchard-Fields, 2015):

1. Nature or nurture? How do both genetics (nature) and life experiences (nurture) influence development? According to the *nature position*, development is largely governed by automatic, genetically predetermined signals in a process known as *maturation*. Just as a flower unfolds in accord with its genetic blueprint, humans crawl before we walk, and walk before we run.

In addition, naturists believe there are **critical periods**, or windows of opportunity, that occur early in life when exposure to certain stimuli or experiences is necessary for proper development. For example, research has shown that when doctors operate on infants who are born with cataracts, a condition in which the eye's lens is cloudy and distorts vision, they're able to see much better than if they're operated on after the age of 8. Other research finds that appropriate social interaction in the first few weeks of life is essential for creating normal cognitive and social development (Berger, 2015; Ekas et al., 2013). As you'll see later in the sad story of Genie, featured in the *Psychological Science*, there appears to be a critical period for cognitive and language development. Due to severe isolation and physical abuse, Genie never progressed beyond simple sentences (Raaska et al., 2013; Rymer, 1993). These and other similar studies provide further evidence for critical periods—at least in the early years.

On the other side of the debate, those who hold the *nurturist position* would argue that development results from environmental forces, learning through personal experiences, and observation of others.

2. Stages or continuity? Some developmental psychologists suggest that development generally occurs in *stages* that are discrete and qualitatively different from one to another, whereas others believe it follows a *continuous pattern*, with gradual, but steady and quantitative (measurable), changes (■ **Figure 9.1**).

3. Stability or change? Which of our traits are stable and present throughout our life span, and what aspects will change? Psychologists who emphasize *stability* hold that measurements of personality taken during childhood are important predictors of adult personality; those who emphasize *change* disagree.

Which of these positions is most correct? Most psychologists do not take a hard line either way. Rather, they prefer an *interactionist perspective* and/or the *biopsychosocial model*. For instance, in the *nature-versus-nurture* debate, psychologists generally agree that development emerges from unique genetic predispositions *and* from experiences in the environment (Cavanaugh & Blanchard-Fields, 2015; Duncan et al., 2014; Plomin et al., 2014).

See multipart **FIGURE 9.1**, Stages versus continuity in development

Infancy   Adulthood

127

Research Approaches

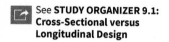
See **STUDY ORGANIZER 9.1:** Cross-Sectional versus Longitudinal Design

To answer these three controversies and other questions, developmental psychologists typically use all the research methods discussed in Chapter 1. To study the entire human life span, they also need two additional techniques—*cross-sectional* and *longitudinal designs*.

The **cross-sectional design** compares groups of people of different ages at a single point in time. For example, one cross-sectional study examined women in three different age groups (ages 22–34, 35–49, and 50–65) to examine whether body weight dissatisfaction changes with age (Siegel, 2010). Unfortunately, female body dissatisfaction appears to be quite stable—and relatively high—across the life span.

On the other hand, a **longitudinal design** takes repeated measures of the same group of people over a long period of time. One interesting longitudinal design study investigated whether peer ratings of personality taken during childhood are better predictors of later adult personality than self-ratings (Martin-Storey et al., 2012). They first asked grade school children in 1976–1978 to rate themselves and their peers on several personality factors, such as likeability, aggression, and social withdrawal. In 1999–2003, the researchers returned and asked the same participants, now in mid-adulthood, to complete a second series of personality tests. As hypothesized, the peer ratings were better than self-ratings in predicting adult personality. Does this finding surprise you? If so, try contacting some of your childhood peers and then compare notes on how you remember one another's personality as children and now as adults.

See **FIGURE 9.2, Reasoning ability using cross-sectional versus longitudinal research**

Now that you have a better idea of these two types of research, if you were a developmental psychologist interested in studying intelligence in adults, which design would you choose—cross-sectional or longitudinal?

Why do the two methods show such different results? Cross-sectional studies sometimes confuse genuine age differences with *cohort effects*—differences that result from specific histories of the age group studied. As shown in the top line in **Figure 9.2**, the 81-year-olds measured by the cross-sectional design have dramatically lower scores than the 25-year-olds. But is this due to aging, or perhaps to broad environmental differences, such as less formal education or poor nutrition? A prime example of possible environmental effects comes from a recent study showing that individuals who enter their teens and early 20s during a recession are less narcissistic relative to those who come of age in more prosperous times (Bianchi, 2014, 2015). For instance, CEOs who were in their teens and early 20s during bad economic times later paid themselves less compared to other top executives.

The key thing to remember is that because different age groups, called *cohorts*, grow up in different historical periods, the results may not apply to people growing up at other times. With the cross-sectional design, age effects and cohort effects are sometimes inextricably tangled.

In contrast, a key advantage of longitudinal research is that researchers can be reasonably confident that any observed changes are the result of time and developmental experiences. However, these studies also have their limits. Given the need to follow participants over many years, longitudinal studies are very expensive in terms of time and money. Some research also suggests that repeated testing over time can influence scores on certain types of abilities, such as cognitive skills (Salthouse, 2014).

One disadvantage shared by both cross-sectional and longitudinal research is the problem of *generalizability*—how well the results from a study conducted on a sample population can be applied ("generalized") to the population at large. Cross-sectional studies only measure behaviors and mental processes at one point in time, which doesn't identify how or when age-related changes may have occurred. For example, if older participants show memory declines, cross-sectional studies can't answer whether those declines appeared suddenly in their 60s, or if they accumulated gradually over several years. Longitudinal research has its own problems with generalizability due to a restricted, self-selected sample. Participants often drop out, lose interest, move away, or even die during extended testing periods.

Look back at Study Organizer 9.1, and note that each method of research has its own comparative strengths and weaknesses. Keep these differences in mind when you read the findings of developmental research in this chapter—and in the popular press.

Before we go on, it's important to point out that modern researchers sometimes combine both cross-sectional and longitudinal designs into one study, known as *cross-sequential research*. For instance, in Chapter 1, we discussed a study that examined whether well-being decreases with age (Sutin et al., 2013). When these researchers examined their combined cross-sectional and longitudinal data from two independent samples taken over 30 years, they initially found that well-being *declined* with age. However, when they then controlled for the fact that older cohorts started out with lower levels of well-being, they found that all the cohorts *increased* rather than decreased in well-being with age. The reversal in findings was explained by the fact that the older group of people had experienced instances of major turmoil in their younger years, including America's Great Depression during the 1930s. This means that this group started out with lower levels of well-being. Sadly, they also maintained these attitudes into their later years, compared to those who grew up during more prosperous times.

Can you see why this combination of two research designs is important? It offers a more accurate and positive view of well-being in old age than what was indicated in either the cross-sectional design or the longitudinal design. It also suggests some troubling possibilities for today's young adults who are entering a stagnant workforce and high unemployment. As the study's authors say, this "economic turmoil may impede [their] psychological, as well as financial, growth even decades after times get better" (Sutin et al., 2013, p. 384).

Q Answer the **Concept Check** questions.

9.2 PHYSICAL DEVELOPMENT

Have you ever compared photos of yourself as a baby, in adolescence, and adulthood? If so, you may be amused and surprised by all the dramatic changes in your physical appearance. But have you stopped to appreciate the incredible underlying process that transforms all of us from birth to death? In this section, we will explore the fascinating processes of physical development from conception through childhood, adolescence, and adulthood.

Prenatal Development

Do you remember being a young child and feeling like it would "take forever to grow up"? Contrary to a child's sense of interminable, unchanging time, the early years of development are characterized by rapid and unparalleled change. In fact, if you continued to develop at the same rapid rate that marked your first two years of life, you would weigh over a thousand pounds and be over 12 feet tall as an adult! Thankfully, physical development slows, yet it is important to note that change continues until the very moment of death.

See multipart **FIGURE 9.3, Conception**

Francis Leroy, Biocosmos/ Photo Researchers

Your prenatal development began at conception, when your biological mother's egg, or ovum, united with your biological father's sperm cell (▣ **Figure 9.3**). At that time, you were a single cell barely 1/175 of an inch in diameter—smaller than the period at the end of this sentence. This new cell, called a *zygote*, then began a process of rapid cell division that resulted in a multimillion-celled infant (you) some nine months later.

The vast changes that occur during the nine months of a full-term pregnancy are usually divided into three stages: the **germinal period**, the **embryonic period**, and the **fetal period** (**Process Diagram 9.1**). Prenatal growth, as well as growth during the first few years after birth, is *proximodistal* (near to far), with the innermost parts of the body developing before the outermost parts. Thus, a fetus's arms develop before its hands and fingers. Development also proceeds *cephalocaudally* (head to tail). Thus, a fetus's head is disproportionately large compared to the lower part of its body. Can you see how these two terms—proximodistal and cephalocaudal—help explain why an infant first lifts its head before then lifting its arms and then its legs?

Process Diagram 9.1

Prenatal Development

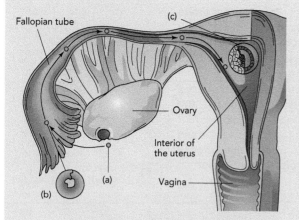

Fallopian tube
(c)
Ovary
Interior of the uterus
(a)
Vagina
(b)

Step 1 Germinal period: From conception to implantation

After discharge from either the left or right ovary (a), the ovum travels to the opening of the fallopian tube.

If fertilization occurs (b), it normally takes place in the first third of the fallopian tube. The fertilized ovum is referred to as a zygote.

When the zygote reaches the uterus, it implants itself in the wall of the uterus (c), and begins to grow tendril-like structures that intertwine with the rich supply of blood vessels located there. After implantation, the organism is known as an *embryo*.

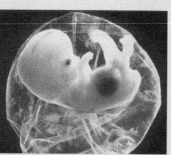

Biophoto Associates/Photo Researchers

Petit Format/Nestle/Photo Researchers

Step 2 Embryonic period: From implantation to eight weeks

At eight weeks, the major organ systems have become well differentiated. Note that at this stage, the head is far larger, and grows at a faster rate, than other parts of the body.

Step 3 Fetal period: From eight weeks to birth

After the eighth week, and until the moment of birth, the embryo is called a *fetus*. At four months, all the actual body parts and organs are established. The remainder of the fetal stage is primarily a time for increased growth and "fine detailing."

Hazards to Prenatal Development

As you recall from Chapter 2, human development begins with the genes we inherit from our parents. However, a new field of research, known as **epigenetics**, studies how non-genetic factors can dramatically affect how (and if) these inherited genes are expressed at each prenatal stage, and throughout our lives (Iakoubov et al., 2015; Li et al., 2014; Vassoler et al., 2014). Unlike simple genetics, which are based on changes in the DNA sequence, these changes in gene expression have other causes, such as age, environment/lifestyle, or disease. (The term "epi" means "above" or "outside of.")

For example, during pregnancy, the *placenta* connects the fetus to the mother's uterus, and serves as the link for delivery of food and excretion of wastes. It also screens out some, but not all, harmful substances. *Epigenetic* environmental hazards such as X-rays, toxic waste, drugs, and diseases can still cross the placental barrier, and negatively affect the child's inherited, genetic potential (**Table 9.2**). These influences generally have the most devastating effects during the first three months of pregnancy, making this a *critical period* in development.

Perhaps the most important—and generally avoidable—danger to a fetus comes from drugs, both legal and illegal. Nicotine and alcohol are major **teratogens**, factors that cause fetal

 See **TABLE 9.2: Sample Prenatal Environmental Conditions That Endanger a Child**

damage or death during prenatal development. Mothers who smoke tobacco or drink alcohol during pregnancy have significantly higher rates of premature births, low-birth-weight infants, and fetal deaths. Their children also show increased behavior and cognitive problems (Doulatram et al., 2015; Mason & Zhou, 2015; Popova et al., 2014).

See **FIGURE 9.4, Fetal alcohol spectrum disorders (FASD)**

Andy Levin/Science Source

As you can see in ▣ **Figure 9.4**, heavy maternal drinking may lead to a cluster of serious abnormalities called *fetal alcohol spectrum disorders (FASD)*. The most severe form of this disorder is known as *fetal alcohol syndrome (FAS)*. Recent research suggests that alcohol may leave chemical marks on DNA that abnormally turn off or on specific genes (Mason & Zhou, 2015). Tobacco might have a similar epigenetic effect. For example, children whose mothers smoked during pregnancy are more likely to be obese as adolescents, perhaps because in-utero exposure to nicotine changes a part of the brain that increases a preference for fatty foods (Haghighi et al., 2013, 2014).

The pregnant mother obviously plays a primary role in prenatal development because her nutrition, health, and almost everything she ingests can cross the placental barrier (a better term might be placental sieve). However, the father also plays a role. Epigenetic environmental factors, such as the father's smoking, can pollute the air the mother breathes. Genetically, the father can transmit heritable diseases, and alcohol, opiates, cocaine, various gases, lead, pesticides, and industrial chemicals can all damage sperm (Finegersh et al., 2015; Ji et al., 2013; Vassoler et al., 2014). Research also shows that children of older fathers may be at higher risk of a range of mental difficulties, including attention deficits, bipolar disorder, autism, and schizophrenia (D'Onofrio et al., 2014; McGrath et al., 2014).

Early Childhood Development

Like the prenatal period, early childhood is also a time of rapid physical development. Let's explore three key areas of change in early childhood: *brain*, *motor*, and *sensory/perceptual development*.

Brain Development

Our brain and other parts of the nervous system grow faster than any other part of the body during both prenatal development and the first two years of life, as illustrated in ▣ **Figure 9.5**. This brain development and learning occur primarily because neurons grow in size. Also, the number of dendrites, as well as the extent of their connections, increases (Bornstein et al., 2014; Garrett, 2015; Swaab, 2014).

See multipart **FIGURE 9.5, Brain development**

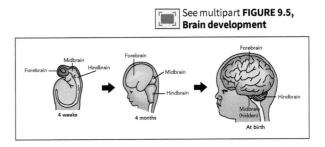

Motor Development

Compared to the hidden, internal changes in brain development, the orderly emergence of active movement skills, known as *motor development*, is easily observed and measured. A newborn's first motor abilities are limited to *reflexes*, or involuntary responses to stimulation (Chapter 2). For example, the rooting reflex occurs when something touches a baby's cheek: The infant will automatically turn its head, open its mouth, and root for a nipple.

In addition to simple reflexes, the infant soon begins to show voluntary control over the movement of various body parts (▣ **Figure 9.6**). Thus, a helpless newborn, who cannot even lift her head, is soon transformed into an active toddler capable of crawling, walking, and climbing. In fact, babies are highly motivated to begin walking because they can move faster than when crawling, and they get better with practice (Adolph & Berger, 2012; Berger, 2015). Note that motor development is largely due to natural maturation, but, like brain development, it can also be affected by environmental influences such as disease and neglect.

See multipart **FIGURE 9.6, Milestones in motor development**

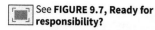
See **FIGURE 9.7, Ready for responsibility?**

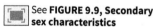
See **FIGURE 9.8, Adolescent growth spurt**

Sergey Novikov/Shutterstock

John Miles/The Image Bank/Getty Images, Inc.

See **FIGURE 9.9, Secondary sex characteristics**

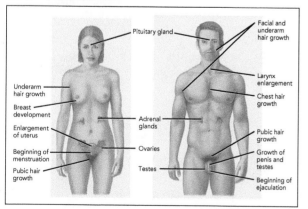

Sensory and Perceptual Development

At birth, and during the final trimester of pregnancy, the developing child's senses are quite advanced (Bardi et al., 2014; Levine & Munsch, 2014). For example, research shows that a newborn infant prefers his or her mother's voice, providing evidence that the developing fetus can hear sounds outside the mother's body (Lee & Kisilevsky, 2014; Von Hofsten, 2013). This raises the interesting possibility of fetal learning, and some have advocated special stimulation for the fetus as a way of increasing intelligence, creativity, and general alertness (Jarvis, 2014; Van de Carr & Lehrer, 1997).

In addition, a newborn can smell most odors and distinguish between sweet, salty, and bitter tastes. Newborns, breast fed by their mothers, also easily recognize the odor of their mother's milk compared to other mother's milk, formula, and other substances (Allam et al., 2010; Nishitani et al., 2009). Similarly, the newborn's sense of touch and pain is highly developed, as evidenced by reactions to circumcision and to heel pricks for blood testing, and by the fact that their pain reactions are lessened by the smell of their own mother's milk (Nishitani et al., 2009; Rodkey & Riddell, 2013; Vinall & Grunau, 2014).

The newborn's sense of vision, however, is poorly developed. At birth, an infant is estimated to have vision between 20/200 and 20/600 (Haith & Benson, 1998). Imagine what the infant's visual life is like: The level of detail you see at 200 or 600 feet (if you have 20/20 vision) is what an infant sees at 20 feet. Within the first few months, vision quickly improves, and by 6 months it is 20/100 or better. At 2 years, visual acuity is nearly at the adult level of 20/20 (Courage & Adams, 1990).

Adolescence

Adolescence is the loosely defined psychological period of development between childhood and adulthood, commonly associated with puberty. In the United States, it roughly corresponds to the teenage years. However, the concept of adolescence varies greatly across cultures (■ **Figure 9.7**).

As we'll see in the next section on cognitive development, our brains continue to change during adolescence—particularly the frontal lobes. For now, let's focus on the physical changes. Do you recall how changes in height and weight, breast development and menstruation for girls, and a deepening voice and beard growth for boys, were such important milestones for you and your adolescent peers? These are all primary signs of **puberty**, the period of adolescence when a person becomes capable of reproduction. They're also a clear biological signal of the end of childhood.

The clearest and most dramatic physical sign of puberty is the *growth spurt*, which is characterized by rapid increases in height, weight, and skeletal growth (■ **Figure 9.8**) and by significant changes in reproductive structures and sexual characteristics. Maturation and hormone secretion cause rapid development of the ovaries, uterus, and vagina, and the onset of menstruation (*menarche*) in the adolescent female. In the adolescent male, the testes, scrotum, and penis develop, and he experiences his first ejaculation (*spermarche*). The testes and ovaries produce hormones that lead to the development of secondary sex characteristics, such as the growth of pubic hair, deepening of the voice and growth of facial hair in men, and growth of breasts in women (■**Figure 9.9**).

Have you ever wondered why teenagers seem to sleep so much? Researchers have found that puberty is triggered by changes in the brain, including the release of certain hormones, which occur only during periods of *deep sleep* (D'Ambrosio & Redline, 2014; Shaw et al., 2012). This finding suggests that getting adequate, deep (slow-wave) sleep (see Chapter 5) during adolescence is an essential part of activating the reproductive system. Can you see why the increasing number of sleep problems in adolescents is a cause for concern, and why parents should actually be encouraging "oversleeping" in their teenagers?

Adulthood

When does adulthood begin? In Western cultures, many define it as beginning after high-school or college graduation, whereas others mark it as when we get our first stable job and become self-sufficient. Adulthood is typically divided into at least three periods: young adulthood (ages 20–45), middle adulthood (ages 45–60), and late adulthood (ages 60 to death).

Emerging/Young Adulthood

Although young adulthood is generally considered to begin at age 20, many developmental psychologists have added a new term, **emerging adulthood**, to refer to the time from the end of adolescence to the young-adult stage, approximately ages 18–25. This time period is found primarily in modern cultures, and is characterized by the search for a stable job, self-sufficiency, and/or marriage and parenthood, along with five distinguishing features (Arnett, 2000, 2015; Munsey, 2006; Newman & Newman, 2015):

- *Identity exploration*—young people decide who they are and what they want out of life.
- *Instability*—a time marked by multiple changes in residence and relationships.
- *Self-focus*—freed from social obligations and commitments to others, young people at this stage are focused on what they want and need before constraints of marriage, children, and career.
- *Feeling in-between*—although taking responsibility for themselves, they still feel in-between adolescence and adulthood.
- *Age of possibilities*—a time of optimism and belief that their lives will be better than their parents.

In contrast to these rather unique cognitive, life style experiences, physical changes during emerging/young adulthood are rather minimal. Some individuals experience modest physical increases in height and muscular development, and most of us find this to be a time of maximum strength, sharp senses, and overall stamina. However, a decline in strength and speed becomes noticeable in the 30s, and our hearing starts to decline after around age 18.

Middle Adulthood

Many physical changes during young adulthood happen so slowly that most people don't notice them until they enter their late 30s or early 40s. For example, around the age of 40, we first experience difficulty in seeing things close up and after dark, a thinning and graying of our hair, wrinkling of our skin, and a gradual loss in height coupled with weight gain (Landsberg et al., 2013; Saxon et al., 2014).

For women between ages 45–55, *menopause*, the cessation of the menstrual cycle, is the second most important life milestone in physical development. The decreased production of estrogen (the dominant female hormone) produces certain physical changes, including decreases in some types of cognitive and memory skills (Doty et al., 2015; Hussain et al., 2014; Pines, 2014). However, the popular belief that menopause (or "the change of life") causes serious psychological mood swings is not supported by current research. In fact, younger women are more likely to report irritability and mood swings, whereas women at midlife generally report positive reactions to aging, and the end of the menstrual cycle. They're also less likely to have negative experiences, such as headaches (Sievert et al., 2007; Sugar et al., 2014).

In contrast to women, men experience a more gradual decline in hormone levels, and most men can father children until their seventies or eighties. Physical changes such as unexpected weight gain, decline in sexual responsiveness, loss of muscle strength, and graying or loss of hair may lead some men (and women as well) to feel depressed and to question their life progress. They often see these alterations as a biological signal of aging and mortality. Such physical and psychological changes in men are generally referred to as the *male climacteric* (or *andropause*). However, the popular belief that almost all men (and some women) go through a deeply disruptive midlife crisis, experiencing serious dissatisfaction with their work and personal relationships, is largely a myth.

Late Adulthood

After middle age, most physical changes in development are gradual and occur in the heart and arteries, and in the sensory receptors. Cardiac output (the volume of blood pumped by the heart each minute) decreases, whereas blood pressure increases, due to the thickening and stiffening of arterial walls. Visual acuity and depth perception decline, hearing acuity lessens (especially for high-frequency sounds), smell sensitivity decreases, and some decline in cognitive and memory skills occurs (Dupuis et al., 2015; Fletcher & Rapp, 2013; Newman & Newman, 2015).

Biological Theories of Aging

Why do we go through so many physical changes? What causes us to age and die? Setting aside aging and deaths resulting from disease, abuse, or neglect, known as *secondary aging*, let's focus on *primary aging* (gradual, inevitable age-related changes in physical and mental processes).

According to *cellular-clock theory*, primary aging is genetically controlled. Once the ovum is fertilized, the program for aging and death is set and begins to run. Researcher Leonard Hayflick (1965, 1996) found that human cells seem to have a built-in life span. After about 100 doublings of laboratory-cultured cells, they cease to divide. Based on this limited number of cell divisions, Hayflick suggests that we humans have a maximum life span of about 120 years—we reach the *Hayflick limit*. Why? One answer may be that small structures on the tips of our chromosomes, called *telomeres*, shorten each time a cell divides. After about 100 replications, the telomeres are too short and the cells can no longer divide (Broer et al., 2013; Hayashi et al., 2015; Rode et al., 2015).

The second major explanation of primary aging is *wear-and-tear theory*. Like any machine, repeated use and abuse of our organs and cell tissues cause our human bodies to simply wear out over time.

See **VIRTUAL FIELD TRIP: Dying with Dignity**

Q   Answer the **Concept Check** questions.

9.3 COGNITIVE DEVELOPMENT

Just as our bodies and physical abilities change over the life span, our way of knowing and perceiving the world also grows and changes. Jean Piaget [pee-ah-ZHAY] provided some of the first demonstrations of how a child's thinking and reasoning abilities differ from that of an adult's (Piaget, 1952). He showed that an infant begins at a cognitively "primitive" level, and that intellectual growth progresses in distinct stages, motivated by an innate need to know.

To appreciate Piaget's contributions, we need to consider three major concepts: *schemas, assimilation*, and *accommodation*. **Schemas** are the cognitive frameworks formed through interactions with objects and events. We use them like an architect's drawings or a builder's blueprints to organize our knowledge and to understand the world around us. For most of us, a common, shared schema for a car would likely be "a moving object with wheels and seats for passengers." However, we also develop unique schemas based on differing life experiences.

In the first few weeks of life, the infant apparently has several *schemas* based on the innate reflexes of sucking, grasping, and so on. These schemas are primarily motor and may be little more than stimulus-and-response mechanisms (for example, the nipple is presented, and the baby sucks). Soon, other schemas emerge. The infant develops a more detailed schema for eating solid food, a different schema for the concepts of "mother" and "father," and so on.

Assimilation and *accommodation* are the two major processes by which schemas grow and change over time. **Assimilation** is the process of absorbing new information into existing schemas. For instance, infants use their sucking schema not only in sucking nipples, but also in sucking blankets and fingers. In **accommodation**, existing ideas are modified to fit new information. Accommodation generally occurs when new information or stimuli cannot be assimilated (▣ **Figure 9.10**).

See **FIGURE 9.10,** Accommodation

Similarly, if you meet and become friends with someone through an online chat room, you may be surprised by your awkward interactions with them when you later meet face to face. This discomfort may result from the mental work involved in readjusting, or accommodating, your earlier schemas to match the new reality.

Stages of Cognitive Development

According to Piaget, all children go through approximately the same four stages of cognitive development, regardless of the culture in which they live (**Process Diagram 9.2**). Piaget also believed that these stages cannot be skipped because skills acquired at earlier stages are essential to mastery at later stages (Berger, 2015).

Sensorimotor Stage

The **sensorimotor stage** lasts from birth until "significant" language acquisition (about age 2). During this time, children explore the world and develop their schemas primarily through their senses and motor activities—hence the term *sensorimotor*. One important concept that infants develop during the sensorimotor stage is **object permanence**—an

This Process Diagram contains essential information NOT found elsewhere in the text, which is likely to appear on quizzes and exams. Be sure to study it CAREFULLY!

Process Diagram 9.2

Piaget's Four Stages of Cognitive Development

Step ❶ Sensorimotor stage (birth to age 2)	**Limits** • Lacks "significant" language and object permanence (understanding that things continue to exist even when not seen, heard, or felt) **Abilities** • Uses senses and motor skills to explore and develop cognitively **Example** • Children at this stage like to explore and play with their food.	Jeff R Clow/Flickr/Getty Images
Step ❷ Preoperational stage (ages 2 to 7)	**Limits** • Cannot perform mental "operations" (lacks reversibility and conservation) • Egocentric thinking (inability to consider another's point of view) • Animistic thinking (believing all things are living) **Abilities** • Has significant language and thinks symbolically **Example** • Children at this stage often believe the moon follows them.	© Igor Demchenkov/ iStockphoto

Step ③ Concrete operational stage (ages 7 to 11)	**Limits** • Cannot think abstractly and hypothetically • Thinking tied to concrete, tangible objects and events **Abilities** • Can perform "operations" on concrete objects • Understands conservation (realizing that changes in shape or appearance can be reversed) • Less egocentric **Example** • Children at this stage begin to question the existence of Santa.	 Hill Street Studios/Blend Images/Getty Images, Inc.
Step ④ Formal operational stage (ages 11 and over)	**Limits** • Adolescent egocentrism at the beginning of this stage, with related problems (imaginary audience, personal fable) **Abilities** • Can think abstractly and hypothetically **Example** • Children at this stage generally show great concern for physical appearance.	 © Adrian Burke/Corbis Images

 See multipart **FIGURE 9.11, Object permanence**

Doug Goodman/Photo Researchers/Getty Images

understanding that objects continue to exist even when they cannot be seen, heard, or touched (■**Figure 9.11**).

Preoperational Stage

During the **preoperational stage** (roughly ages 2 to 7), language advances significantly, and the child begins to think symbolically—using symbols, such as words, to represent concepts. Three other qualities characterize this stage: *cannot perform mental operations, egocentrism,* and *animism.*

1. *Cannot perform mental operations* Piaget labeled this period "preoperational" because the child lacks *operations,* meaning the ability to imagine an action and mentally reverse it. For instance, a preoperational child would not understand that milk poured into a tall, thin glass is not "more" than the same amount poured into short, wide glass. This is because they lack reversibility, and the concept of **conservation**—the principle that certain characteristics (such as volume) stay the same, even though appearances may change.

2. *Egocentrism* Children at this stage are **egocentric**, which refers to the preoperational child's limited ability to distinguish between his or her own perspective and someone else's.

Egocentrism is not the same as "selfishness." Preschoolers who move in front of you to get a better view of the TV, or repeatedly ask questions while you are talking on the telephone, are not being selfish. They are demonstrating their natural limits and egocentric thought processes. Children in this stage naively assume that others see, hear, feel, and think exactly as they do. Consider the following telephone conversation between a 3-year-old, who is at home, and her mother, who is at work:

MOTHER: Emma, is that you?
EMMA: (Nods silently.)
MOTHER: Emma, is Daddy there? May I speak to him?
EMMA: (Twice nods silently.)

Egocentric preoperational children fail to understand that the phone caller cannot see their nodding head. Charming as this is, preoperational children's egocentrism also sometimes leads them to believe their "bad thoughts" caused their sibling or parent to get sick or that their misbehavior caused their parents' marital problems. Because they think the world centers on them, they often cannot separate reality from what goes on inside their own head.

3. *Animistic thinking* During this stage, children generally believe objects, such as the moon, trees, clouds, and bars of soap, have motives, feelings, and intentions (for example, "the moon follows me when I walk," "dark clouds are angry" and "soap sinks to the bottom of the bathtub because it is tired"). *Animism* refers to the belief that all things are living (or animated).

Concrete Operational Stage

At approximately age 7, children enter the **concrete operational stage**. During this time, many important thinking skills emerge. However, as the name implies, thinking tends to be limited to *concrete*, tangible objects and events. As most parents know, children now stop believing in Santa Claus because they logically conclude that one man can't deliver presents to everyone in one night.

Because they understand the concept of *reversibility*, concrete operational children also can now successfully perform "operations." They recognize that certain physical attributes (such as volume) remain unchanged, although the outward appearance is changed.

Formal Operational Stage

The final period in Piaget's theory, the **formal operational stage**, typically begins around age 11. In this stage, children begin to apply their operations to abstract concepts in addition to concrete objects. They also become capable of hypothetical thinking ("What if?"), which allows systematic formulation and testing of concepts.

For example, before filling out applications for part-time jobs, adolescents may think about possible conflicts with school and friends, the number of hours they want to work, and the kind of work for which they are qualified. Formal operational thinking also allows the adolescent to construct a well-reasoned argument based on hypothetical concepts and logical processes. Consider the following argument:

1. If you hit a glass with a feather, the glass will break.

2. You hit the glass with a feather.

What is the logical conclusion? The correct answer, "The glass will break," is contrary to fact and direct experience. Therefore, the child in the concrete operational stage would have difficulty with this task, whereas the formal operational thinker understands that this problem is about abstractions that need not correspond to the real world.

Along with the benefits of this cognitive style come several problems. Adolescents in the early stages of the formal operational period demonstrate a type of egocentrism different from that of the preoperational child. Adolescents certainly recognize that others have unique thoughts and perspectives. However, they may fail to differentiate between what they are thinking and what others are thinking. Can you see how this might explain why when adolescents get a new haircut or fail to make the sports team, they may be overly concerned about how others will react? Instead of considering that everyone is equally wrapped up in his or her own appearance, concerns, and plans, they tend to believe that they are the center of others' thoughts and attentions. David Elkind (1967, 2007) referred to this as the *imaginary audience*.

In sharp, ironic contrast to believing that others are always watching and evaluating them (the imaginary audience), Piaget also contended that adolescents typically believe they are special and unique. They alone are having insights or difficulties that no one else understands or experiences. Sadly, these feelings of special uniqueness, known as the *personal fable*, are associated with several forms of risk taking, such as engaging in sexual intercourse without protection, driving dangerously, indoor tanning, and experimenting with drugs (Banerjee et al., 2015; Landicho et al., 2014). Adolescents apparently recognize the dangers of risky activities, but they believe the rules and statistics just don't apply to them.

In sum, the imaginary audience apparently results from an inability to differentiate the self from others, whereas the personal fable may be a product of differentiating too much. Thankfully, these two forms of adolescent egocentrism tend to decrease during later stages of the formal operational period.

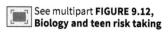
See multipart **FIGURE 9.12,**
Biology and teen risk taking

© Radka Linkova/Alamy Inc.

Brain, Cognition, and Behavior in Adolescence

Rather than simply describing and labeling the adolescent's exaggerated self-consciousness, feelings of special uniqueness, and risky behaviors as the result of the imaginary audience and personal fable, psychologists now believe these effects may be largely due to their less-than-fully-developed frontal lobes (Casey et al., 2014; Pokhrel et al., 2013). In contrast to the rapid synaptic growth experienced in the earlier years, the adolescent's brain actively destroys (prunes) unneeded connections (**Figure 9.12**). Although it may seem counterintuitive, this pruning actually improves brain functioning by making the remaining connections between neurons more efficient. Nevertheless, full maturity of the frontal lobes is not accomplished until the mid-twenties.

Brain, Cognition, and Behavior in Adulthood

The public and most researchers long believed aging was inevitably accompanied by declining cognitive abilities and widespread death of neurons in the brain. Although this decline does happen with degenerative disorders like Alzheimer's disease, it is no longer believed to be an inevitable part of normal aging (Hillier & Barrow, 2011; Whitbourne & Whitbourne, 2014). It's also important to remember that age-related cognitive problems are not on a continuum with Alzheimer's disease. That is, normal forgetfulness does not mean that serious dementia is around the corner.

Aging does seem to take its toll on the *speed* of information processing (Chapter 7). Decreased speed of processing may reflect problems with *encoding* (putting information into long-term storage) and *retrieval* (getting information out of storage). If memory is like a filing system, older people may have more filing cabinets, and it may take them longer to initially file and later retrieve information.

Although mental speed declines with age, general information processing and much of memory ability are largely unaffected by the aging process (Carey, 2014; Ramscar et al., 2014; Whitbourne & Whitbourne, 2014). Despite their concerns about "keeping up with 18-year-olds," older returning students often do as well or better than their younger counterparts in college classes.

This superior performance by older adult students may be due, in part, to their generally greater academic motivation, but it also reflects the importance of prior knowledge. Cognitive psychologists have clearly demonstrated that the more people know, the easier it is for them to lay down new memories (Goldstein, 2014; Matlin, 2016). Older students, for instance, generally find this chapter on development easier to master than younger students. Their interactions with children and greater, accumulated knowledge about life changes create a framework upon which to hang new information.

In short, the more you know, the more you can learn. Thus, more education and having an intellectually challenging life may help you stay mentally sharp in your later years—another good reason for going to college and life-long learning (Branco et al., 2014; Huang & Zhou, 2013; Sobral et al., 2015).

See **TUTORIAL VIDEO:**
Attitudes Toward Aging

Ageism Versus the Age-Related Positivity Effect

Television, magazines, movies, and advertisements generally portray aging as a time of balding and graying hair, sagging body parts, poor vision, hearing loss, and, of course, no sex life. Can you see how our personal fears of aging and death, combined with these negative media portrayals, contribute to our society's widespread **ageism**—prejudice and discrimination based on physical age (Lamont et al., 2015; West, 2015)? Sadly, a recent study found that older adults who reported discrimination based on their age had significantly lower physical and emotional health, and greater declines in health, than those who did not report such discrimination (Sutin et al., 2015). The good news is that advertisers have noted the large number of aging baby boomers, and are now producing ads with a more positive and accurate portrayal of aging as a time of vigor, interest, and productivity (■**Figure 9.13**).

Further positive news comes from research showing that as we age, our overall well-being tends to increase—contrary to what negative stereotypes about aging might predict (Kern et al., 2014; Sutin et al., 2013). This increased happiness and well-being is often attributed to the **age-related positivity effect**, showing that older adults tend to prefer positive over negative information (Carstensen, 1993, 2006; Reed et al., 2014).

See multipart **FIGURE 9.13, Achievement in later years**

Mark Wilson/Getty Images

Vygotsky Versus Piaget

As influential as Piaget's account of cognitive development has been, there are other important theories, and criticisms of Piaget, to consider. For example, Russian psychologist Lev Vygotsky emphasized the sociocultural influences on a child's cognitive development, rather than Piaget's focus on internal schemas (Vygotsky, 1962). According to Vygotsky, children construct knowledge through their culture, language, and collaborative social interactions with more experienced thinkers (Mahn & John-Steiner, 2013; Scott, 2015; Yasnitsky, 2015). Unlike Piaget, Vygotsky believed that adults play an important instructor role in development, and that this instruction is particularly helpful when it falls within a child's **zone of proximal development (ZPD)** (■**Figure 9.14**).

Having briefly discussed Vygotsky's alternative theory, let's consider two major criticisms of Piaget—*underestimated abilities* and *underestimated genetic and cultural influences*. As you've just seen, Piaget believed that infancy and early childhood were a time of extreme egocentrism, in which children have little or no understanding of the perspective of others. However, later research finds that empathy develops at a relatively young age (■**Figure 9.15**). Even newborn babies tend to cry in response to the cry of another baby (Diego & Jones, 2007; Geangu et al., 2010). Also, preschoolers adapt their speech by using shorter, simpler expressions when talking to 2-year-olds as compared to talking with adults.

Piaget's model, like other stage theories, has also been criticized for not sufficiently taking into account genetic and cultural differences (Newman & Newman, 2015; Shweder, 2011). During Piaget's time, the genetic influences on cognitive abilities were poorly understood, but as in the case of epigenetics, there has been a rapid explosion of information in this field in the last few years. In addition, formal education and specific cultural experiences can significantly affect cognitive development. Consider the following example from a researcher attempting to test the formal operational skills of a farmer in Liberia (Scribner, 1977):

RESEARCHER: All Kpelle men are rice farmers. Mr. Smith is not a rice farmer. Is he a Kpelle man?

KPELLE FARMER: I don't know the man. I have not laid eyes on the man myself.

Instead of reasoning in the "logical" way of Piaget's formal operational stage, the Kpelle farmer reasoned according to his specific cultural and educational training, which apparently emphasized personal knowledge. Not knowing Mr. Smith,

See **FIGURE 9.14, Vygotsky's zone of proximal development (ZPD)**

Upper limit (tasks beyond reach at present)

Zone of proximal development (ZPD) (tasks achievable with guidance)

Lower limit (tasks achieved without help)

© omgimages/iStockphoto

See **FIGURE 9.15, Are preoperational children always egocentric?**

Ermolaev Alexander/Shutterstock

the Kpelle farmer did not feel qualified to comment on him. Thus, Piaget's theory may have underestimated the effect of culture on a person's cognitive functioning.

Despite criticisms, Piaget's contributions to psychology are enormous. As one scholar put it, "assessing the impact of Piaget on developmental psychology is like assessing the impact of Shakespeare on English literature or Aristotle on philosophy—impossible" (Beilin, 1992, p. 191).

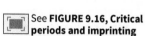
Answer the **Concept Check** questions.

9.4 ADDITIONAL FACTORS IN DEVELOPMENT

In addition to physical and cognitive development, developmental psychologists study the way social and emotional factors affect development over the life span. We begin with a focus on *attachment* and *parenting styles*.

Attachment

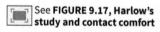
See **FIGURE 9.16, Critical periods and imprinting**

An infant arrives in the world with a multitude of behaviors that encourage a strong emotional bond of **attachment** with special others. Although research has focused on the attachment between mother and child, infants also form attachment bonds with fathers, grandparents, and other caregivers.

In studying attachment behavior, researchers are often divided along the lines of the now familiar nature-versus-nurture debate. Those who advocate the nurture position suggest attachment results from a child's interactions and experiences with his or her environment. In contrast, the nativist, or innate, position cites John Bowlby's work (1969, 2000). He proposed that newborn infants are biologically equipped with verbal and nonverbal behaviors (such as crying, clinging, smiling) and with "following" behaviors (such as crawling and walking after the caregiver) that elicit instinctive nurturing responses from the caregiver. Early studies of **imprinting**, forming attachments during critical periods, further support the biological argument for attachment (■ **Figure 9.16**).

See **FIGURE 9.17, Harlow's study and contact comfort**

Studies have found numerous benefits to a child's good attachment, including lower levels of aggressive behavior, fewer sleep problems, and less social withdrawal (Ding et al., 2014). But as you've seen with the sad case of Genie, some children never form appropriate, loving attachments. What happens to these children? Researchers have investigated this question in two ways: They have looked at children and adults who spent their early years in institutions without the stimulation and love of a regular caregiver, as well as those who lived at home but were physically isolated under abusive conditions.

Tragically, infants raised in impersonal or abusive surroundings suffer from a number of problems. They seldom cry, coo, or babble; they become rigid when picked up; and they have few language skills. As for their social-emotional development, they tend to form shallow or anxious relationships. Some appear forlorn, withdrawn, and uninterested in their caretakers, whereas others seem insatiable in their need for affection. They also tend to show intellectual, physical, and perceptual deficiencies, along with increased susceptibility to infection, and neurotic "rocking" and isolation behaviors. There are even cases where healthy babies who were well fed and kept in clean diapers—but seldom held or stimulated—actually died from lack of attachment (Bowlby, 2000; Duniec & Raz, 2011; Spitz & Wolf, 1946). Some research suggests that childhood emotional abuse and neglect is as harmful, in terms of long-term mental problems, as physical and sexual abuse (Spinazzola et al., 2014).

Touch

Harry Harlow and his colleagues (1950, 1971) also investigated the variables that might affect attachment. They created two types of wire-framed surrogate (substitute) "mother" monkeys: one covered by soft terry cloth and one left uncovered (■ **Figure 9.17**). The infant monkeys were fed by either the cloth or the wire mother, but they otherwise had access to both mothers. The researchers found that regardless of which surrogate was feeding them, the infant monkeys overwhelmingly preferred the soft, cloth surrogate–even when the wire surrogate was the one providing the food.

Thanks in part to Harlow's research, psychologists discovered that *contact comfort*, the pleasurable, tactile sensations provided by a soft and cuddly "parent," is a powerful contributor to attachment ( **Figure 9.18**). Further support comes from a study described in Chapter 4, which shows that touch is an essential part of normal human infant development. This is why hospitals now encourage premature babies to receive "kangaroo care," in which babies have immediate and repeated skin-to-skin contact with a parent (Head, 2014; Metgud & Honap, 2015).

See **FIGURE 9.18, The power of touch**

Marili Forastieri/Digital Vision/Getty Images

Styles of Attachment

Although physical contact between caregiver and child appears to be an innate, biological part of attachment, Mary Ainsworth and her colleagues (1967, 1978) discovered several interesting differences in the type and level of human attachment (**Figure 9.19**). For example, infants with a secure attachment style generally had caregivers who were sensitive and responsive to their signals of distress, happiness, and fatigue. In contrast, anxious/avoidant infants had caregivers who were aloof and distant, and anxious/ambivalent infants had inconsistent caregivers, who alternated between strong affection and indifference. Caregivers of disorganized/disoriented infants tended to be abusive or neglectful (Ainsworth, 1967; Ainsworth et al., 1978; Zeanah & Gleason, 2015).

As a critical thinker, can you offer additional explanations for attachment, other than differences in caregivers? What about differences in the infants themselves? Researchers have found that the temperament of the child does play an important role. An infant who is highly anxious and avoidant might not accept or respond to a caregiver's attempts to comfort and soothe. In addition, children and their parents share genetic tendencies and attachment patterns may reflect these shared genes. Finally, critics have suggested that Ainsworth's research does not account for cultural variations, such as cultures that encourage infants to develop attachments to multiple caregivers (Rothbaum et al., 2007; van IJzendoorn & Bakermans-Kranenburg, 2010).

See **TUTORIAL VIDEO: The Strange Situation: Measuring Attachment**

See multipart **FIGURE 9.19, Research on infant attachment**

1. The baby plays while the mother is nearby.
2. A stranger enters the room, speaks to the mother, and approaches the child.
3. The mother leaves and the stranger stays in the room with an unhappy baby.
4. The mother returns and the stranger leaves.
5. The baby is reunited with the mother.

Attachment Styles in Adulthood

In addition to finding varying levels of infant attachment to parents, researchers have also examined adult attachment patterns independent of their earlier infant patterns, with several interesting and/or troublesome results. For example, a secure attachment pattern is associated with higher subjective well-being (SWB), whereas adolescents and young adult couples with avoidant and anxious attachment patterns, show more depressive symptoms (Desrosiers et al., 2014; Galinha et al., 2014). Another study found an association between pathological jealousy and the anxious/ambivalent style of attachment (Costa et al., 2015).

Researchers also looked at how varying types of attachment as infants might shape our later adult styles of romantic love (Fraley & Roisman, 2015; Salzman et al., 2014; Sprecher & Fehr, 2011). If we developed a secure, anxious/ambivalent, anxious/avoidant, or disorganized/disoriented style as infants, we tend to follow these same patterns in our adult approach to intimacy and affection. For example, young adults who experienced either unresponsive or overintrusive parenting during childhood are more likely to avoid committed romantic relationships as adults (Dekel & Farber, 2012).

See **TUTORIAL VIDEO: Attachment Through the Lifespan**

Parenting Styles

How much of our personality comes from the way our parents treat us as we're growing up? Researchers since the 1920s have studied the effects of different methods of childrearing on children's behavior, development, and mental health. For example, one interesting study found that teenagers whose parents used a controlling style, such as withholding love or

See **VIRTUAL FIELD TRIP: A Guide to Parenting**

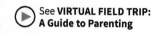 See **TABLE 9.3: Parenting Styles**

creating feelings of guilt-later have more difficulty working out conflicts with friends and romantic partners (Oudekerk et al., 2015).

Other studies by Diana Baumrind (1980, 2013) found that parenting styles could be reliably divided into four broad patterns—*permissive-neglectful, permissive-indulgent, authoritarian*, and *authoritative*—which can be differentiated by their degree of *control/ demandingness* (C) and *warmth/responsiveness* (W) (■ **Table 9.3**).

Do fathers differ from mothers in their parenting style? Until recently, the father's role in discipline and child care was largely ignored. But fathers in Western countries have begun to take a more active role in child-rearing, and there has been a corresponding increase in research. From these studies, we now know that children do best with authoritative dads, and with fathers who are absorbed with, excited about, and responsive to their newborns. Researchers also find few differences in the way children form attachments to either parent (Lopez & Hsu, 2002; Talitwala, 2007). In general, fathers are just as responsive, nurturing, and competent as mothers when they assume child-care responsibilities.

Moral Development

In Europe, a cancer-ridden woman was near death, but an expensive drug existed that might save her. The woman's husband, Heinz, begged the druggist to sell the drug at a lower price, or to let him pay later, but he refused. Heinz became desperate and broke into the druggist's store and stole it. (Adapted from Kohlberg, 1964, pp. 18–19)

Was Heinz right to steal the drug? What do you consider moral behavior? Is morality "in the eye of the beholder," or are there universal truths and principles? Whatever your answer, your ability to think, reason, and respond to Heinz's dilemma demonstrates another type of development that is very important to psychology—*morality*.

One of the most influential researchers in moral development was Lawrence Kohlberg (1927–1987). He presented what he called "moral stories" like the Heinz dilemma to people of all ages, not to see whether they judged Heinz right or wrong but to examine the reasons they gave for their decisions. On the basis of his findings, Kohlberg (1964, 1984) developed a model of moral development, with three broad levels and six distinct stages (**Process Diagram 9.3**). Individuals at each stage and level may or may not support Heinz's stealing of the drug, but Kohlberg believed that looking at their moral reasoning was important because it correlated with their moral behavior.

Kohlberg's ideas have led to considerable research on how we think about moral issues But his theories have also been the focus of three major areas of criticism:

1. *Moral reasoning versus behavior* Are people who achieve higher stages on Kohlberg's scale really more moral than others? Or do they just "talk a good game"? Some researchers have shown that a person's sense of moral identity, meaning the use of moral principles to define oneself, is often a good predictor of his or her behavior in real-world situations (Johnston et al., 2013; Stets & Carter, 2012). But others have found that situational factors are better predictors of moral behavior (Bandura, 1986, 2008; Frimer et al., 2014; Noval & Stahl, 2015). For example, research participants are more likely to steal when they are told the money comes from a large company rather than from individuals (Greenberg, 2002). And both men and women will tell more sexual lies during casual relationships than during close relationships (Williams, 2001).

2. *Cultural differences* Studies confirm that children from a variety of cultures generally follow Kohlberg's model and progress sequentially from his first level, the *preconventional*, to his second, the *conventional* (Rest et al., 1999; Snarey, 1995). However, other studies find differences among cultures (Csordas, 2014; Vozzola, 2014). For example, comparisons of responses to Heinz's moral dilemma show that Europeans and Americans tend to consider whether they like or identify with the victim in questions of morality. In contrast, Hindu Indians consider social responsibility and personal concerns two separate issues (Miller & Bersoff, 1998). Researchers suggest that the difference reflects the Indians' broader sense of social responsibility.

Furthermore, in India, Papua New Guinea, and China, as well as in Israeli kibbutzim, people don't choose between the rights of the individual and the rights of society (as the top levels of Kohlberg's model require). Instead, most people seek a compromise solution that

Process Diagram 9.3

Kohlberg's Stages of Moral Development

Step 1: Preconventional Level Morality is based on rewards, punishment, and exchange of favors

Step 2: Conventional Level Moral judgments are based on compliance with the rules and values of society

Step 3: Postconventional Level Individuals develop personal standards for right and wrong.

POSTCONVENTIONAL LEVEL

CONVENTIONAL LEVEL

PRECONVENTIONAL LEVEL

(Stages 1 and 2—birth to adolescence) Moral judgment is *self-centered*. What is right is what one can get away with, or what is personally satisfying. Moral understanding is based on rewards, punishments, and the exchange of favors.

① Punishment-obedience orientation

Focus is on self-interest—obedience to authority and avoidance of punishment. Because children at this stage have difficulty considering another's point of view, they also ignore people's intentions.

② Instrumental-exchange orientation

Children become aware of others' perspectives, but their morality is based on reciprocity—an equal exchange of favors.

(Stages 3 and 4—adolescence to young adulthood) Moral reasoning is *other-centered*. Conventional societal rules are accepted because they help ensure the social order

③ Good-child orientation

Primary moral concern is being nice and gaining approval; judges others by their intentions—"His heart was in the right place."

④ Law-and-order orientation

Morality based on a larger perspective—societal laws. Understanding that if everyone violated laws, even with good intentions there would be chaos.

(Stages 5 and 6—adulthood) Moral judgments based on *personal standards for right and wrong*. Morality also defined in terms of abstract principles and values that apply to all situations and societies

⑤ Social-contact orientation

Appreciation for the underlying purposes served by laws Societal laws are obeyed because of the "social contract," but they can be morally disobeyed if they fail to express the will of the majority or fail to maximize social welfare.

⑥ Universal-ethics orientation

"Right" is determined by universal ethical principles (e.g., nonviolence, human dignity, freedom) that moral authorities might view as compelling or fair. These principles apply whether or not they conform to existing laws.

Sources: Based on Kohlberg, L. "Stage and Sequence: The Cognitive Developmental Approach to Socialization," in D.A. Goslin, The handbook of socialization theory and research. Chicago: Rand McNally, 1969, p. 376 (Table 6.2).

See **FIGURE 9.20**, Gilligan versus Kohlberg

Jonathan Nourok/Stone/Getty Images

accommodates both interests. Thus, Kohlberg's standard for judging the highest level of morality (the postconventional) may be more applicable to cultures that value individualism over community and interpersonal relationships.

3. *Possible gender bias* Researcher Carol Gilligan criticized Kohlberg's model because on his scale, women often tend to be classified at a lower level of moral reasoning than men (▣ **Figure 9.20**). Gilligan suggested that this difference occurred because Kohlberg's theory emphasizes values more often held by men, such as rationality and independence, while deemphasizing common female values, such as concern for others and belonging (Gilligan, 1977, 1993; Kracher & Marble, 2008). However, most follow-up studies of Gilligan's specific theory, have found few, if any, gender differences (Friesdorf et al., 2015; Gibbs, 2014).

Personality Development

As an infant, did you lie quietly and seem oblivious to loud noises? Or did you tend to kick and scream and respond immediately to every sound? Did you respond warmly to people, or did you fuss, fret, and withdraw? Your answers to these questions help determine what developmental psychologists call your **temperament**, an individual's innate disposition, behavioral style, and characteristic emotional response.

Thomas and Chess's Temperament Theory

One of the earliest and most influential theories regarding temperament came from the work of psychiatrists Alexander Thomas and Stella Chess (Thomas & Chess, 1977, 1987). Thomas and Chess found that approximately 65 percent of the babies they observed could be reliably separated into three categories:

1. *Easy children* These infants were happy most of the time, relaxed and agreeable, and adjusted easily to new situations (approximately 40 percent).

2. *Difficult children* Infants in this group were moody, easily frustrated, tense, and overreactive to most situations (approximately 10 percent).

3. *Slow-to-warm-up children* These infants showed mild responses, were somewhat shy and withdrawn, and needed time to adjust to new experiences or people (approximately 15 percent).

Follow-up studies have found that certain aspects of these temperament styles tend to be consistent and enduring throughout childhood and even adulthood (Bates et al., 2014; Sayal et al., 2014). That is not to say every shy, cautious infant ends up a shy adult. Many events take place between infancy and adulthood that shape an individual's development. Moreover, culture can also influence infant temperament. One recent study found that Dutch babies tend to be happier and easier to soothe, whereas babies born in the United States are typically more active and vocal (Sung et al., 2015). These temperament differences are thought to reflect cultural differences in parenting styles and values.

One of the most influential factors in early personality development is *goodness of fit* between a child's nature, parental behaviors, and the social and environmental setting (Mahoney, 2011; Seifer et al., 2014). For example, a slow-to-warm-up child does best if allowed time to adjust to new situations. Similarly, a difficult child thrives in a structured, understanding environment but not in an inconsistent, intolerant home. Alexander Thomas, the pioneer of temperament research, thinks parents should work with their child's temperament rather than trying to change it. Can you see how this idea of goodness of fit is yet another example of how nature and nurture interact?

⏵ See **TUTORIAL VIDEO:** Erikson's Psychosocial Theory

Erikson's Psychosocial Theory

Like Piaget and Kohlberg, Erik Erikson developed a stage theory of development. He identified eight **psychosocial stages** of social development, each stage marked by a "psychosocial" crisis or conflict that must be successfully resolved for proper future development (**Process Diagram 9.4**).

The name for each psychosocial stage reflects the specific crisis encountered at that stage and two possible outcomes. The crisis or task of most young adults is *intimacy*

Process Diagram 9.4

Erikson's Eight Stages of Psychosocial Development

Stage 1
Trust versus mistrust (birth–age 1)

Infants learn to *trust* or *mistrust* their caregivers and the world based on whether or not their needs—such as food, affection, safety—are met.

Stage 2
Autonomy versus shame and doubt (ages 1–3)

Toddlers start to assert their sense of independence (*autonomy*). If caregivers encourage this self-sufficiency, the toddler will learn to be independent versus feeling *shame* and *doubt*.

Stage 3
Initiative versus guilt (ages 3–6)

Preschoolers learn to *initiate* activities, which develops their self-confidence and a sense of social responsibility. If not, they feel irresponsible, anxious, and *guilty*.

Stage 4
Industry versus inferiority (ages 6–12)

Elementary school-aged children, who succeed in learning new, productive life skills, develop a sense of pride and competence (*industry*). Those who fail to develop these skills feel inadequate and unproductive (*inferior*).

Stage 5
Identity versus role confusion (ages 12–20)

Adolescents develop a coherent and stable self-definition (identity) by exploring many roles and deciding who or what they want to be in terms of career, attitudes, etc. Failure to resolve this **identity crisis** may lead to apathy, withdrawal and/or *role confusion*.

Stage 6
Intimacy versus isolation (early adulthood)

Young adults form lasting, meaningful relationships, which help them develop a sense of connectedness and *intimacy* with others. If not, they become psychologically *isolated*.

Stage 7
Generativity versus stagnation (middle adulthood)

The challenge for middle-aged adults is in nurturing the young, and making contributions to society through their work, family, or community activities. Failing to meet this challenge leads to self-indulgence and a sense of *stagnation*.

Stage 8
Ego integrity versus despair (late adulthood)

During this stage, older adults reflect on their past. If this reflection reveals a life well-spent, the person experiences self-acceptance and satisfaction (*ego integrity*). If not, he or she experiences regret and deep dissatisfaction (*despair*).

versus isolation. This age group's developmental challenge is establishing deep, meaningful relations with others. Those who don't meet this challenge risk social isolation. Erikson believed that the more successfully we overcome each psychosocial crisis, the better chance we have to develop in a healthy manner (Erikson, 1950).

Many psychologists agree with Erikson's general idea that psychosocial crises, which are based on interpersonal and environmental interactions, do contribute to personality development (Marcia & Josselson, 2013; Svetina, 2014). However, Erikson also has his critics (Arnett, 2015; Spano et al., 2010). Some evidence now suggests that people may approach these valued life tasks at different ages, and potentially in different orders. Furthermore, identity development often continues through the late 20s, and is not simply a task approached by adolescents (Carlsson et al., 2015). In addition, Erikson's psychosocial stages are difficult to test scientifically, and the labels he used to describe the eight stages may not be entirely appropriate cross-culturally. For example, in individualistic cultures, *autonomy* is highly preferable to *shame and doubt*. But in collectivist cultures, the preferred resolution might be *dependence* or *merging relations* (Berry et al., 2011).

Despite their limits, Erikson's stages have greatly contributed to the study of North American and European psychosocial development. Also, by suggesting that development continues past adolescence, Erikson's theory has encouraged ongoing research and theory development across the life span.

Final Note

As you've discovered in this chapter, our physical, cognitive, moral, and psychosocial development continue throughout life. And we firmly believe that, with this increased knowledge and understanding, you'll be better equipped to deal with life's inevitable, changes, challenges, and difficulties. You have our best wishes for a happy, successful, and satisfied life.

Q Answer the **Concept Check** questions.

WP LS Go to your WileyPLUS Learning Space course for video episodes, examples, art, tables, Concept Checks, practice, and other pedagogical resources that will help you succeed in this course.

MOTIVATION AND EMOTION

 Go to your WileyPLUS Learning Space course for video episodes, examples, art, tables, Concept Checks, practice, and other pedagogical resources that will help you succeed in this course.

10.1 THEORIES OF MOTIVATION

Years of research on motivation has created six major theories, which fall into three general categories—*biological*, *psychological*, and *biopsychosocial*. While studying these theories, try to identify which theory best explains your personal behaviors, such as going to college or choosing a lifetime partner. This type of personal focus will not only improve your exam performance but also may lead to increased self-knowledge and personal motivation!

See **STUDY ORGANIZER 10.1: Six Major Theories of Motivation**

Biological Theories

Many theories of **motivation** focus on inborn biological processes that control behavior. Among these biologically oriented theories are *instinct*, *drive-reduction*, and *arousal* theories.

One of the earliest researchers, William McDougall (1908), proposed that humans had numerous instincts, such as repulsion, curiosity, and self-assertiveness. Other researchers later added their favorite instincts, and by the 1920s, the list of recognized instincts had become impossibly long. One researcher found listings for more than 10,000 human instincts (Bernard, 1924).

See multipart **FIGURE 10.1, Do humans have instincts?**

In addition, the label *instinct* led to unscientific, circular explanations—"men are aggressive because they are instinctively aggressive" or "women are maternal because they have a natural maternal instinct." However, in recent years, a branch of biology called sociobiology (Weingart et al., 2013; Wilson, 2013) has revived the case for **instincts** when strictly defined as a *fixed, unlearned response patterns found in almost all members of a species* (■ **Figure 10.1**).

In the 1930s, the concepts of drive and drive reduction began to replace the theory of instincts. According to **drive-reduction theory**, when biological needs such as for food, water, and oxygen are unmet, a state of tension known as a *drive* is created (Hull, 1952). The organism is then motivated to reduce that drive. Drive-reduction theory is based largely on the biological concept of **homeostasis**, a term that literally means "standing still" and describes the body's natural tendency to maintain a state of internal balance (**Process Diagram 10.1**).

© imaginary_nl/iStockphoto

This Process Diagram contains essential information NOT found elsewhere in the text, which is likely to appear on quizzes and exams. Be sure to study it CAREFULLY!

Process Diagram 10.1

Drive-Reduction Theory

When we are hungry or thirsty, the disruption of our normal state of equilibrium creates a drive that motivates us to search for food or water. Once action is taken, and the need is satisfied, homeostasis is restored, and our motivation decreases.

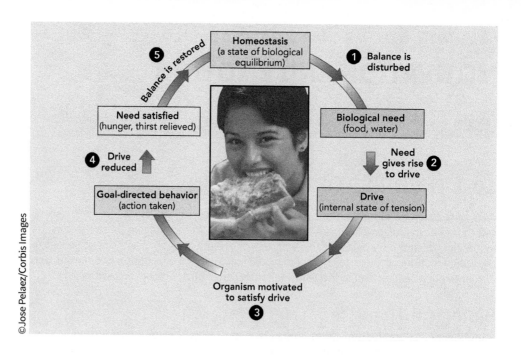

©Jose Pelaez/Corbis Images

See **FIGURE 10.2, Optimal level of arousal**

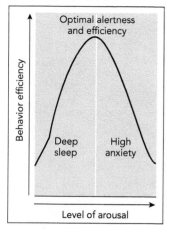

In addition to having obvious biological needs, humans and other animals are innately curious and require a certain amount of novelty and complexity from the environment.

According to **optimal-arousal theory**, organisms are motivated to achieve and maintain an optimal level of arousal that maximizes their performance. Both too much and too little arousal diminish performance (**Figure 10.2**). The desired amount of arousal also may vary from person to person.

Interestingly, according to the **Yerkes-Dodson law**, our performance on difficult tasks is optimal at relatively low arousal states, whereas the reverse is true at high arousal states (see **Figure 10.3**). Can you see how this information could be helpful in your everyday life? Have you ever "blanked out" during a very important exam and couldn't answer questions that you thought you knew? This is probably due to your overarousal. And it helps explain why *over-learning* is so important—especially if you're someone who suffers from test anxiety. As we discussed in Chapter 1, you need to study and practice completing the self-tests within this text, until the information is firmly locked in place. On the other hand, if you find yourself getting bored and distracted while studying, do something to raise your arousal level, such as drinking coffee, taking a walk, and/or reminding yourself how important it is to do well on the exam.

In addition to the personal applications of the optimal arousal theory, can you see how understanding the differing levels of sensation seeking might also apply to your everyday life? Research suggests that four distinct factors characterize sensation seeking (LaBrie et al., 2014; Lauriola et al., 2014; Zuckerman & Aluja, 2015).

See **FIGURE 10.3, The Yerkes-Dodson law**

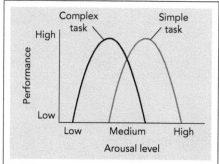

1. Thrill and adventure seeking (skydiving, driving fast, or traveling to an unusual, "off the beaten path" location)

2. Experience seeking (unusual friends, exotic foods or restaurants, drug experimentation)

3. Disinhibition ("letting loose")

4. Susceptibility to boredom (lower tolerance for repetition and sameness)

Can you see how being very high or very low in sensation seeking might cause problems in relationships with two individuals who score toward opposite extremes? This is true not just between partners or spouses, but also between parent and child or therapist and client. There also might be job difficulties for high-sensation seekers in routine clerical or assembly-line jobs, and for low-sensation seekers in highly challenging and variable occupations.

Psychological Theories

Instinct and drive-reduction theories explain some motivations, but why do we continue to eat after our biological need has been completely satisfied? Why do some of us work overtime when our salary is sufficient to meet all basic biological needs? These questions are best answered by psychosocial theories that emphasize incentives and cognition.

Unlike drive-reduction theory, which states that internal factors *push* people in certain directions, **incentive theory** maintains that external stimuli *pull* people toward desirable goals or away from undesirable ones. Most of us initially eat because our hunger "pushes" us (drive-reduction theory). But the sight of apple pie or ice cream too often "pulls" us toward continued eating (incentive theory).

What about using rewards to increase motivation? Researchers interested in increasing good grades, adherence to school rules, and regular school attendance found that paying middle-school students $2.00 daily for each goal met was enough incentive to produce higher reading test scores, especially for boys and those with disciplinary problems in the past (Fryer, 2010). But as you'll discover later in this chapter, this type of *extrinsic reward* (paying students for achieved goals) may create problems of its own.

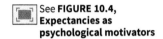

See **FIGURE 10.4, Expectancies as psychological motivators**

Other psychological explanations, such as *cognitive theories*, suggest that motivation is directly affected by our *attributions*, which involve the ways in which we interpret or think about our own and others' actions. Imagine that you receive a high grade on a test in your psychology course. You can interpret that grade in several ways: You earned it because you really studied, you "lucked out" because the test was easy, or the textbook was exceptionally interesting and (our preference!). Not surprisingly, people who attribute their successes to personal ability and effort tend to work harder toward their goals than people who attribute their successes to luck or external factors (Gorges & Göke, 2015; Weiner, 1972, 1982; Zhou & Urhahne, 2013).

Expectancies, or what we believe or assume will happen, are also important to motivation (Durik et al., 2015; Kim & Pekrun, 2014; Tamir et al., 2015). (See **Figure 10.4**). If you anticipate that you will receive a promotion at work, you're more likely to work overtime for no pay than if you do not expect a promotion.

Don Smetzer/PhotoEdit

Biopsychosocial Theories

As we've just shown, psychological research generally emphasizes either biological or psychosocial factors (nature or nurture). But biopsychosocial factors almost always provide the best explanation. One researcher who believed in biopsychosocial factors as predictors of motivation was Abraham Maslow (1954, 1999). Maslow believed we all have numerous biological AND psychoscial needs that compete for fulfillment, but that some needs are more important than others. For example, your need for food and shelter is generally more important than your need for good grades.

Maslow's **hierarchy of needs** prioritizes human needs, starting with survival needs (which must be met before others) at the bottom. In contrast, social and esteem needs are near the top (**Figure 10.5**).

See **FIGURE 10.5, Maslow's hierarchy of needs**

Maslow's hierarchy of needs seems intuitively correct: A starving person would first look for food, then love and friendship, and then self-esteem. This prioritizing, and the highest, elusive level of **self-actualization**, the inborn drive to develop all our talents and capabilities, are important factors in motivation. And Maslow's work has had a major impact on psychology, economics, and other related fields (Henwood et al., 2015; Hsu, 2016; van Lenthe et al., 2015). For example, standard marketing texts often use Maslow's hierarchy of needs to imply that brand consumption is a natural, driving force in shopping behaviors. Ironically, Maslow's work and humanistic ideals would emphasize less, not more, consumption (Hackley, 2007).

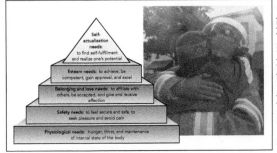

Nam Y Huh/©AP/Wide World Photos

Q Answer the **Concept Check** questions.

Maslow's critics argue that parts of Maslow's theory are poorly researched and biased toward Western preferences for individualism. Furthermore, his theory presupposes that the lower needs must be satisfied before someone can achieve self-actualization, but people sometimes seek to satisfy higher-level needs even when their lower-level needs have not been met (Cullen & Gotell, 2002; Kress et al., 2011; Neher, 1991). As you know, protestors all over the world have used starvation as a way to protest unfair laws and political situations.

10.2 MOTIVATION AND BEHAVIOR

Why do people put themselves in dangerous situations? Why do salmon swim upstream to spawn? Behavior results from many motives. We discuss the need for sleep in Chapter 5, and we look at aggression, altruism, and interpersonal attraction in Chapter 14. Here, we focus on hunger, eating, achievement, and sexuality. Then we turn to a discussion of how different kinds of motivation affect our intrinsic interests and performance.

Hunger and Eating

What motivates hunger? Is it your growling stomach? Or is it the sight of a juicy hamburger or the smell of a freshly baked cinnamon roll?

The Stomach

Early hunger researchers believed that the stomach controlled hunger, contracting to send hunger signals when it was empty. Today, we know it's more complicated. As dieters who drink lots of water to keep their stomachs feeling full have been disappointed to discover, sensory input from an empty stomach is not essential for feeling hungry. In fact, humans and nonhuman animals without stomachs continue to experience hunger.

However, there is a connection between the stomach and feeling hungry. Receptors in the stomach and intestines detect levels of nutrients, and specialized pressure receptors in the stomach walls signal feelings of either emptiness or *satiety* (fullness). The stomach and other parts of the gastrointestinal tract also release chemical signals that play a role in hunger (Gluck et al., 2014; Harmon-Jones & Harmon-Jones, 2015; Hellström, 2013).

Biochemistry

Like the stomach, the brain and other parts of the body produce numerous neurotransmitters, hormones, enzymes, and other chemicals that affect hunger and satiety (Bunnett, 2014; van Avesaat et al., 2015; Williams, 2014). Research in this area is complex because of the large number of known (and unknown) bodily chemicals and the interactions among them. It's unlikely that any one chemical controls our hunger and eating. Other internal factors, such as *thermogenesis*—the heat generated in response to food ingestion—also play a role (Hofmann et al., 2014; Hudson et al., 2015; Williams, 2014).

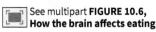

See multipart **FIGURE 10.6, How the brain affects eating**

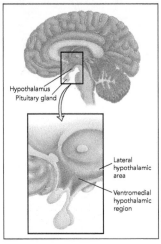

Hypothalamus
Pituitary gland

Lateral hypothalamic area

Ventromedial hypothalamic region

See **TUTORIAL VIDEO, Hormones and Hunger**

The Brain

In addition to its chemical signals, particular brain structures also influence hunger and eating. Let's look at the *hypothalamus*, which helps regulate eating, drinking, and body temperature.

Early research suggested that one area of the hypothalamus, the lateral hypothalamus (LH), stimulates eating, while another area, the ventromedial hypothalamus (VMH), creates feelings of satiation, signaling the animal to stop eating. When the VMH area was destroyed in rats, researchers found that the rats overate to the point of extreme obesity (■ **Figure 10.6**). In contrast, when the LH area was destroyed, the animals starved to death if they were not force-fed.

Later research, however, showed that the LH and VMH areas are not simple on–off switches for eating. Lesions (damage) to the VMH not only lead ultimately to severe weight gain, but the lesions also make animals picky eaters that reject a wide variety of foods. Can you see how this picky eating doesn't match the idea that VMH lesioned rats overeat because they aren't satiated and just can't stop eating? Furthermore, normal rats force-fed to become overweight also become picky eaters. Today, researchers know that the hypothalamus plays an important role in hunger and eating, but it is not the brain's "eating center." In fact, hunger and eating, like virtually all other behaviors, are influenced by numerous neural circuits that run throughout the brain (Berman et al., 2013; Seeley & Berridge, 2015; Williams, 2014).

Psychosocial Factors

The internal motivations for hunger we've discussed (the stomach, biochemistry, and the brain) are powerful. But *psychosocial factors*—spying a luscious dessert or a McDonald's billboard, or even simply noticing that it's almost lunchtime—can be equally important *stimulus cues* for hunger and eating. In fact, researchers have found that simply looking at pictures of high-fat foods, such as hamburgers, cookies, and cake, can stimulate parts of the brain in charge of appetite, thereby increasing feelings of hunger and cravings for sweet and salty foods (Luo et al., 2014; Schüz et al., 2015).

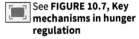

See **FIGURE 10.7, Key mechanisms in hunger regulation**

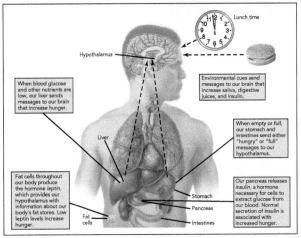

Another important psychosocial influence on when, what, where, and why we eat is *cultural conditioning*. Americans, for example, tend to eat dinner at around 6 p.m., whereas people in Spain and South America tend to eat around 10 p.m. When it comes to *what* we eat, have you ever eaten rat, dog, or horse meat? If you are a typical American, this might sound repulsive to you, yet most Hindus in India would feel a similar revulsion at the thought of eating meat from cows.

In sum, numerous biological and psychosocial factors operate in the regulation of hunger and eating (**Figure 10.7**), and researchers are still struggling to discover and explain how all these processes work together.

Eating Problems and Disorders

The same biopsychosocial forces that explain hunger and eating also play a role in four serious eating disorders: *obesity*, *anorexia nervosa*, *bulimia nervosa*, and *binge-eating disorder*.

Obesity

Imagine yourself being born and living your life on another planet, but you could receive all the Earth's normal television channels. Given that almost all the television stars, newscasters, and commercial spokespeople you've ever seen are very thin, would you wonder why there are so many ads promoting weight loss? How would you explain the recent news that obesity has reached epidemic proportions in the United States and other developed nations?

Obviously, there is a large gap between the select few appearing on television and the real world. In fact, more than one-third of adults in the United States are considered to be overweight, another third are considered to be medically obese, and in 2013, obesity was officially classified as a disease in the DSM-5 (American Psychiatric Association, 2013; Maggi et al., 2015; Reed, 2015). (See Chapter 12 for a discussion of the DSM-5.)

What is **obesity**? The most widely used measure of weight status is *body mass index (BMI)*, which is a single numerical value that calculates height in relation to weight. Sadly, obesity is one of our greatest health threats because of its significant contribution to serious illnesses, like heart disease, diabetes, stroke and certain cancers (American Heart Association, 2013; Arca, 2015; Miller & Brooks-Gunn, 2015). In addition, each year billions of dollars are spent treating serious and life-threatening medical problems related to obesity, with consumers spending billions more on largely ineffective weight-loss products and services.

See **VIRTUAL FIELD TRIP: Surgical Weight Loss**

Controlling weight is a particularly difficult task for people in the United States. We are among the most sedentary people of all, and we've become accustomed to "supersized" cheeseburgers, "Big Gulp" drinks, and huge servings of dessert (Almiron-Roig et al., 2015; Fast et al., 2015; Reitzel et al., 2014). We've also learned that we should eat three meals a day, whether we're hungry or

See **FIGURE 10.8, A fattening environment**

not; that "tasty" food requires lots of salt, sugar, and fat; and that food is an essential part of the workplace and all social gatherings (**Figure 10.8**).

However, we all know some people who can seemingly eat anything they want and still not add pounds. This may be a result of their ability to burn calories more effectively in the process of thermogenesis, a higher metabolic rate, or other factors. Adoption and twin studies confirm that genetics also plays a role (Pérusse et al., 2014; van Dongen et al., 2015; Zhou et al., 2015). Unfortunately, identifying the genes for obesity is difficult. Researchers have isolated a large number of genes that contribute to normal and abnormal weight (Albuquerque et al., 2015; Butler et al., 2015; van Dijk et al., 2015).

The good news is that one of these identified genes may provide a potential genetic explanation for why some people overeat and run a greater risk for obesity. For example, recent research finds that people who carry variants of the FTO gene don't feel full after eating, and, therefore, overeat because they have higher blood levels of ghrelin—a known hunger-producing hormone (Hess & Bruning, 2014; King et al., 2015; van Name et al., 2015). But the researchers cautioned that more research is needed, and that human appetite and obesity are undoubtedly more complex than a single hormone. In addition, this focus on genes should not encourage people to feel helpless against obesity. Ghrelin also can be reduced by exercise and a high-protein diet (Bailey et al., 2015; Martins et al., 2014; Williams, 2013).

Eating Disorders

The major eating disorders—*anorexia nervosa, bulimia nervosa*, and *disorder*—are found in all ethnicities, socioeconomic classes, and both sexes. However they are more common in women (American Psychiatric Association, 2013; Bohon, 2015; Himes et al., 2015). **Anorexia nervosa** is characterized by an obsessive fear of obesity, a need for control, the use of dangerous weight-loss measures, and a body image that is so distorted that even a skeletal, emaciated body is perceived as fat. The resulting extreme malnutrition often leads to osteoporosis, bone fractures, interruption of menstruation in women, and loss of brain tissue. A significant percentage of individuals with anorexia nervosa ultimately die of the disorder (Nakai et al., 2014; Robinson & Nicholls, 2015; Saito et al., 2014).

Occasionally, a person suffering from anorexia nervosa succumbs to the desire to eat and gorges on food, then vomits or takes laxatives. However, this type of bingeing and purging is more characteristic of **bulimia nervosa**. Individuals with bulimia go on recurrent eating binges and then purge by self-induced vomiting or the use of laxatives. They often show impulsivity in other areas, sometimes engaging in excessive shopping, alcohol abuse, or petty shoplifting (Buckholdt et al., 2015; Pearson et al., 2015; Slane et al., 2014). The vomiting associated with bulimia nervosa causes severe damage to the teeth, throat, and stomach. It also leads to cardiac arrhythmia, metabolic deficiencies, and serious digestive disorders.

Note that bulimia is similar to, but not the same as, **binge-eating disorder**. This disorder involves recurrent episodes of consuming large amounts of food in a discrete period of time, while feeling a lack of control over eating. However, the individual does not try to purge (Amianto et al., 2015). Individuals with binge-eating disorder generally eat more rapidly than normal, eat until they are uncomfortably full, and eat when not feeling physically hungry. They also tend to eat alone because of embarrassment at the large quantities they are consuming, and they feel disgusted, depressed, or very guilty after bingeing.

There are many suspected causes of anorexia nervosa, bulimia nervosa, and binge-eating disorder. Some theories focus on physical causes, such as hypothalamic disorders, low levels of various neurotransmitters, and genetic or hormonal disorders. Other theories emphasize psychosocial factors, such as a need for perfection, a perceived loss of control, teasing about body weight, destructive thought patterns, depression, dysfunctional families, distorted body image, and emotional or sexual abuse (American Psychiatric Association, 2013; Brauhardt et al., 2014; Schneider, 2015).

Culture and ethnicity also play important roles in eating disorders (Lähteenmäki et al., 2014; Smart & Tsong, 2014). For instance, Blacks report fewer overall eating disorders and greater satisfaction with their bodies, but the specific problem of binge-eating disorder occurs across all ethnicities and sexes in the U.S. (Bruns & Carter, 2015; Capodilupo & Kim, 2014; Watson et al., 2013). Regardless of the causes, it is important to recognize the symptoms of anorexia, bulimia, and binge-eating disorder (■ **Table 10.1**) and to seek therapy if the symptoms apply to you. The key point to remember is that all eating disorders are serious and chronic conditions that require treatment. In fact, some studies find that they have the highest mortality rates of all mental illnesses (Goldberg et al., 2015; Zerwas et al., 2015).

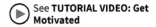
See **TABLE 10.1, Symptoms of Three Common Eating Disorders**

Achievement

Do you wonder what motivates Olympic athletes to work so hard for a gold medal? What about someone like Oprah Winfrey, famous television star, thriving businesswoman, and generous philanthropist? Or Mark Zuckerberg, cofounder of Facebook? What drives some people to high achievement?

One key to understanding what motivates high-achieving individuals lies in what psychologist Henry Murray (1938) identified as a high need for achievement (nAch), or **achievement motivation**. One of the earliest tests for achievement motivation was devised by Christiana Morgan and Henry Murray (1935). Using a series of ambiguous pictures called the *Thematic Apperception Test* (TAT), these researchers asked participants to make up a story about each picture. Their responses were scored for different motivational themes, including achievement. Since that time, other researchers have developed several questionnaire measures of achievement.

See **TUTORIAL VIDEO: Get Motivated**

Several traits distinguish people who have this high nAch (Baneshi et al., 2015; Hsu, 2016; McClelland, 1958, 1993).

- **Preference for moderately difficult tasks** People high in nAch avoid tasks that are too easy because they offer little challenge or satisfaction. They also avoid extremely difficult tasks because the probability of success is too low.

- **Competitiveness** High-achievement-oriented people are more attracted to careers and tasks that involve competition and an opportunity to excel.

- **Preference for clear goals with competent feedback** High-achievement-oriented people tend to prefer tasks with clear outcomes and situations in which they can receive feedback on their performance. They also prefer criticism from a harsh, but competent, evaluator to criticism from one who is friendlier, but less competent.

- **Self-regulation and personal responsibility** High-achievement-oriented people purposefully control their thoughts and behaviors to attain their goals. In addition, they prefer being personally responsible for a project, so that they can feel satisfied when the task is well done.

See **FIGURE 10.9, Future achievers**

- **Mental toughness and persistence** High-achievement-oriented people have a mindset that allows them to persevere through difficult circumstances. It includes attributes like sacrifice and self-denial, which help them overcome obstacles. Such traits also maintain concentration and motivation when things aren't going well.

- **More accomplished** People who have high nAch scores generally do better than others on exams, earn better grades in school, and excel in their chosen professions.

Achievement orientation appears to be largely learned in early childhood, primarily through interactions with parents (■ **Figure 10.9**). Highly motivated children tend to have parents who encourage independence and frequently reward successes (Kim, 2015; Tao & Hong, 2014; Wigfield et al., 2015). In fact, children's motivation for academic success—along with their study skills—are better predictors of long-term math achievement than IQ (Murayama et al., 2012). Our cultural values also affect achievement needs (Baker, 2014; Greenfield & Quiroz, 2013).

Dave & Les Jacobs/Blend/Getty Images

Extrinsic Versus Intrinsic Motivation

Have you ever noticed that for all the money and glory they receive, professional athletes often don't look like they're enjoying themselves very much? What's the problem? Why don't they appreciate how lucky they are to be able to make a living by playing games?

One way psychologists attempt to answer questions about motivation is by distinguishing between **extrinsic motivation**, based on external rewards or punishments, and **intrinsic motivation**, based on internal, personal satisfaction from a task or activity (Deci & Moller, 2005; Ryan & Deci, 2013). When we perform a task for no ulterior purpose, we use internal, personal reasons ("I like it"; "It's fun"). But when extrinsic rewards are added, the explanation shifts to external, impersonal reasons ("I did it for the money"; "I did it to please my parents"). This shift often decreases enjoyment and hampers performance. This is as true for professional athletes as it is for anyone else.

A classic experiment demonstrating this effect was conducted with preschool children who liked to draw (Lepper et al., 1973). These researchers found that children who were given paper and markers, and promised a reward for their drawings, were subsequently less interested in drawing than children who were not given a reward, or who were given an unexpected reward for their pictures when they were done. Likewise, a decade long study of over 10,000 West Point cadets found that those who were motivated to pursue a military career for internal reasons, such as personal ambition, were more likely to receive early career promotions than those who attended a military academy for external reasons, such as family expectations (Wrzesniewski et al., 2014).

As it turns out, however, there is considerable controversy over individual differences, and under what conditions giving external, extrinsic rewards increases or decreases motivation (Deci & Ryan, 1985, 2012; Koo et al., 2015; Pope & Harvey, 2015). Furthermore, research shows that not all extrinsic motivation is bad. A study of elementary school students, who were simply mailed books weekly during the summer, or mailed books along with a reading incentive, or assigned to a control group with no books or incentives, found that students who were initially more motivated to read were also more responsive to incentives (Guryan et al., 2015). As you can see in ▣ **Figure 10.10**, extrinsic rewards, with "no strings attached," can actually increase motivation.

How does this apply to you and your everyday life? As a college student facing many high-stakes exams, have you noticed how often professors try to motivate their students with "scare tactics," such as frequently reminding you of how your overall GPA and/or scores on certain exams may be critical for entry into desirable jobs or for admittance to graduate programs? Does this type of extrinsic motivation help or hurt your motivation? One study found that when instructors use extrinsic consequences, such as fear tactics, as motivational tools, their students' intrinsic motivation and exam scores decrease (Putwain & Remedios, 2014). In fact, fear of failure may be one of the greatest detriments to intrinsic motivation (Covington & Müeller, 2001; Ma et al., 2014; Martin & Marsh, 2006).

What should teachers and students do instead? Rather than emphasizing high exam scores or overall GPA, researchers recommend focusing on specific behaviors required to avoid failure and attain success. In other words, as a student you can focus on improving your overall study techniques and test taking skills. See again the *Strategies for Student Success* at the end of Chapter 1. For additional help, check with your professor and/or your college counseling center.

See **FIGURE 10.10, How extrinsic rewards can sometimes be motivating**

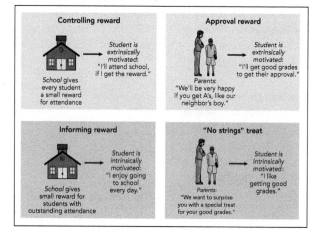

Sexuality

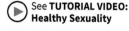
See **TUTORIAL VIDEO: Healthy Sexuality**

Obviously, there is strong motivation to engage in sexual behavior: It's essential for the survival of our species, and it's also pleasurable. But *sexuality* includes much more than reproduction. For most humans (and some other animals), a sexual relationship fulfills many needs, including the need for connection, intimacy, pleasure, and the release of sexual tension. Given the large role sex plays throughout our lives, can you predict how the quality of the first sexual experience might affect later relationships?

William Masters and Virginia Johnson (1966) were the first to conduct laboratory studies on what happens to the human body during sexual activity. They attached recording devices to male and female volunteers and monitored or filmed their physical responses as they moved from nonarousal to orgasm and back to nonarousal. They labeled the bodily changes during this series of events a **sexual response cycle** (■ **Figure 10.11**). Researchers have further characterized differences between sexual response patterns in men and women (■ **Figure 10.12**).

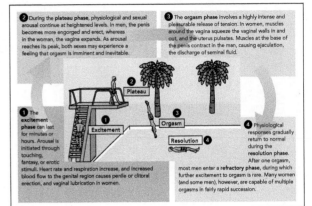

See **FIGURE 10.11**, Masters and Johnson's view of the sexual response cycle

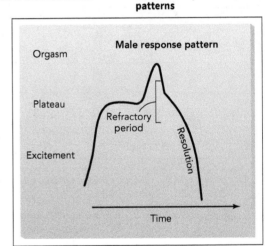

❶ The **excitement phase** can last for minutes or hours. Arousal is initiated through touching, fantasy, or erotic stimuli. Heart rate and respiration increase, and increased blood flow to the genital region causes penile or clitoral erection, and vaginal lubrication in women.

❷ During the **plateau phase**, physiological and sexual arousal continue at heightened levels. In men, the penis becomes more engorged and erect, whereas in the woman, the vagina expands. As arousal reaches its peak, both sexes may experience a feeling that orgasm is imminent and inevitable.

❸ The **orgasm phase** involves a highly intense and pleasurable release of tension. In women, muscles around the vagina squeeze the vaginal walls in and out, and the uterus pulsates. Muscles at the base of the penis contract in the man, causing ejaculation, the discharge of seminal fluid.

❹ Physiological responses gradually return to normal during the **resolution phase**. After one orgasm, most men enter a **refractory phase**, during which further excitement to orgasm is rare. Many women (and some men), however, are capable of multiple orgasms in fairly rapid succession.

Sexual Orientation

Of course, an important part of sexuality is the question of our **sexual orientation**, meaning to whom we are erotically and emotionally attracted. What leads some people to be sexually interested in members of their own sex, the other sex, or both sexes? The highly controversial "causes" of human sexual orientation are poorly understood. However, most studies suggest that genetics and biology play the dominant role (Burri et al., 2015; Camperio Ciani, 2014; LeVay, 2003, 2012). For example, studies on identical twins find that if one identical twin is gay, the second twin is also gay 48 to 65% of the time, whereas the rate for fraternal twins is 26 to 30% (Ando et al., 2013; Långström et al., 2010; Moutinho et al., 2011).

Can you see why research that points to biological "causes" for sexual orientation is so highly controversial? If orientation is not a matter of choice, then society can no longer demand that gays and lesbians change. A biological foundation also challenges some of our most enduring myths and misconceptions about sexual orientation (see ■ **Table 10.2**). Unfortunately, these FALSE beliefs often contribute to **sexual prejudice**, which is a negative attitude directed toward an individual because of his or her sexual orientation, and many gays, lesbians, and transgender people experience serious verbal and physical attacks, disrupted family and peer relationships, and high rates of anxiety, depression, and suicide (Bidell, 2014; Duncan et al., 2014; Herek et al., 2015).

See **TUTORIAL VIDEO: The Sexual Response Cycle**

See multipart **FIGURE 10.12**, **Differences between male and female sexual response patterns**

Male response pattern

Orgasm

Plateau

Refractory period

Resolution

Excitement

Time

See **TABLE 10.2**, Sexual Orientation Myths

Q Answer the **Concept Check** questions.

10.3 COMPONENTS AND THEORIES OF EMOTION

Emotions play an important role in our lives. They color our dreams, memories, and perceptions. When they are disordered, they contribute significantly to psychological problems (Bowers et al., 2014; Charles & Robinette, 2015; Manstead & Parkinson, 2015). But what do we really mean by the term **emotion**? In everyday usage, we use it to describe feeling states; we feel "thrilled" when our political candidate wins an election, "dejected" when our sports team loses, and "miserable" when our loved ones reject us. Obviously, what you and I mean by these terms, or what we ourselves experience with different emotions, can vary greatly among individuals.

Three Components of Emotion

Psychologists define and study emotion according to three basic components—*biological*, *cognitive*, and *behavioral*. See ■ **Figure 10.13**.

See **TUTORIAL VIDEO: Emotional Intelligence**

Biological (Arousal) Component

Internal physical changes occur in our bodies whenever we experience an emotion. Imagine walking alone on a dark street and having someone jump from behind a stack of boxes and

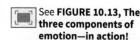
See **FIGURE 10.13, The three components of emotion—in action!**

David J. Phillip/AP Images

See **FIGURE 10.14, The limbic system's role in emotion**

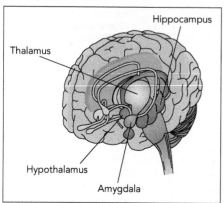

Hippocampus

Thalamus

Hypothalamus

Amygdala

See **FIGURE 10.15, Fast and slow pathways for fear**

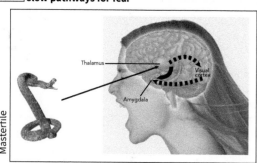

Masterfile

start running toward you. How would you respond? Like most people, you would probably interpret the situation as threatening and would run. Your predominant emotion, fear, would inspire several physiological reactions, such as increased heart rate and blood pressure, perspiration, and goose bumps (piloerection).

Our emotional experiences appear to result from important interactions between several areas of the brain, particularly the *cerebral cortex* and *limbic system* (Grimm et al., 2014; Kemp et al., 2015; Schulze et al., 2015). As we discuss in Chapter 2, the cerebral cortex, the outermost layer of the brain, serves as our body's ultimate control and information-processing center, our ability to recognize and regulate our emotions.

Studies of the limbic system, located in the innermost part of the brain, have shown that one area, the **amygdala**, plays a key role in emotion—especially fear (**Figure 10.14**). It sends signals to the other areas of the brain, causing increased heart rate and other physiological reactions. Interestingly, children with high levels of anxiety have larger amygdalas, as well as stronger connections between the amygdala and other parts of the brain (Qin et al., 2014).

Have you noticed that your own emotional arousal often occurs without your conscious awareness? According to psychologist Joseph LeDoux (1996, 2014), when the *thalamus* (our brain's sensory switchboard) receives sensory inputs, it sends separate messages up to the cortex, which "thinks" about the stimulus, and to the amygdala, which immediately activates the body's alarm system (**Figure 10.15**). Although this dual pathway occasionally leads to "false alarms," such as when we mistake a stick for a snake, LeDoux believes it is a highly adaptive warning system essential to our survival. He states that "the time saved by the amygdala in acting on the thalamic interpretation, rather than waiting for the cortical input, may be the difference between life and death" (LeDoux, 1996, p. 166).

As important as the brain is to emotion, it is the *autonomic nervous system* (ANS) that produces the obvious signs of arousal (Chapter 2). These largely automatic responses result from interconnections between the ANS and various glands and muscles (**Figure 10.16**).

Cognitive (Thinking) Component

Emotional reactions are very individual: What you experience as intensely pleasurable may be boring or aversive to another. To study the cognitive (thought) component of emotions, psychologists typically use self-report techniques, such as paper-and-pencil tests, surveys, and interviews. However, people are sometimes unable or unwilling to accurately remember or describe their emotional states. For these reasons, our cognitions about our own and others' emotions are difficult to measure scientifically. This is why many researchers supplement participants' reports on their emotional experiences with methods that assess such experiences indirectly (e.g., measuring physiological responses such as heart rate, pupil dilation, blood flow).

Behavioral (Expressive) Component

Emotional expression is a powerful form of communication, and facial expressions may be our most important form of emotional communication. Researchers have developed sensitive techniques to measure subtleties of feeling, and to differentiate honest expressions from fake ones. Perhaps most interesting is the difference between the *social smile* and the *Duchenne smile* (named after French anatomist Duchenne de Boulogne, who first described it in 1862). As you can see in **Figure 10.17**, in a false, social smile, our voluntary cheek muscles are pulled back, but our eyes are unsmiling. Smiles of real pleasure, on the other hand, use the muscles not only around the cheeks but also around the eyes.

Now that you know how to recognize a real smile versus a fake smile, let's consider other forms of nonverbal communication. Imagine yourself as a job interviewer. Your first applicant greets you with a big smile, full eye contact, a firm handshake, and an erect, open posture.

The second applicant offers a similar big smile and full eye contact, but provides a weak handshake, and slouches. Whom do you think you will hire?

See **FIGURE 10.16, Emotion and the autonomic nervous system (ANS)**

Like most people, you're probably much less likely to hire the second applicant due to his or her "mixed messages." Psychologist Albert Mehrabian suggests that when we're communicating feelings or attitudes and our verbal and nonverbal dimensions don't match, the receiver trusts the predominant form of communication, which is about 93% nonverbal and consists of the facial expression and the way the words are said, rather than the literal meaning of the words (Mehrabian, 1968, 2007).

Unfortunately, Mehrabian's research is often overgeneralized, and many people misquote him as saying that "over 90% of communication is nonverbal." Clearly, if a police officer says, "Put your hands up," his or her verbal words might carry 100% of the meaning. However, when we're confronted with a mismatch between verbal and nonverbal communication, it is safe to say that we pay far more attention to the nonverbal because we believe it more often tells us what someone is really thinking or feeling. The importance of nonverbal communication, particularly facial expressions, is further illustrated by the popularity of smileys and other emoticons in our everyday e-mail and text messages.

Sympathetic		Parasympathetic
Pupils dilated	Eyes	Pupils constricted
Decreased saliva	Mouth	Increased saliva
Vessels constricted (skin cold and clammy)	Skin	Vessels dilated (normal blood flow)
Respiration increased	Lungs	Respiration normal
Increased heart rate	Heart	Decreased heart rate
Increased epinephrine and norepinephrine	Adrenal glands	Decreased epinephrine and norepinephrine
Decreased motility	Digestion	Increased motility

See multipart **FIGURE 10.17, Duchenne smile**

Courtesy Karen Huffman

Three Major Theories of Emotion

Researchers generally agree that emotion has biological, cognitive, and behavioral components, but there is less agreement about *how* we become emotional (Laird & Lacasse, 2014; Manstead & Parkinson, 2015; Reisenzein, 2015). The major competing theories are the *James-Lange*, *Cannon-Bard*, and *Schachter and Singer's two-factor theories* (**Process Diagram 10.2**).

This Process Diagram contains essential information NOT found elsewhere in the text, which is likely to appear on quizzes and exams. Be sure to study it CAREFULLY!

Process Diagram 10.2

Comparing Three Major Theories of Emotion

After perceiving an environmental stimulus, such as a dangerous snake, we experience the emotion of fear. However, there are three competing theories attempting to explain how and why this happens:

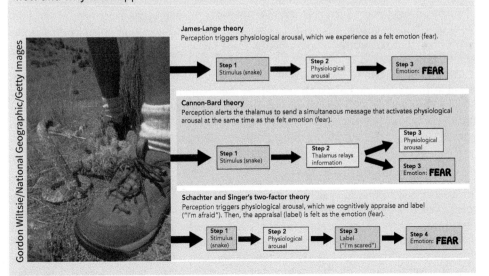

Gordon Wiltsie/National Geographic/Getty Images

James-Lange theory
Perception triggers physiological arousal, which we experience as a felt emotion (fear).

Step 1 Stimulus (snake) → Step 2 Physiological arousal → Step 3 Emotion: **FEAR**

Cannon-Bard theory
Perception alerts the thalamus to send a simultaneous message that activates physiological arousal at the same time as the felt emotion (fear).

Step 1 Stimulus (snake) → Step 2 Thalamus relays information → Step 3 Physiological arousal / Step 3 Emotion: **FEAR**

Schachter and Singer's two-factor theory
Perception triggers physiological arousal, which we cognitively appraise and label ("I'm afraid"). Then, the appraisal (label) is felt as the emotion (fear).

Step 1 Stimulus (snake) → Step 2 Physiological arousal → Step 3 Label ("I'm scared") → Step 4 Emotion: **FEAR**

157

Imagine that you're walking in the desert, and suddenly see a coiled snake on the path next to you. What emotion would you experience? Most people would say they would be very afraid. But why? Common sense tells us that our hearts pound and we tremble when we're afraid, and that we cry when we're sad. But according to the **James-Lange theory**, felt emotions begin with physiological arousal of the ANS. This arousal (a pounding heart, breathlessness, trembling all over) then causes us to experience the emotion we call "fear." Contrary to popular opinion, James wrote: "We feel sorry because we cry, angry because we strike, afraid because we tremble" (James, 1890, pp. 449–450).

In contrast, the **Cannon-Bard theory** proposes that arousal and emotion occur separately but simultaneously. Following perception of an emotion-provoking stimulus, the thalamus (a subcortical brain structure) sends two simultaneous messages: one to the ANS, which causes physiological arousal, and one to the brain's cortex, which causes awareness of the felt emotion.

Finally, Schachter and Singer's **two-factor theory** suggests that our emotions start with physiological arousal, followed by a conscious, cognitive appraisal. We then look to external cues from the environment, and from others around us, to find a label and explanation for the arousal. Therefore, if we cry at a wedding, we label our emotion as joy or happiness. If we cry at a funeral, we label the emotion as sadness.

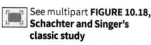
See multipart **FIGURE 10.18, Schachter and Singer's classic study**

In their classic study demonstrating this effect, Schachter and Singer (1962) gave research participants injections of epinephrine (adrenaline), a hormone/neurotransmitter that produces feelings of arousal, or saline shots (a placebo) and then exposed the participants to either a happy or angry confederate (◫ **Figure 10.18**). The way participants responded suggested that arousal could be labeled happiness or anger, depending on the context. Thus, Schachter and Singer's research demonstrated that emotion is determined by two factors: physiological arousal and cognitive appraisal (labeling).

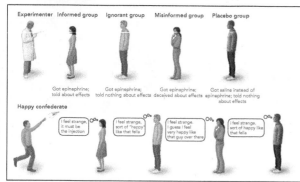

Schachter and Singer's two-factor theory may have important practical implications. Depending on the cues present in our environment, we apparently can interpret exactly the same feelings of arousal in very different ways. If you're shy or afraid of public speaking, try interpreting your feelings of nervousness as the result of too much coffee or the heating in the room. Numerous studies on this type of *misattribution of arousal* have had both positive and negative outcomes (Greenaway et al., 2015; Jouffre, 2015; Webber et al., 2015).

A common negative example might be when we're frustrated by a heavy traffic jam, but yell at our friends or family for a small infraction. We've misattributed the actual cause of arousal from the traffic jam to other people.

Before going on, we need to add one more approach that helps expand our understanding of our emotional reactions. According to the **facial-feedback hypothesis**, movements of our facial muscles produce, and may even intensify, our subjective experience of emotion. More specifically, sensory input (such as seeing a snake) is first routed to subcortical areas of the brain that activate facial movements. These facial changes then initiate and intensify emotions (Adelmann & Zajonc, 1989; Lamer et al., 2015; Vrticka et al., 2013).

Researchers in one study asked participants to maintain a smile while they were engaging in a stress-inducing task, such as keeping their hand in a bucket of very cold ice water (Kraft & Pressman, 2012). Compared to participants who held their face in a neutral position, those who smiled had lower heart rates, showing that smiling can help reduce the experience of stress—further supporting the facial-feedback hypothesis.

Evaluating Theories of Emotion

Which theory is correct? As you may imagine, each theory has its limits. The *James–Lange theory* fails to acknowledge that physiological arousal can occur without emotional experience (e.g., when we exercise). This theory also requires a distinctly different pattern of arousal for each emotion. Otherwise, how do we know whether we are sad, happy, or mad? Positron emission tomography (PET) scans of the brain do show subtle differences and overall physical arousal with basic emotions, such as happiness, fear, and anger (Levenson,

1992, 2007; Werner et al., 2007). But most people are not aware of these slight variations. Thus, there must be other explanations for how we experience emotion.

The *Cannon–Bard theory* (that arousal and emotions occur simultaneously and that all emotions are physiologically similar) has received some experimental support. Recall from our earlier discussion that victims of spinal cord damage still experience emotions—often more intensely than before their injuries. Instead of the thalamus, however, other research shows that it is the limbic system, hypothalamus, and prefrontal cortex that are activated in emotional experience (Junque, 2015; LeDoux, 2007; Schulze et al., 2015).

As mentioned earlier, research on the *facial-feedback hypothesis* has found a distinctive physiological response for basic emotions such as fear, sadness, and anger—thus partially confirming James–Lange's initial position. Facial feedback does seem to contribute to the intensity of our subjective emotional experience and our overall moods. Thus, if you want to change a bad mood or intensify a particularly good emotion, adopt the appropriate facial expression. Try smiling when you're sad and expanding your smiles when you're happy.

Finally, Schachter and Singer's *two-factor theory* emphasizes the importance of cognitive labels in emotions. But research shows that some neural pathways involved in emotion by-pass the cortex and go directly to the limbic system. Recall our earlier example of jumping at the sight of a supposed snake and then a second later using the cortex to interpret what it was. This and other evidence suggest that emotions can take place without conscious cognitive processes. Thus, emotion is not simply the labeling of arousal.

In sum, certain basic emotions are associated with subtle differences in arousal. These differences can be produced by changes in facial expressions or by organs controlling the autonomic nervous system. In addition, "simple" emotions (fear and anger) do not initially require conscious cognitive processes. This allows a quick, automatic emotional response that can later be modified by cortical processes. On the other hand, "complex" emotions (jealousy, grief, depression, embarrassment, love) seem to require more extensive cognitive processes.

Q Answer the **Concept Check** questions.

10.4 EMOTION AND BEHAVIOR

As you recall from Chapter 1 and throughout this text, critical thinking is a core part of psychological science and a major goal of this text. In this section, we will use our critical thinking skills to explore two special and sometimes controversial topics—how evolution and culture affect our emotions, and the use of polygraph tests for lie detection.

Evolution and Culture

Early research indicated that people in all cultures express, recognize, and experience several basic emotions, such as happiness, surprise, anger, sadness, fear, and disgust, in essentially the same way (Darwin, 1872; Ekman & Keltner, 1997; Zhang et al., 2015). However, more recent research suggests that this previous list of six basic emotions could be combined and reduced to just four: happy, sad, afraid/surprised, and angry/disgusted (Jack et al., 2014). Regardless of the exact number, researchers tend to agree that across cultures, the facial expression of emotions, such as a smile, is recognized by all as a sign of pleasure, whereas a frown is recognized as a sign of displeasure.

From an evolutionary perspective, the idea of universal facial expressions makes adaptive sense because such expressions signal others about our current emotional state (Awasthi & Mandal, 2015; Ekman & Keltner, 1997; Hwang & Matsumoto, 2015). Charles Darwin first advanced the evolutionary theory of emotion in 1872. He proposed that expression of emotions evolved in different species as a part of survival and natural selection. Fear helps animals avoid danger, whereas expressions of anger and aggression are useful when fighting for mates or resources. Modern evolutionary theory suggests that basic emotions originate in the *limbic system*. Given that higher brain areas (like the cortex) developed later than the subcortical limbic system, evolutionary theory proposes that basic emotions evolved before thought.

Studies with infants provide further support for an evolutionary basis for emotions. For example, infants only a few hours old show distinct expressions of emotion that closely

See **FIGURE 10.19, Cultural differences in emotional expressions**

Behrouz Mehri/AFP/Getty Images

match adult facial expressions, and by the age of 7 months, they can reliably interpret and recognize emotional information across both face and voice (Cole & Moore, 2015; Jessen & Grossman, 2015; Meltzoff & Moore, 1977, 1994). And all infants, even those who are born deaf and/or blind, show similar facial expressions in similar situations (Denmark et al., 2014; Field et al., 1982; Gelder et al., 2006). Interestingly, studies have found that families may have characteristic facial expressions that are shared even by family members who have been blind from birth (Peleg et al., 2006). This collective evidence points to a strong biological, evolutionary basis for emotional expression and decoding.

In contrast to the evolutionary approach, other researchers suggest that emotions are much more complex than originally thought (Gendron et al., 2014; Hsu, 2016; Hwang & Matsumoto, 2015). As you can see in ▣ **Figure 10.19**, we may all share similar facial expressions for some emotions, but each culture has its own *display rules* governing how, when, and where to express them (Allen et al., 2014; Ekman, 1993, 2004; Hess & Hareli, 2015). In addition to culture, gender, family background, norms, and individual differences also affect our emotions and their expression (▣ **Figure 10.20**).

See **FIGURE 10.20, Can you identify these emotions?**

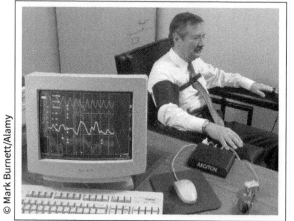

© Ocean/Corbis

The Polygraph as a Lie Detector

We've discussed the three major theories of emotion, and the way emotions are affected by culture and evolution. Now we turn our attention to another issue in emotion and behavior, which is also one of the hottest, and most controversial, topics in emotion research—the **polygraph**.

The polygraph is a machine that measures your physiological arousal (such as heart rate and blood pressure) to detect emotional arousal, which supposedly reflects whether or not you are lying. Traditional polygraph tests are based on the assumption that when people lie, they feel stressed, and that this stress can be measured. As you can see in ▣ **Figure 10.21**, during a polygraph test, multiple (poly) signals from special sensors assess four major indicators of stress and autonomic arousal: heart rate (pulse), blood pressure, respiration (breathing) rate, and perspiration (or skin conductivity). If the subject's bodily responses significantly increase when responding to key questions, the examiner will infer that the subject is lying (Granhag et al., 2015; Tomash & Reed, 2013).

See multipart **FIGURE 10.21, Polygraph testing**

© Mark Burnett/Alamy

Can you imagine what problems might be associated with the polygraph? First, many people become stressed even when telling the truth, whereas others can conceal their stress, and remain calm when deliberately lying. Second, emotions also cause physiological arousal, and a polygraph cannot tell which emotion is being felt (anxiety, irritation, excitement, or any other emotion). For this reason, some have suggested the polygraph should be relabeled as an "arousal detector." The third, and perhaps most important, problem is one of questionable accuracy. In fact, people can be trained to beat a lie detector (Kaste, 2015; Wollan, 2015). When asked a general, control question ("Where do you live?"), subjects wishing to mislead the examiner can artificially raise their arousal levels by imagining their worst fears (being burned to death or buried alive). Then when asked relevant/guilty knowledge questions ("Did you rob the bank?"), they can calm themselves by practicing meditation tricks (imagining themselves relaxing on a beach).

In response to these and other problems, countless research hours and millions of dollars have been spent on new and improved lie-detection techniques (Kluger & Masters, 2006). Perhaps the most promising is the use of brain scans, like *functional magnetic resonance imaging (fMRI)* (Farah et al., 2014; Jiang et al., 2015). Unfortunately, each of these new lie-detection techniques has potential problems. Furthermore, researchers have questioned their reliability and validity, while civil libertarians and judicial scholars debate their ethics and legality (Kaste, 2015; Lilienfeld et al., 2015; Roskey, 2013). For these reasons, most courts do not accept polygraph test results, laws have been passed to restrict their use, and we should remain skeptical about drawing conclusions about their ability to detect guilt or innocence (Granhag et al., 2015; Handler et al., 2013; Tomash & Reed, 2013).

Final Note

Earlier in the chapter we focused on understanding motivation, and ended with an *Applying Psychology* on how to increase it.

Q Answer the **Concept Check** questions.

WP LS Go to your WileyPLUS Learning Space course for video episodes, examples, art, tables, Concept Checks, practice, and other pedagogical resources that will help you succeed in this course.

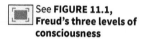

11

Reading for

PERSONALITY

WP LS Go to your WileyPLUS Learning Space course for video episodes, examples, art, tables, Concept Checks, practice, and other pedagogical resources that will help you succeed in this course.

11.1 PSYCHOANALYTIC/PSYCHODYNAMIC THEORIES

Before discussing the following psychoanalytic/psychodynamic theories, we need to provide a basic definition of **personality** as *a unique and relatively stable pattern of thoughts, feelings, and actions*. In other words, it describes how we are different from other people and what patterns of behavior are typical of us. We might qualify as an "extrovert," if we're talkative and outgoing most of the time. Or we may be described as "conscientious" if we're responsible and self-disciplined most of the time. (Keep in mind that *personality* is not the same as *character*, which refers to our ethics, morals, values, and integrity.)

See **FIGURE 11.1,** Freud's three levels of consciousness

Freud's Psychoanalytic Theory

One of the earliest theories of personality was Sigmund Freud's psychoanalytic perspective, which emphasized unconscious processes and unresolved past conflicts. Working from about 1890 until he died in 1939, Freud developed a theory of personality that has been one of the most influential—and controversial—theories in all of science (Barenbaum & Winter, 2013; Carducci, 2015; Cordón, 2012). Let's examine some of Freud's most basic and debatable concepts.

Levels of Consciousness

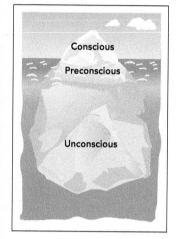

Freud called the mind the "psyche" and asserted that it contains three *levels of consciousness*, or awareness: the conscious, the preconscious, and the **unconscious** (⊞ **Figure 11.1**). For Freud, the unconscious is all-important because it reportedly stores our most primitive, instinctual impulses, plus anxiety-laden thoughts and memories, which are normally blocked (*repressed*) from our personal awareness. However, it supposedly still has an enormous impact on our behavior—like the hidden part of the iceberg that sunk the ocean liner *Titanic*.

Because many of our unconscious thoughts and motives are unacceptable and threatening, Freud believed that they are normally *repressed* (held out of awareness)—unless they are unintentionally revealed by dreams or slips of the tongue, later called *Freudian slips* (⊞ **Figure 11.2**).

Freud believed that most psychological disorders originate from repressed memories or sexual and aggressive instincts hidden in the unconscious. To treat these disorders, he developed a form of therapy called *psychoanalysis* (Chapter 13).

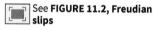

See **FIGURE 11.2, Freudian slips**

"Good morning, beheaded—uh, I mean beloved."

Personality Structure

In addition to proposing that the mind functions at three levels of consciousness, Freud also believed personality was composed of three mental structures: the *id*, *ego*, and *superego* (⊞ **Figure 11.3**).

According to Freud, the **id** is made up of primitive, biological instincts and urges. It is immature, impulsive, and irrational. The id is also completely unconscious and serves as the reservoir of mental energy. When its primitive drives build up, the id seeks immediate gratification to relieve the tension—a concept known as the *pleasure principle*. In other words, the id is like a newborn baby. It wants what it wants, when it wants it!

Freud further believed that the second part of the psyche, the **ego**, is responsible for decision-making, and controlling the potentially destructive energy of the id. In Freud's system, the ego corresponds to the *self*—our conscious identity of ourselves as persons.

One of the ego's tasks is to channel and release the id's energy in ways that are compatible with the external world. Thus, the ego is responsible for delaying gratification when necessary. Contrary to the id's pleasure principle, the ego operates on the *reality principle* because it can understand and deal with objects and events in the real world.

The final part of the psyche to develop is the **superego**, which acts as an internal judge using a set of ethical rules for behavior. The superego develops from internalized parental and societal standards. It constantly strives for perfection and is therefore as unrealistic as the id. Some Freudian followers have suggested that the superego operates on the *morality principle* because violating its rules results in feelings of guilt (Carducci, 2015; Castelloe, 2013).

See **FIGURE 11.3**, Freud's personality structure

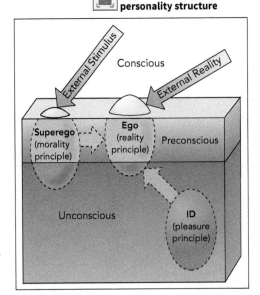

Defense Mechanisms

When the ego fails to satisfy both the id and the superego, anxiety slips into our conscious awareness. Because anxiety is uncomfortable, we avoid it through **defense mechanisms** (■ **Figure 11.4**). In addition to *intellectualization* and *rationalization* (discussed in Figure 11.4), Freud identified several other defense mechanisms (■ **Table 11.1**).

See **TABLE 11.1**, Sample Psychological Defense Mechanisms

See multipart **FIGURE 11.4**, Why do we use defense mechanisms?

Psychosexual Stages of Personality Development

Although defense mechanisms are now an accepted part of modern psychology, other Freudian ideas are more controversial (Breger, 2014). For example, according to Freud, strong biological urges residing within the id push all children through five universal **psychosexual stages** (**Process Diagram 11.1**). The term *psychosexual* reflects Freud's belief that personality development occurs in five stages named for the sexual pleasure it produces in one part of the body more than others. For instance, the oral phase is named for the mouth, which is the key erogenous zone during infancy.

At each psychosexual stage, the id's impulses and social demands come into conflict. Therefore, if a child's needs are not met, or are overindulged, at one particular stage, the child supposedly may *fixate,* and a part of his or her personality will remain stuck at that stage. Freud believed most individuals successfully pass through each of the five stages. But during stressful times, they may return (or *regress*) to an earlier stage in which prior needs were badly frustrated or overgratified.

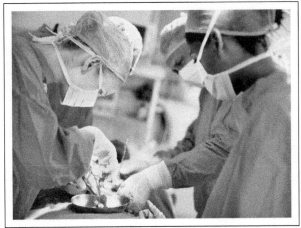

© kupicoo/iStockphoto

1. Oral stage (birth to 18 months) During this period, an infant receives satisfaction through sucking, eating, biting, and so on. Because the infant is highly dependent on parents and other caregivers to provide opportunities for oral gratification, fixation at this stage can easily occur. If the mother overindulges her infant's oral needs, the child may fixate and as an adult become gullible ("swallowing" anything), dependent, and passive. The underindulged child, however, will develop into an aggressive, sadistic person who exploits others. According to Freud, orally fixated adults may also orient their life around their mouth—overeating, becoming an alcoholic, smoking, or talking a great deal.

2. Anal stage (18 to 36 months) Once the child becomes a toddler, his or her erogenous zone shifts to the anus. The child supposedly receives satisfaction by having and retaining bowel movements. Because this is also the time when most parents begin toilet training, the child's

Process Diagram 11.1

Freud's Five Psychosexual Stages of Development

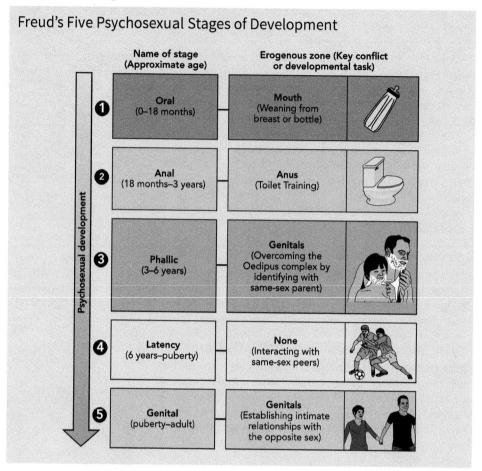

	Name of stage (Approximate age)	Erogenous zone (Key conflict or developmental task)	
①	**Oral** (0–18 months)	**Mouth** (Weaning from breast or bottle)	
②	**Anal** (18 months–3 years)	**Anus** (Toilet Training)	
③	**Phallic** (3–6 years)	**Genitals** (Overcoming the Oedipus complex by identifying with same-sex parent)	
④	**Latency** (6 years–puberty)	**None** (Interacting with same-sex peers)	
⑤	**Genital** (puberty–adult)	**Genitals** (Establishing intimate relationships with the opposite sex)	

Psychosexual development

desire to control his or her own bowel movements often leads to strong conflict. Adults who are fixated at this stage, in Freud's view, may display an *anal-retentive* personality and be highly controlled and compulsively neat. Or they may be very messy, disorderly, rebellious, and destructive—the *anal-expulsive* personality.

3. Phallic stage (3–6 years) During the *phallic stage*, the major center of pleasure is the genitals. Masturbation and "playing doctor" with other children are common during this time. According to Freud, a 3- to 6-year-old boy also develops an unconscious sexual longing for his mother and jealousy and hatred for the rival father. This attraction creates a conflict Freud called the **Oedipus complex**, named after Oedipus, the legendary Greek king who unwittingly killed his father and married his mother. The young boy reportedly experiences guilt and fear of punishment from the rival father, perhaps by castration. The anxiety this produces is supposedly repressed into the unconscious, which leads to the development of the superego. The boy then identifies with his father and adopts the male gender role. If this stage is not resolved completely or positively, the boy supposedly grows up resenting his father and generalizes this feeling to all authority figures.

What happens with little girls? Because a girl does not have a penis, she does not fear castration and fails to fully complete this stage and move on to successful identification with her mother. According to Freud, she develops *penis envy* and fails to develop an adequate superego, which Freud believed resulted in women being morally inferior to men. (You are undoubtedly surprised or outraged by this statement, but remember that

Freud was a product of his times. Sexism was common at this point in history. And most modern psychodynamic theorists reject Freud's notion of penis envy, as we will see in the next section.)

4. Latency period (6 years to puberty) Following the phallic stage, children supposedly repress sexual thoughts and engage in nonsexual activities, such as developing social and intellectual skills. The task of this stage is to develop successful interactions with same-sex peers and refine appropriate gender roles.

5. Genital stage (puberty to adulthood) With the beginning of adolescence, the genitals are again erogenous zones. Adolescents seek to fulfill their sexual desires through emotional attachment to members of the opposite sex. Unsuccessful outcomes at this stage lead to participation in sexual relationships based only on lustful desires, not on respect and commitment.

Psychodynamic/Neo-Freudian Theories

Some initial followers of Freud expanded on his work to include their own beliefs, ideas, and theories. They became known as *neo-Freudians*.

Alfred Adler (1870–1937) was the first to leave Freud's inner circle. Instead of seeing behavior as motivated by unconscious forces, Adler believed it is purposeful and goal directed. According to his *individual psychology*, we are motivated by our goals in life—especially our goals of obtaining security and overcoming feelings of inferiority (Carlson & Englar-Carlson, 2013).

See **FIGURE 11.5, An upside to feelings of inferiority?**

Adler believed that almost everyone suffers from an *inferiority complex*, which he defined as deep feelings of inadequacy and incompetence that arise from our feelings of helplessness as infants (Adler, 1924, 1998). According to Adler, these early feelings result in a "will-to-power" that can take one of two paths. It can lead children to strive to develop superiority over others through dominance, aggression, or expressions of envy. Or, on a more positive note, it can encourage them to develop their full potential and creativity, and to gain mastery and control in their lives (▣**Figure 11.5**).

© Hero Images/Corbis

Another early Freud follower turned dissenter, Carl Jung [Yoong], developed *analytical psychology*. Like Freud, Jung (1875–1961) emphasized unconscious processes, but he believed that the unconscious contains positive and spiritual motives, as well as sexual and aggressive forces.

See **FIGURE 11.6, Archetypes in the collective unconscious?**

Jung also thought that we have two forms of the unconscious mind. The *personal unconscious* is created from our individual experiences, whereas the **collective unconscious** is identical in each person and is inherited from our common ancestral past (Jung, 1946, 1969). The collective unconscious contains "blueprints" or models for human personality that are strongly influenced by universal memories and images and patterns, called **archetypes**, that have symbolic meaning for all people (▣**Figure 11.6**).

Because of archetypal patterns in the collective unconscious, we supposedly perceive and react in certain predictable ways. Jung claimed that both males and females have patterns for feminine aspects of personality—*anima*—and masculine aspects of personality—*animus*—which allow us to express both masculine and feminine personality traits, and to understand the opposite sex.

Like Adler and Jung, psychoanalyst Karen Horney [HORN-eye] was an influential follower of Freud who later came to reject major aspects of Freudian theory. She is credited with having developed a creative blend of Freudian, Adlerian, and Jungian concepts, along with the first feminist critique of Freud's approach (Horney, 1939, 1945). Horney emphasized women's positive traits and suggested that most of Freud's

Roger Wood/Corbis

ideas about female personality reflected male bias and misunderstanding. For example, she proposed that women's everyday experience with social inferiority led to power envy, not to Freud's idea of biological *penis envy*.

Horney also emphasized that personality development depends largely on social relationships—particularly on the relationship between parent and child. She believed that when a child's needs are not met by nurturing parents, the child may develop lasting feelings of helplessness and insecurity. The way people respond to this so-called *basic anxiety*, in Horney's view, sets the stage for later adult psychological health. She believed that everyone copes with this basic anxiety in one of three ways—we move toward, away from, or against other people—and that psychological health requires a balance among these three styles.

In sum, Horney proposed that our adult personalities are shaped by our childhood relationship with our parents—not by fixation or regression at some stage of psychosexual development, as Freud argued.

Evaluating Psychoanalytic Theories

See **FIGURE 11.7, Oral fixation or simple self-soothing?**

© PonyWang/Stockphoto

Q Answer the **Concept Check** questions.

Before going on, let's consider the major criticisms of Freud's psychoanalytic theories (Carducci, 2015; Gagnepain et al., 2014; Koocher et al., 2014):

- **Inadequate empirical support** Many psychoanalytic concepts—such as the psychosexual stages—cannot be empirically tested.
- **Overemphasis on sexuality, biology, and unconscious forces** Modern psychologists believe Freud underestimated the role of learning and culture in shaping personality (■ **Figure 11.7**).
- **Sexism** Many psychologists, beginning with Karen Horney, reject Freud's theories as derogatory toward women.

In contrast to the claim of inadequate empirical support, numerous modern studies have supported certain Freudian concepts, like defense mechanisms and the fact that a lot of our information processing occurs outside our conscious awareness (*automatic processing*, Chapter 5, and *implicit memories*, Chapter 7). In addition, a study found that people who identify as having a heterosexual orientation, but show a strong sexual attraction to same-sex people in psychological tests, tend to have more homophobic attitudes, and higher levels of hostility toward gay people (Weinstein et al., 2012). Can you see how Freud's theory might suggest that these negative attitudes and beliefs spring from unconscious repression of same-sex desires? Moreover, many contemporary clinicians still value Freud's insights about childhood experiences and unconscious influences on personality development (Gelso et al., 2014; Sand, 2014; Schimmel, 2014).

Today there are few Freudian purists. Instead, modern psychodynamic theorists and psychoanalysts tend to place less emphasis on sexual instincts and more on sociocultural influences on personality development.

11.2 TRAIT THEORIES

Think for a moment about the key personality characteristics of your best friend. You might say: *He's a great guy who's a lot of fun to be with. But I sometimes get tired of his constant jokes and pranks. On the other hand, he does listen well and will be serious when I need him to be.*

When describing another's personality, we generally use terms that refer to that person's most frequent and typical characteristics ("fun," "constant jokes and pranks," "listens well"). These unique and defining characteristics are the foundation for the *trait approach*, which seeks to discover what characteristics form the core of human personality.

Early Trait Theorists

An early study of dictionary terms found almost 4,500 words that described personality **traits**, which are relatively stable dispositions that describe an individual's characteristic way of thinking, feeling, and acting (Allport & Odbert, 1936). Faced with this enormous list, Gordon Allport (1937) believed that the best way to understand personality was to arrange a person's unique personality traits into a hierarchy, with the most pervasive or important traits at the top.

Later psychologists reduced the list of possible personality traits using a statistical technique called *factor analysis*, in which large arrays of data are grouped into more basic units (factors). Raymond Cattell (1950, 1990) condensed the list of traits to 16 source traits. Hans Eysenck (1967, 1990) reduced the list even further. He described personality as a relationship among three basic types of traits: *extroversion–introversion*, *neuroticism* (the tendency toward insecurity, anxiety, guilt, and moodiness), and *psychoticism* (being out of touch with reality).

Modern Trait Theory

Factor analysis was also used to develop the most promising modern trait theory, the **five-factor model (FFM)** (Costa & McCrae, 2011; McCrae & Costa, 2013). A handy way to remember this model is to note that the first letters of the five words spell *ocean* (■ **Figure 11.8**).

See **FIGURE 11.8, The five-factor model (FFM)**

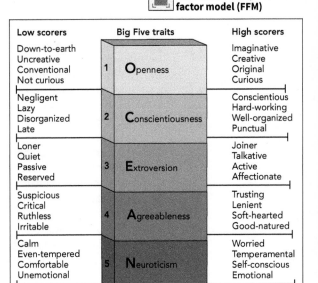

Combining previous research findings and the long list of possible personality traits, researchers discovered that these five traits came up repeatedly, even when different tests were used:

O **Openness** People who rate high on this factor are original, imaginative, curious, open to new ideas, artistic, and interested in cultural pursuits. For instance, if you score high on this dimension, you probably enjoy talking with people with sharply different opinions because you realize that what the other person is saying may have value. Low scorers tend to be conventional, down-to-earth, narrower in their interests, and not artistic.

C **Conscientiousness** This factor ranges from responsible, self-disciplined, organized, and achieving at the high end to irresponsible, careless, impulsive, lazy, and undependable at the other. If you are often late to class or social events, and commonly misplace things, you may be low on conscientiousness.

E **Extroversion** This factor contrasts people who are sociable, outgoing, talkative, fun loving, and affectionate at the high end, with introverted individuals who tend to be withdrawn, quiet, passive, and reserved at the low end. If you dislike being the center of attention and enjoy your time alone, you're probably more introverted. The reverse is true for someone who is extroverted.

A **Agreeableness** Individuals who score high in this factor are good-natured, warm, gentle, cooperative, trusting, and helpful, whereas low scorers are irritable, argumentative, ruthless, suspicious, uncooperative, and vindictive. Given that almost everyone likes to think of themselves as agreeable, it's difficult to accurately assess ourselves on this dimension. You may want to take the five factor online self-test mentioned in Figure 11.8.

N **Neuroticism (*or emotional stability*)** People who score high in neuroticism are emotionally unstable and prone to insecurity, anxiety, guilt, worry, and moodiness. People at the other end are emotionally stable, calm, even-tempered, easygoing, and relaxed. Again, most of us like to believe we're emotionally stable, so it may be helpful to take the self-grading, online quiz mentioned in Figure 11.8.

This five-factor model (FFM) has led to numerous research follow-ups and intriguing insights about personality (Connelly et al., 2014; DeYoung et al., 2014; Giluk & Postlethwaite, 2015). For example, a meta-analysis of 20 studies, involving over 17,000 students, found that conscientiousness, agreeableness, and openness were all positively correlated with grade point average (GPA) (Vedel, 2014). Based on your own college experience, can you see why this makes intuitive sense? How would you explain why extroversion does not positively correlate with GPA?

What about nonhuman animals? Do you think they have distinct personalities? If so, do they also have specific personality traits that might be associated with a longer life span?

Do Nonhuman Animals Have Unique Personalities?

Pet owners have long believed that their dogs and cats have unique personalities, and a growing body of research tends to support these beliefs (Cote et al., 2014; Cussen & Mench, 2014; Gosling & John, 1999).

Dog lovers might be interested in knowing that when 78 dogs of all shapes and sizes were rated by both owners and strangers, a strong correlation was found on traits such as affection versus aggression, anxiety versus calmness, and intelligence versus stupidity. Researchers also found that personalities vary widely within a breed, which means that not all pit bulls are aggressive, and not all Labrador retrievers are affectionate (Gosling et al., 2004).

In addition, research has discovered that chimpanzees have distinct types of personality traits that are quite similar to those described by the five-factor model (FFM) of human personality (Latzman et al., 2015). To discover whether particular personality traits were associated with a longer life expectancy in nonhuman animals, researchers studied 298 gorillas in zoos and sanctuaries across North America. Using standardized measures similar to the FFM, they asked zoo keepers, volunteers, researchers, and caretakers who knew the gorillas well to score each gorilla's personality. With these scorings, they reliably identified four distinct personality traits: *dominance, extroversion, neuroticism*, and *agreeableness* (Weiss et al., 2013).

Next, the researchers examined the association between levels of each of these personality traits and life expectancy. They found that gorillas scoring high on extroversion, which included behaviors such as sociability, activity, play, and curiosity, lived longer lives. This link was found in both male and female gorillas and across all the different types of environments in which this research was conducted.

What might explain this link? One possibility is that extroverted apes—just like extroverted people—develop stronger social networks, which helps increase survival and reduce stress. Can you think of other possible explanations?

Evaluating Trait Theories

The five-factor model (FFM) is the first model to achieve the major goal of trait theory—to describe and organize personality characteristics using the smallest number of traits. There also is strong cross-cultural research support for the FFM. Psychologist David and his colleagues (Buss, 1989, 2008) surveyed more than 10,000 men and women from 37 countries and found a surprising level of agreement in the characteristics that men and women value in a mate (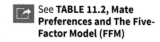 **Table 11.2**). Note that both sexes tend to prefer mates with traits that closely match the FFM—dependability (conscientiousness), emotional stability (low neuroticism), pleasing disposition (agreeableness), and sociability (extroversion) to the alternatives.

See **TABLE 11.2, Mate Preferences and The Five-Factor Model (FFM)**

Why is there such a high degree of shared preferences for certain personality traits? Scientists suggest that these traits may provide an evolutionary advantage. For instance, people who are conscientious have better health, which is clearly advantageous (Israel et al., 2014). The evolutionary advantage is also confirmed by cross-cultural studies, and comparative studies with dogs, chimpanzees, and other highly social species (e.g., Carlo et al., 2014; Gosling, 2008; Valchev et al., 2014).

In addition to having strong cross-cultural support, trait theories, like the FFM, allow us to predict real-life preferences and behaviors, such as our political attitudes, beliefs, and voting preferences, and even how much time we spend on Facebook (Hart et al., 2015; Mondak & Canache, 2014). Interestingly, FFM traits, such as conscientiousness and openness to experience, are stronger predictors of academic success and educational attainment than intelligence (Krapohl et al., 2014; Poropat, 2014). As you might expect, personality may even affect your career choice and job satisfaction.

In short, trait theories, particularly the FFM, have proven to be successful at *describing* personality and predicting certain behaviors and preferences. However, critics argue that they fail to consider situational determinants of personality or to offer sufficient *explanations* for why people develop specific traits (Chamorro-Premuzic, 2011; Cheung et al., 2011; Furguson et al., 2011). And although trait theories have documented fairly high levels of personality stability over the life span, they have failed to identify which characteristics last a lifetime and which are most likely to change (Carlo et al., 2014; Hosie et al., 2014; McCrae, 2011).

Q | Answer the **Concept Check** questions.

11.3 HUMANISTIC THEORIES

Humanistic theories of personality emphasize each person's internal feelings, thoughts, and sense of basic worth. In contrast to Freud's generally negative view of human nature, humanists believe that people are naturally good (or, at worst, neutral), and that we possess a natural tendency toward **self-actualization**, the inborn drive to realize our full potential by developing all our talents and capabilities.

According to this view, our personality and behavior depend on how we perceive and interpret the world, not on traits, unconscious impulses, or rewards and punishments. Humanistic psychology was developed largely by Carl Rogers and Abraham Maslow.

Rogers's Theory

To psychologist Carl Rogers (1902–1987), the most important component of personality is our **self-concept**, the way we see and feel about ourselves, which develops from our interactions with significant others, and our personal life experiences. Rogers emphasized that mental health and adjustment reflect the degree of overlap (congruence) between our perceived real and ideal selves. This self-perception is relatively stable over time and develops from our life experiences, particularly the feedback and perception of others.

See **TUTORIAL VIDEO:** Applying Rogerian Techniques to Your Life

Why do some people develop negative self-concepts and poor mental health? Rogers believed that such outcomes generally result from early childhood experiences with parents and other adults who make their love and acceptance contingent on behaving in certain ways and expressing only certain feelings. Imagine being a child who is repeatedly told that your naturally occurring negative feelings and behaviors (which we all have) are totally unacceptable and unlovable. Can you see how your self-concept may become distorted? And why as an adult you might develop a shy, avoidant personality, always doubting the love and approval of others because they don't know "the real person hiding inside"?

See **FIGURE 11.9, Unconditional positive regard**

To help children develop their fullest personality and life potential, Rogers cautioned that adults need to provide **unconditional positive regard**—love and acceptance with no "strings" (contingencies) attached (Feeney & Collins, 2015; Ray et al., 2014; Schneider et al., 2015). Interestingly, parents who engage in *responsive caregiving*, a form of unconditional positive regard, also tend to show this same pattern of behavior toward their spouses, which in turn leads to higher levels of relationship satisfaction (Millings et al., 2013). This suggests that unconditional positive regard is important in all types of relationships.

This is *not* to say that adults must approve of everything a child does. Rogers emphasizes that we must separate the value of the person from his or her behaviors—encouraging the person's innate positive nature, while discouraging destructive or hostile behaviors. Humanistic psychologists in general suggest that both children and adults must control their behavior so they can develop a healthy self-concept and satisfying relationships with others (□ **Figure 11.9**).

David Laurens/PhotoAlto/Corbis

Parker Eshelman/The Columbia Daily Tribune/AP Photos

Maslow's Theory

Like Rogers, Abraham Maslow believed there is a basic goodness to human nature and a natural tendency toward *self-actualization.* Maslow saw personality development as a natural progression from lower to higher levels—a basic *hierarchy of needs* (Chapter 10). As newborns, we focus on physiological needs like hunger and thirst, and then as we grow and develop, we move on through four higher levels (■ **Figure 11.10**). Surveys from 123 countries found that people from around the world do share a focus on the same basic needs, and when those needs are met, they report higher levels of happiness (Tay & Diener, 2011).

According to Maslow, self-actualization is the inborn drive to develop all our talents and capacities. It requires understanding our own potential, accepting ourselves and others as unique individuals, and taking a problem-centered approach to life (Maslow, 1970). Self-actualization is an ongoing process of growth rather than an end product or accomplishment.

Maslow believed that only a few, rare individuals, such as Albert Einstein, Mahatma Gandhi, and Eleanor Roosevelt, become fully self-actualized. However, he saw self-actualization as part of every person's basic hierarchy of needs. (See Chapter 10 for more information about Maslow's theory.)

Evaluating Humanistic Theories

Humanistic psychology was extremely popular during the 1960s and 1970s. It was seen as a refreshing new perspective on personality after the negative determinism of the psychoanalytic approach and the mechanical nature of learning theories (Chapter 6). Although this early popularity has declined, humanistic theories have provided valuable insights that are useful for personal growth and self-understanding. They also play a major role in contemporary counseling and psychotherapy, as well as in modern childrearing, education, and managerial practices (Angus et al., 2015; DeRobertis, 2013; Schneider et al., 2015).

However, humanistic theories have also been criticized (Berger, 2015; Henwood et al., 2014; Nolan, 2012) for the following:

1. Naïve assumptions Some critics suggest that humanistic theories are unduly optimistic and overlook the negative aspects of human nature. How would humanists explain immoral politicians, the spread of terrorism, and humankind's ongoing history of war and murder?

2. Poor testability and inadequate evidence Like many psychoanalytic terms and concepts, humanistic concepts such as unconditional positive regard and self-actualization are difficult to define operationally and to test scientifically.

3. Narrowness Like trait theories, humanistic theories have been criticized for merely describing personality rather than explaining it. They ask, where does the motivation for self-actualization come from? To say that it is an "inborn drive" doesn't satisfy those who favor using experimental research and scientific standards to study personality.

Q Answer the **Concept Check** questions.

11.4 SOCIAL-COGNITIVE THEORIES

As you've just seen, psychoanalytic/psychodynamic, trait, and humanistic theories all tend to focus on internal, personal factors in personality development. In contrast, today's *social-cognitive* theories emphasize the influence of our *social* interpersonal interactions with the environment, along with our *cognitions*—our thoughts, feelings, expectations, and values.

Bandura's and Rotter's Approaches

Albert Bandura (also discussed in Chapter 6) has played a major role in reintroducing thought processes into personality theory. Cognition, or thought, is central to his concept of **self-efficacy**, which is our expectations of success—or in other words, our self-confidence (Bandura, 1997, 2015).

According to Bandura, if you have a strong sense of self-efficacy, you believe you can generally succeed and reach your goals, regardless of past failures and current obstacles. Your self-efficacy will, in turn, affect which challenges you choose to accept and the effort you expend in reaching goals. A study found that children who assisted their parents with meal preparation reported higher self-efficacy for selecting and eating healthy foods (Chu et al., 2013).

How does self-efficacy affect personality? Bandura sees personality as being shaped by **reciprocal determinism**, which suggests that the *person* (his or her personality, thoughts, expectations, etc.), *environment*, and *behavior* all interact and influence one another (◼ **Figure 11.11**). Using Bandura's concept of self-efficacy, can you see how your own beliefs will affect how others respond to you, and thereby influence your chance for success? Your belief ("I can succeed") will affect behaviors ("I'll work hard and ask for a promotion"), which in turn will affect the environment ("My employer recognized my efforts and promoted me").

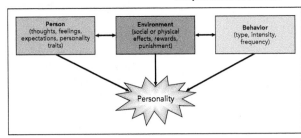
See **FIGURE 11.11, Bandura's theory of reciprocal determinism**

Julian Rotter's theory is similar to Bandura's in that it suggests that learning experiences create *cognitive expectancies* that guide behavior and influence the environment (Rotter, 1954, 1990). According to Rotter, your behavior or personality is determined by (1) what you expect to happen following a specific action and (2) the reinforcement value attached to specific outcomes.

See **FIGURE 11.12, Locus of control and achievement**

To understand your personality and behavior, Rotter would use personality tests that measure your internal versus external *locus of control* (Chapter 3). Rotter's tests ask volunteers to respond to statements such as, "People get ahead in this world primarily by luck and connections rather than by hard work and perseverance," and "When someone doesn't like you, there is little you can do about it." As you may suspect, people with an *external locus of control* think the environment and external forces have primary control over their lives, whereas people with an *internal locus of control* think they can personally control events in their lives through their own efforts (◼ **Figure 11.12**).

"We're encouraging people to become involved in their own rescue."

Evaluating Social-Cognitive Theories

The social-cognitive perspective holds several attractions. First, it offers testable, objective hypotheses and operationally defined terms, and it relies on empirical data. Second, social-cognitive theories emphasize the role of cognitive processes in personality, and the interplay between personality characteristics and situational forces.

On the other hand, critics argue that social-cognitive theories focus too much on situational influences. They also suggest that this approach fails to adequately acknowledge the stability of personality, as well as sociocultural, emotional, and biological influences (Ahmetoglu & Chamorro-Premuzic, 2013; Berger, 2015; Cea & Barnes, 2015).

Q Answer the **Concept Check** questions.

11.5 BIOLOGICAL THEORIES

In this section, we explore how biological factors influence our personalities. We conclude with a discussion of how all theories of personality ultimately interact in the *biopsychosocial model*.

Three Major Contributors to Personality

Hans Eysenck, the trait theorist mentioned earlier in the chapter, was one of the first to propose that personality traits are biologically based—at least in part. And modern research supports the belief that certain brain structures, neurochemistry, and genetics all may contribute to some personality characteristics. For example, how do we decide in the real world which risks are worth taking and which are not? Modern research using functional magnetic resonance imaging (fMRI) and other brain mapping techniques has identified specific areas of the brain that correlate with impulsiveness, and areas that differ between people with risk-averse and those with risk-seeking personalities (Laricchiuta et al., 2014; Schilling et al., 2014).

Earlier research also found that increased electroencephalographic (EEG) activity in the left frontal lobes of the brain is associated with sociability (or extroversion), whereas greater EEG activity in the right frontal lobes is associated with shyness or introversion (Fishman & Ng, 2013; Tellegen, 1985).

A major limitation of research on brain structures and personality is the difficulty of identifying which structures are uniquely connected with particular personality traits. Neurochemistry seems to offer more precise data on how biology influences personality. For instance, sensation seeking (Chapter 10) has consistently been linked with levels of monoamine oxidase (MAO), an enzyme that regulates levels of neurotransmitters such as dopamine (García et al., 2014; Zuckerman, 1994, 2004, 2014). Dopamine also seems to be correlated with addictive personality traits, novelty seeking, and extroversion (Blum et al., 2013; Harris et al., 2015; Schilling et al., 2014).

How can neurochemistry have such effects? Studies suggest that high-sensation seekers and extroverts tend to experience less physical arousal than introverts from the same stimulus (Fishman & Ng, 2013; Munoz & Anastassiou-Hadjicharalambous, 2011). Extroverts' low arousal apparently motivates them to seek out situations that will elevate their arousal. Moreover, it is

See multipart **FIGURE 11.13, Multiple influences on personality**

believed that a higher arousal threshold is genetically transmitted. In other words, personality traits like sensation seeking and extroversion may be inherited. The genes we inherit may also predict how nice we are, how we parent, and even possible criminal behaviors, including arrest records (Armstrong et al., 2014; Klahr & Burt, 2014).

Rosanne Olson/Stone/Getty Images

Evaluating Biological Theories

Modern research in biological theories has provided exciting insights, and established clear links between some personality traits and various brain areas, neurotransmitters, and/or genetics. However, it's important to keep in mind that personality traits are never the result of a single biological process. Studies do show a strong inherited basis for personality, but researchers are careful not to overemphasize genetics (Allan et al., 2014; Latzman et al., 2015; Turkheimer et al., 2014). Some believe the importance of the unshared environment—aspects of the environment that differ from one individual to another, even within a family—has been overlooked. Others fear that research on "genetic determinism" could be misused to "prove" that a particular ethnic or racial group is inferior, that male dominance is natural, or that social progress is impossible.

 Answer the **Concept Check** questions.

As you can see, each personality theory offers different insights into how a person develops the distinctive set of characteristics we call "personality." That's why, instead of adhering to any one theory, many psychologists believe in the *biopsychosocial approach*, or the idea that several factors—biological, psychological, and social—overlap in their contributions to personality (■ **Figure 11.13**).

11.6 PERSONALITY ASSESSMENT

Numerous unscientific methods have been used over the decades to assess personality. Even today, some people consult fortune-tellers, horoscope columns in the newspaper, tarot cards, and fortune cookies in Chinese restaurants. But scientific research has provided much more reliable and valid methods for measuring personality (Berger, 2015; Dana, 2014).

Clinical and counseling psychologists, psychiatrists, and other helping professionals use these modern methods to help with the diagnosis of patients, and to assess their progress in therapy. Personality assessments can be grouped into a few broad categories: *interviews*, *observations*, *objective tests*, and *projective tests*.

Interviews and Observation

We all use informal "interviews" to get to know other people. When first meeting someone, we usually ask about his or her job, academic interests, family, or hobbies. Psychologists also use interviews. In an unstructured format, interviewers get impressions and pursue hunches, or let the interviewee expand on information that promises to disclose personality characteristics. In structured interviews, the interviewer asks specific questions in order to evaluate the interviewee's responses more objectively and compare them with others' responses.

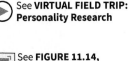 See **VIRTUAL FIELD TRIP: Personality Research**

In addition to conducting interviews, psychologists also assess personality by directly and methodically observing behavior. They look for examples of specific behaviors and fol-

See **FIGURE 11.14, Behavioral observation**

low a careful set of evaluation guidelines. For instance, a psychologist might arrange to observe a troubled client's interactions with his or her family. Does the client become agitated by the presence of certain family members and not others? Does he or she become passive and withdrawn when asked a direct question? Through careful observation, the psychologist gains valuable insights into the client's personality, as well as family dynamics (■ **Figure 11.14**).

David De Lossy/Getty Images

Objective Tests

Objective personality tests, or inventories, are the most widely used method of assessing personality, for two reasons: They can be administered to a large number of people relatively quickly, and they can be evaluated in a standardized fashion. Some objective tests measure one specific personality trait, such as sensation seeking (Chapter 12) or locus of control. However, psychologists in clinical, counseling, and industrial settings often wish to assess a range of personality traits. To do so, they generally use multi-trait, or *multiphasic*, inventories.

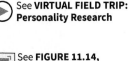 See **TUTORIAL VIDEO: Exploring Your Personality**

The most widely studied and clinically used multitrait test is the **Minnesota Multiphasic Personality Inventory (MMPI)**—or its revisions, the MMPI-2 and MMPI-2-RF (Bagby et al., 2014; Butcher, 2000, 2011; Finn et al., 2014). The latest version, the MMPI-2-RF, consists of 388 statements, condensed from the 567 items on the original MMPI. Participants respond with *True*, *False*, or *Cannot Say*. The following are examples of the kinds of statements found on the MMPI:

My stomach frequently bothers me.

I have enemies who really wish to harm me.

I sometimes hear things that other people can't hear.

I would like to be a mechanic.

I have never indulged in any unusual sex practices.

Did you notice that some of these questions are about very unusual, abnormal behavior? Although there are many "normal" questions on the full MMPI, the test is designed primarily to help clinical and counseling psychologists diagnose psychological disorders. However, it's also sometimes used in hiring decisions and forensic settings.

Other objective personality measures are less focused on psychological disorders. A good example is the NEO Personality Inventory–Revised, which assesses the dimensions of the five-factor model.

Note that personality tests like the MMPI are often confused with *career inventories*, or vocational interest tests. Career counselors use these latter tests (along with aptitude and achievement tests) to help people identify occupations and careers that match their unique traits, skills, and interests.

Projective Tests

See multipart **FIGURE 11.15, Sample projective tests**

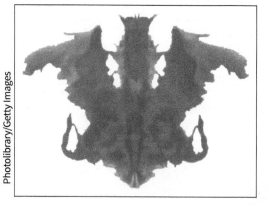

Photolibrary/Getty Images

Unlike objective tests, **projective tests** use a standard set of ambiguous stimuli that people can perceive in many ways. When you listen to a piece of music or look at a picture, you might say that the music is sad, or that the people in the picture look happy—but not everyone would have the same interpretation. Some psychologists believe that these different interpretations reveal important things about each individual's personality.

As the name implies, projective tests are meant to allow each person to project his or her own unconscious conflicts, psychological defenses, motives, and personality traits onto the test materials. Because respondents may be unable or unwilling to express their true feelings if asked directly, the ambiguous stimuli reportedly provide an indirect "psychological X-ray" of important unconscious processes (Hogan, 2013). As depicted in ▣ **Figure 11.15**, the **Rorschach Inkblot Test** and **Thematic Apperception Test (TAT)** are two of the most widely used projective tests (Choca, 2013; Silverstein, 2013).

Are Personality Measurements Accurate?

Let's evaluate the strengths and the challenges of each of the four methods of personality assessment: *interviews*, *observation*, *objective tests*, and *projective tests*.

Interviews and Observations

Both interviews and observations can provide valuable insights into personality, but they are time-consuming and expensive. Furthermore, raters of personality tests frequently disagree in their evaluations of the same individuals. Interviews and observations also take place in unnatural settings, and, in fact, the very presence of an observer can alter a participant's behavior. For example, can you recall a time in which you were nervous in a job interview, and maybe didn't act quite the same as you would in a more relaxed setting?

 See **TUTORIAL VIDEO: Measuring Personality**

Objective Tests

Tests like the MMPI-2 provide specific, objective information about a broad range of personality traits in a relatively short period. However, they are also subject to at least three major criticisms:

1. Deliberate deception and social desirability bias Some items on personality inventories are easy to see through, so respondents may fake particular personality traits. In addition, some respondents want to look good and will answer questions in ways that they perceive are socially desirable. For instance, people might try to come across as less hostile, and more kind to others in their responses to some test items.

To avoid these problems, the MMPI-2 has built-in validity scales. In addition, personality researchers and some businesses avoid self-reports. Instead, they rely on other people, such as friends or coworkers, to rate an individual's personality, as well as how their personality influences their work performance (Connelly & Hülsheger, 2012). In fact, three meta-analyses involving over 44,000 participants found that ratings from others were better predictors of actual behavior, academic achievement, and job performance than those based on self-reports (Connelly & Ones, 2010).

2. Diagnostic difficulties In addition to the problems with all self-report inventories, when they are used for diagnosis, overlapping items sometimes make it difficult to pinpoint a disorder (Ben-Porath, 2013; Hogan, 2013; Hunsley et al., 2015). Clients with severe disorders also sometimes score within the normal range, and normal clients sometimes score within the elevated range (Borghans et al., 2011; Morey, 2013). Furthermore, one study found that computer-based tests were more accurate at determining someone's personality than those made by humans (Youyou et al., 2015).

3. Cultural bias and inappropriate use Some critics think the standards for "normalcy" on objective tests fail to recognize the impact of culture (Dana, 2014; Geisinger & McCormick, 2013; Malgady et al., 2014). Research examining personality traits in members of the Tsimané culture, a community of foragers and farmers in Bolivia with relatively little contact with the outside world, reveals two distinct dimensions of personality—prosociality and industriousness—instead of the more widely accepted five personality traits (Gurven et al., 2013).

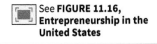 See **FIGURE 11.16, Entrepreneurship in the United States**

Cultural differences in personality may also be seen in different parts of a single country (Rentfrow, 2014). A study of over half a million people in the United States revealed regional differences in personality traits linked with entrepreneurial activity, defined as business-creation and self-employment rates (Obschonka et al., 2013). As you can see in ◼ **Figure 11.16**, certain regions are much more entrepreneurial than others. Why? The authors of the study suggested that the higher scores in the West, for example, might be a reflection of America's historical migration patterns of people moving from the East into the West (or from outside of America). They cite other research (e.g., Rentfrow et al., 2008) that suggests this selective migration may have had a lasting effect on personality due to the heritability of personality traits, and the passing on of norms and values within the regions.

What do you think? How would you explain the differences? Can you see the overall value of expanding our study of personality from just looking at differences between individuals to examining regional differences, and how this expansion might increase our understanding of how personality is formed and its potential applications?

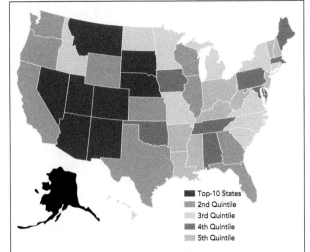

Top-10 States
2nd Quintile
3rd Quintile
4th Quintile
5th Quintile

Source: Obschonka, M., et. al. (2013). The regional distribution and correlates of an entrepreneurshipprone personality profile in the United States, Germany, and the United Kingdom: A socioecological perspective. Journal of Personality and *Social Psychology, Vol 105*(1), Jul 2013, 104–122. Copyright ©2013 by the American Psychological Association.

Projective Tests

Although projective tests are extremely time-consuming to administer and interpret, their proponents suggest that because the method is unstructured, respondents may be more willing to talk honestly about sensitive topics. Critics point out several limitations with projective tests, including problems with their *reliability* and *validity* (Hartmann & Hartmann, 2014; Hunsley et al., 2015; Koocher et al., 2014). [Recall from Chapter 8 that reliability (the consistency of test results) and validity (whether the test actually measures what it was designed to measure) are the two most important measures of a good test.]

As you can see, each of these methods of personality assessment has limits, which is why psychologists typically combine the results from various sources to create a fuller picture of any individual's personality. However, you're unlikely to have access to this type of professional analysis, so what's the most important take-home message? Beware of pop-psych books and pop-culture personality quizzes in magazines and websites! They may be entertaining, but they're rarely based on standardized testing or scientific research of any kind, and you should never base important decisions on their input.

Finally, throughout this text, we have emphasized the value of critical thinking, which also is useful in evaluating personality tests.

Q Answer the **Concept Check** questions.

WP LS Go to your WileyPLUS Learning Space course for video episodes, examples, art, tables, Concept Checks, practice, and other pedagogical resources that will help you succeed in this course.

Reading for

PSYCHOLOGICAL DISORDERS

12.1 STUDYING PSYCHOLOGICAL DISORDERS

Most people agree that neither the artist who stays awake for 72 hours finishing a painting, nor the shooter who kills 20 young school children is behaving normally. But what exactly is "normal"? How do we distinguish between eccentricity in the first case, and abnormal behavior in the second?

Identifying and Explaining Psychological Disorders

See multipart **FIGURE 12.1, Four criteria for identifying abnormal behavior**

Corbis Flirt/Alamy

See **TUTORIAL VIDEO: Myths About Mental Illness**

See **FIGURE 12.2, Witchcraft or mental illness?**

Corbis Images

As you can see, it's difficult to distinguish normal from abnormal behavior, and psychologists have struggled to create a precise definition. However, mental health professionals generally agree that **abnormal behavior** (or psychopathology) can be identified as the patterns of behaviors, thoughts, or emotions considered pathological (diseased or disordered) for one or more of these four reasons: *deviance, dysfunction, distress,* and *danger* (■ **Figure 12.1**). Keep in mind that abnormal behavior, like intelligence or creativity, is not composed of two discrete categories—"normal" and "abnormal." Instead, mental health lies along a continuum, with people being unusually healthy at one end, and extremely disturbed at the other (Kring et al., 2014; Sue et al., 2016). As you consider the deviance and danger criteria, remember that judgments of what is deviant vary historically and cross-culturally, and that the public generally overestimates the danger posed by the mentally ill.

What causes abnormal behavior? Historically, evil spirits and witchcraft have been blamed (King, 2015; Shiraev, 2015; Walsh et al., 2014). For example, during the Stone Age, it was believed that abnormal behavior stemmed from demonic possession. The prescribed "therapy" for such behavior was to bore a hole in the skull so the evil spirit could escape, a process known as *trephining*. During the European Middle Ages, a troubled person was sometimes treated with *exorcism*, an effort to drive the devil out by praying, fasting, making noise, and beating the suffering victim. During the fifteenth century, as the Renaissance got under way, many believed that some individuals chose to consort with the devil. These supposed witches were often tortured, imprisoned for life, or executed (■ **Figure 12.2**).

As the Middle Ages ended, special mental hospitals called *asylums* began to appear in Europe. Initially designed to provide quiet retreats for the victims, and to protect society, the asylums unfortunately became overcrowded, inhumane prisons (Radhika et al., 2015; Shiraev, 2015).

Improvements in the conditions of these asylums came in 1792, when Philippe Pinel, a French physician, was placed in charge of a Parisian asylum. Believing that the inmates' behavior was caused by underlying physical illness, Pinel insisted that they be unshackled and removed from their dark, unheated cells. Many inmates improved so dramatically that they could be released. Pinel's actions reflected the ideals of the modern **medical model**, which assumes that diseases (including mental illness) have physical causes that can be diagnosed, treated, and possibly cured and prevented. This medical model is the foundation of the branch of medicine, known as **psychiatry**, which deals with the diagnosis, treatment, and prevention of mental disorders.

Although the medical model brought about a significant improvement in the treatment of mental illness, when we label people "mentally ill," we may create new problems. One of the most outspoken critics, psychiatrist Thomas Szasz proposed that this medical model encourages people to believe they have little or no responsibility for their actions. Furthermore, Szasz contended that mental illness is a myth used to label individuals who are peculiar or offensive to others, and that these labels can become self-perpetuating (Haldipur, 2013; Szasz, 1960, 2012). In other words, the label may become permanent because the sufferer accepts the label, and begins to behave according to his or her diagnosed disorder.

In addition to potential problems for the identified person, the public also may develop negative attitudes about those labeled as mentally ill, and about mental illness in general. This is particularly common following intensive media coverage of mass shootings. For example, many Americans were shocked and deeply saddened on June 17, 2015, when 21-year-old Dylann Roof, shot and killed nine people who were attending a church prayer service in Charleston, South Carolina. Roof later confessed to the crime, saying that he murdered the Black church members because wanted to ignite a race war.

Following the shootings, the news media, and almost all of the candidates for the 2016 U.S. presidential election, expressed their shock and condolences, and then focused on issues of race and gun control. However, many also said that mental illness was a motivating factor, and a common thread for mass shootings (Gonyea & Montanaro, 2015; Lysiak, 2015). Can you see how this type of intensive media coverage increases the myths, misconceptions, and exaggerated fears of mental illness? To make matters worse, the media seldom, if ever, mentions the fact that people with mental illness are more often self-destructive, or the victims of violence—rather than the perpetrators (McGinty et al., 2013; Metzl & MacLeish, 2015).

Despite these potential problems and dangers, the medical model—and the concept of mental illness—remains a founding principle of psychiatry. Psychology, in contrast, offers a multifaceted approach to explaining abnormal behavior.

See **STUDY ORGANIZER 12.1: Seven Psychological Perspectives on Psychological Disorders**

Classifying Psychological Disorders

Given the potential problems with labeling mental illness, why do we do it? Without a clear, reliable system for classifying the wide range of psychological disorders, scientific research on them would be almost impossible, and communication among mental health professionals would be seriously impaired. Fortunately, mental health specialists share a uniform classification system, and the manual that includes this classification of psychological disorders is known as the ***Diagnostic and Statistical Manual of Mental Disorders (DSM)*** (American Psychiatric Association, 2013). This text has been updated and revised several times, and the latest, fifth edition, was published in 2013.

Each revision of the *DSM* has expanded the list of disorders, and changed the descriptions and categories to reflect the latest in scientific research (Black & Grant, 2014; Roberts & Louie, 2015; Sharp, 2015). For example, in previous editions of the *DSM*, the term **neurosis** reflected Freud's belief that all neurotic conditions arise from unconscious conflicts (Chapter 11). As modern research has diminished Freud's influence on the field, conditions that were previously grouped under the heading *neurosis* have been formally studied and redistributed as separate categories. Unlike neurosis, the term **psychosis**, which is commonly defined as a *loss of contact with reality*, is still listed in the current edition of the *DSM* because it remains useful for distinguishing the most severe psychological disorders, such as schizophrenia.

See **FIGURE 12.3, The insanity plea—guilty of a crime or mentally ill?**

What about the term *insanity*? Where does it fit in? **Insanity** is a legal term indicating that a person cannot be held responsible for his or her actions, or is incompetent to manage his or her own affairs, because of mental illness. The definition rests primarily on a person's inability to tell right from wrong (■ **Figure 12.3**).

Eddie Ray Routh Chris Kyle

© STRINGER/Reuter/Corbis

Chris Haston/NBC/NBCU Photo Bank/Getty Images, Inc.

See **TABLE 12.1,
Subcategories of
Psychological Disorders in
the DSM-5**

Explaining the DSM

The *DSM* identifies and describes the symptoms of approximately 400 disorders, which are grouped into 22 categories (⬛ **Table 12.1**). We focus on only the first 7 in this chapter (categories 8–14 are discussed in other chapters; and 15–22 are beyond the scope of this text). Note that the 22 categories are based on groups that share common, key features. For example, generalized anxiety disorder, panic disorder, and phobias are all classified under "anxiety disorders" because their main symptom is excessive anxiety. Also remember that there is considerable overlap among the categories, and that people may be diagnosed with more than one disorder at a time, a condition referred to as **comorbidity**.

Critiques of the DSM

As mentioned earlier, classification and diagnosis of mental disorders are essential to scientific study. Without a system such as the *DSM*, we could not effectively identify and diagnose the wide variety of disorders, predict their future courses, or suggest appropriate treatment. The *DSM* also facilitates communication among professionals, those suffering from psychological disorders, and their families. In addition, it serves as a valuable educational tool.

Unfortunately, the *DSM* does have several limitations and potential problems (Aragona, 2015; Bornstein, 2015; Vandeleur et al., 2015). For example, critics suggest that it overemphasizes the medical model, and that it may be casting too wide a net and *overdiagnosing*. Given that insurance companies compensate physicians and psychologists only if each client is assigned a specific *DSM* code number, can you see how compilers of the *DSM* may be encouraged to add more categories and diagnoses?

In addition, the *DSM* has been criticized for having a potential *cultural bias*. It does provide a culture-specific section and a glossary of culture-bound syndromes, such as *amok* (Indonesia), *genital retraction syndrome* (Asia), and *windigo psychosis* (Canadian Indians), which we discuss later in this chapter. However, the overall classification still reflects a Western European and U.S. perspective (Chang & Kwon, 2014; Hsu, 2016; Jacob, 2014).

Perhaps the most troubling criticism of the *DSM* is the potentially dangerous power of diagnostic labels. Consider the classic and controversial study conducted by David Rosenhan (1973) in which he and seven other adults, none with psychological disorders, presented themselves at several hospital admissions offices complaining of hearing voices (a classic symptom of schizophrenia). Aside from making this single false complaint, and providing false names and occupations, the researchers answered all questions truthfully. Not surprisingly, given their reported symptom, they were all diagnosed with psychological disorders, and admitted to the hospital. Once there, these pseudo-patients stopped reporting any symptoms and behaved as they normally would, yet none were ever recognized by hospital staff as phony. All eight of these pseudo-patients were eventually released, after an average stay of 19 days. However, all but one were assigned a label on their permanent record of "schizophrenia in remission."

What do you think about this study? Did you believe it demonstrated the inherent dangers and "stickiness" of all forms of labels? This particular study has been criticized, but few doubt that the stigma, prejudice, and discrimination surrounding mental illness often create lifetime career and social barriers for those who are already struggling with the psychological disorder itself.

Q Answer the
Concept Check questions.

12.2 ANXIETY DISORDERS

Have you ever faced a very important exam, job interview, or first date and broken out in a cold sweat, felt your heart pounding, and had trouble breathing? If so, you have some understanding of anxiety. But when the experiences and symptoms of fear and anxiety become disabling (uncontrollable and disrupting), mental health professionals may diagnose it as an **anxiety disorder**. Fortunately, anxiety disorders are also among the easiest to treat, and offer some of the best chances for recovery (Chapter 13).

Describing Anxiety Disorders

See **STUDY ORGANIZER 12.2:**
Anxiety Disorders

In this section, we discuss three anxiety disorders: *generalized anxiety disorder* (GAD), *panic disorder*, and *phobias*. Although we cover these disorders separately, two or more often occur together (Dibbets et al., 2015; Iwamasa & Regan, 2014; Zinbarg et al., 2015).

Generalized Anxiety Disorder

Sufferers of **generalized anxiety disorder (GAD)** experience persistent, uncontrollable, and free-floating, nonspecified anxiety. The fears and anxiety are referred to as "free-floating" because they're unrelated to any specific threat, which also explains why the anxiety is referred to as *generalized* (Louie & Roberts, 2015; Szkodny & Newman, 2014). However, the anxiety is chronic, uncontrollable, and lasts at least six months. Because of persistent muscle tension and autonomic fear reactions, people with this disorder may develop headaches, heart palpitations, dizziness, and insomnia, making it even harder to cope with normal, daily activities. The disorder affects twice as many women as men (Horwath & Gould, 2011).

Panic Disorder

Most of us have experienced feelings of intense panic, such as after narrowly missing a potentially fatal traffic collision. However, people with **panic disorder** endure repeated, sudden onsets of intense terror, and inexplicable *panic attacks*. Symptoms include severe heart palpitations, trembling, dizziness, difficulty breathing, and feelings of impending doom. The reactions are so extreme that many sufferers believe they are having a heart attack. Panic disorder is diagnosed when several apparently spontaneous panic attacks lead to a persistent concern about future attacks. A common complication of panic disorder is agoraphobia, discussed in the next section (Petrowski et al., 2014; Schmidt et al., 2014).

Phobias

Like the previously mentioned feelings of panic, most of us also share a common fear of spiders, sharks, and snakes. However, people who suffer from **phobias** experience a persistent and intense, irrational fear, and avoidance of a specific object, activity, or situation. Their fears are so disabling that they significantly interfere with their daily life. Although the person recognizes that the level of fear is irrational, the experience is still one of overwhelming anxiety, and a full-blown panic attack may follow. The fifth edition of the *DSM* divides phobia into separate categories: agoraphobia, specific phobias, and social anxiety disorder (social phobia) (Black & Grant, 2014; Louie & Roberts, 2015).

People with *agoraphobia* restrict their normal activities because they fear having a panic attack in crowded, enclosed, or wide-open places where they would be unable to receive help in an emergency. In severe cases, people with agoraphobia may refuse to leave the safety of their homes.

A *specific phobia* is a fear of a specific object or situation, such as needles, rats, spiders, or heights. Claustrophobia (fear of closed spaces) and acrophobia (fear of heights) are the specific phobias most often treated by therapists. People with specific phobias generally recognize that their fears are excessive. However, they're unable to control their anxiety, and will go to great lengths to avoid the feared stimulus.

One factor that helps explain why people with specific phobias have such extreme reactions is that they perceive the feared objects as larger than they actually are. Researchers in one study asked people with spider phobias to stand beside a glass tank and observe five different tarantulas (Vasey et al., 2012). They were then asked to estimate the size of each tarantula by drawing a line on an index card to illustrate its length. Those who reported experiencing the highest levels of fear while standing beside the tank drew the longest lines.

See **FIGURE 12.4, Anxiety disorders and the biopsychosocial model**

People with *social anxiety disorder* (or *social phobia*) are irrationally fearful of embarrassing themselves in social situations. Fear of public speaking and of eating in public are the two most common social phobias. Concerns over public scrutiny and potential humiliation may become so pervasive that normal life is impossible (Beidel et al., 2014; Bruce & Heimberg, 2014; Wild & Clark, 2015).

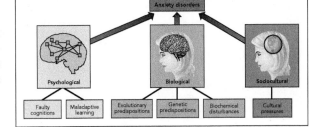

Explaining Anxiety Disorders

Why do people develop anxiety disorders? Research has focused on the roles of psychological, biological, and sociocultural processes (the *biopsychosocial model*) (■ **Figure 12.4**).

Psychological Factors

Studies have identified two major psychological contributions to anxiety disorders:

1. *Faulty cognitive processes* People with anxiety disorders generally have habits of thinking, or cognitive processes, that make them prone to fear. These faulty cognitions, in turn, make them hypervigilant—meaning they constantly scan their environment for signs of danger, and ignore signs of safety. They also tend to magnify ordinary threats and failures, and to be hypersensitive to others' opinions of them (Helbig-Lang et al., 2015; Sue et al., 2016; Wild & Clark, 2015).

2. *Maladaptive learning* In contrast to this cognitive explanation, learning theorists suggest that anxiety disorders result from inadvertent and improper conditioning and social learning (Duits et al., 2015; Kunze et al., 2015; van Meurs et al., 2014). As we discovered in Chapter 6, during classical conditioning, if a stimulus that is originally neutral (such as a harmless spider) becomes paired with a frightening event (a sudden panic attack), it becomes a conditioned stimulus (CS) that elicits a conditioned emotional response (CER)—in this case, a phobia. The person then begins to avoid spiders in order to reduce anxiety (an operant conditioning process known as negative reinforcement) (**Process Diagram 12.1**).

> This Process Diagram contains essential information NOT found elsewhere in the text, which is likely to appear on quizzes and exams. Be sure to study it CAREFULLY!

Process Diagram 12.1

Conditioning and Phobias

Classical conditioning combined with operant conditioning can lead to phobias. Consider the example of Little Albert's classically conditioned fear of rats, discussed in Chapter 6.

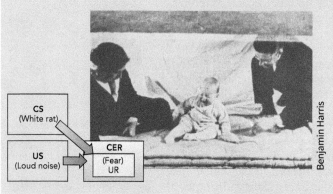

Benjamin Harris

Step ❶ Classical conditioning
By pairing the white rat with a loud noise, experimenters used classical conditioning to train Little Albert to fear rats.

CS
(White rat)

US
(Loud noise)

CER
(Fear)
UR

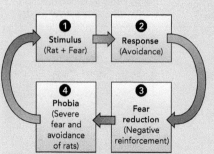

❶ Stimulus (Rat + Fear)

❷ Response (Avoidance)

❸ Fear reduction (Negative reinforcement)

❹ Phobia (Severe fear and avoidance of rats)

Step ❷ Operant conditioning
Later Albert might have learned to avoid rats through operant conditioning. Avoiding rats would reduce Albert's fear, giving unintended reinforcement to his fearful behavior.

© jsteck/iStockphoto

Step ❸ Phobia
Over time, this cycle of fear and avoidance could lead to an intense, irrational fear, known as a phobia.

Some researchers contend that because most people with phobias cannot remember a specific instance that led to their fear, and because frightening experiences do not always trigger phobias, conditioning may not be the only explanation. Social learning theorists propose that some phobias result from modeling and imitation. For example, children whose parents, especially fathers, are afraid of going to the dentist are much more likely to develop a similar fear (Lara-Sacido et al., 2012).

Biological Factors

Some researchers, who take a biological approach, believe phobias reflect an evolutionary predisposition to fear things that were dangerous to our ancestors (Gilbert, 2014; Mineka & Oehlberg, 2008; New & German, 2015). Some people with panic disorder also seem genetically predisposed toward an overreaction of the autonomic nervous system, further supporting the argument for a biological component. In addition, stress and arousal, drugs (e.g., caffeine, nicotine) and even hyperventilation can trigger an attack, all suggesting a biochemical disturbance.

Sociocultural Factors

In addition to psychological and biological components, sociocultural factors can contribute to anxiety. For example, there has been a sharp rise in anxiety disorders in the past 50 years, particularly in Western industrialized countries. Can you see how our fast-paced lives—along with our increased mobility, decreased job security, and less family support—might contribute to anxiety? Unlike the dangers early humans faced in our evolutionary history, today's threats are less identifiable and less immediate. This may, in turn, lead some people to become hypervigilant, and predisposed to anxiety disorders.

Further support for sociocultural influences on anxiety disorders is that they often have dramatically different forms in other cultures. Some Japanese experience a type of social phobia called *taijin kyofusho* (*TKS*), a morbid dread of doing something to embarrass *others*. This disorder is quite different from the Western version of social phobia, which centers on a fear of criticism and self-embarrassment.

Q Answer the **Concept Check** questions.

12.3 DEPRESSIVE AND BIPOLAR DISORDERS

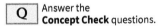
See **FIGURE 12.5, Depressive versus bipolar disorders**

Describing Depressive and Bipolar Disorders

People with **depressive disorders** suffer from sad, empty, or irritable moods, which interfere with their ability to function. We all feel "blue" sometimes—especially following the loss of a job, end of a relationship, or death of a loved one. But victims of depressive disorders are so deeply sad and discouraged that they often have trouble sleeping, are likely to lose (or gain) significant weight, and may feel so fatigued that they cannot go to work or school, or even comb their hair and brush their teeth. Seriously depressed individuals also have trouble concentrating, making decisions, and being social. In addition, they often have a hard time recognizing their common "thinking errors," such as *tunnel vision*, which involves focusing on only certain aspects of a situation (usually the negative parts) and ignoring other interpretations or alternatives. Can you see how this type of depressed thinking would deepen depression, and possibly even lead to suicide (Durbin, 2014; Rasmussen & Aleksandrof, 2015; Sue et al., 2016)?

When depression is *unipolar*, and the depressive episode ends, the person generally returns to a normal emotional level. People with **bipolar disorders**, however, rebound to the opposite state, known as *mania*, which is characterized by unreasonable elation and hyperactivity. (■ **Figure 12.5**).

During a manic episode, the person is overly excited, and easily distracted. In addition, he or she exhibits unrealistically high self-esteem, an inflated sense of importance, and poor judgment—giving away valuable possessions or going on wild spending sprees. The person also is often hyperactive, and may not sleep for days at a time, yet does not become fatigued. Thinking is faster than normal, and can change abruptly to new topics,

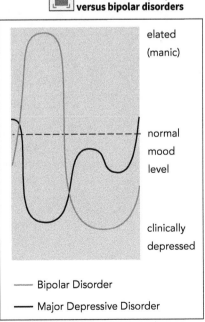
—— Bipolar Disorder
—— Major Depressive Disorder

elated (manic)

normal mood level

clinically depressed

See **VIRTUAL FIELD TRIP: Bipolar Disorder**

showing "rapid flight of ideas." Speech is also rapid ("pressured speech"), making it difficult for others to get a word in edgewise. A manic episode may last a few days or a few months, but ends abruptly. The ensuing depressive episode generally lasts three times as long as the mania (Leigh, 2015; Ray, 2015).

The lifetime risk for bipolar disorder is low—between 0.5 and 1.6%—but it can be one of the most debilitating and lethal disorders, with a suicide rate between 10 and 20% among sufferers (Ketter & Miller, 2015; Moshier & Otto, 2014). As you'll see in the next section, suicide is also a serious risk for anyone suffering from depressive disorders, as well as several other psychological disorders.

Explaining Depressive and Bipolar Disorders

Depressive and bipolar disorders differ in how often they occur, how much they disrupt normal functioning (severity), and how long they last (duration). In this section, we will look at the biological and psychosocial factors that contribute to disorders of mood.

Biological Factors

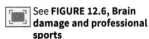
See **FIGURE 12.6, Brain damage and professional sports**

Recent research suggests that structural brain changes may contribute to depressive and bipolar disorders. For example, some professional athletes are at greater risk of developing such disorders as they age, possibly due to brain damage caused by repeated concussions

(Broshek et al., 2015; Yang et al., 2015). Sadly, a study on former NFL players revealed that 41% show cognitive problems and 24% show clinical depression, which may result from neurological changes caused by concussions (Hart et al., 2013). Their brain scans also revealed changes in blood flow within the brain, and abnormalities in various parts of the brain. These changes may contribute to serious depression and increased risk of suicide (**Figure 12.6**).

Other research points to imbalances of the neurotransmitters serotonin, norepinephrine, and dopamine as possible causes of mood disorders (Artigas, 2015; Fakhoury, 2015; Sperner-Unterweger et al., 2014). And both depressive disorders and bipolar disorders are sometimes treated with antidepressants, which affect the amount or functioning of these same neurotransmitters. As an interesting side note, researchers in one study found that people who regularly ate fast food, and commercially produced baked goods (such as croissants and doughnuts), were 51% more likely to develop depression later on. Can you see how this may be the result of the chemicals in such foods leading to physiological changes in the brain and body (Sánchez-Villegas et al., 2011)?

Some biological research also indicates that both depressive and bipolar disorders may be inherited (Bousman et al., 2014; Jacobs et al., 2015; Pandolfo et al., 2015). In contrast, research that takes an evolutionary perspective suggests that moderate depression may be a normal and healthy adaptive response to a very real loss, such as the death of a loved one, which helps us to step back and reassess our goals (Durisko et al., 2015; Neumann & Walter, 2015). Clinical, severe depression may just be an extreme version of this generally adaptive response.

Psychosocial Factors

Psychosocial theories of depression focus on environmental stressors, disturbances in the individual's relationships, life style, or self-concept, and on a history of abuse or assault (Ege et al., 2015; Massing-Schaffer et al., 2015; Raposa et al., 2014). The psychoanalytic explanation sees depression as the result of experiencing a real or imagined loss, which is internalized as guilt, shame, self-hatred and ultimately self-blame. The cognitive perspective explains depression as caused, at least in part, by negative thinking patterns, including a tendency to ruminate, or obsess, about problems (Arora et al., 2015; Yoon et al., 2014). The humanistic school says that depression results when

Kent C. Horner/Getty Images

a person demands perfection of him- or herself, or when positive growth is blocked (McCormack & Joseph, 2014; Short & Thomas, 2015).

In addition, according to the **learned helplessness** model (Seligman, 1975, 2007), depression occurs when people (and other animals) become resigned to the idea that they are helpless to escape from a situation, because of a history of repeated failures in the past. For humans, learned helplessness may be particularly likely to trigger depression if the failure is attributed to causes that are internal ("my own weakness"), stable ("this weakness is long-standing and unchanging"), and global ("this weakness is a problem for me in lots of settings") (Barnum et al., 2013; Smalheiser et al., 2014; Travers et al., 2015).

Keep in mind that suicide is a major danger associated with both depressive disorder and bipolar disorder. Given that some people who suffer from these disorders are so disturbed that they lose contact with reality, they may fail to recognize the danger signs, and when to seek help.

See **TUTORIAL VIDEO: Signs of Suicide**

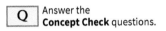
Answer the **Concept Check** questions.

12.4 SCHIZOPHRENIA

Imagine that your 17-year-old son's behavior has changed dramatically over the past few months. He's gone from being actively involved in sports and clubs to suddenly quitting all activities, and refusing to go to school. He now also talks to himself—mumbling and yelling out at times—and no longer regularly showers or washes his hair. Recently he announced, "The voices are telling me to jump out the window" (Kotowski, 2012).

This description is taken from the true case history of a person who suffers from **schizophrenia**. As discussed below, people with schizophrenia have major disturbances in perception (seeing or hearing things that others don't), language (bizarre words and meanings), thought (impaired logic), emotion (exaggerated or blunted), and/or behavior (peculiar movements and social withdrawal). They also may have serious problems caring for themselves, relating to others, and holding a job. The DSM-5 places schizophrenia within the category of "schizophrenic spectrum and other psychotic disorders." Recall that *psychosis* refers to a serious loss of contact with reality. In extreme cases, the illness is so severe that it requires institutional or custodial care.

Tragically, schizophrenia is one of the most widespread and devastating psychological disorders. Approximately 1 out of every 100 people will develop it in his or her lifetime, and approximately half of all people who are admitted to mental hospitals are diagnosed with this disorder (Castle & Buckley, 2015; Gottesman, 1991; Lauriello & Rahman, 2015). Schizophrenia usually emerges between the late teens and the mid-30s, and only rarely prior to adolescence or after age 45. It seems to be equally prevalent in men and women, but it's generally more severe and strikes earlier in men (Castle & Buckley, 2015; Silber, 2014; Zorrilla et al., 2015). It's important to note that people often confuse schizophrenia with dissociative identity disorder, which is sometimes referred to as *multiple personality disorder*.

Characteristics of Schizophrenia

Schizophrenia is a group of disorders characterized by a disturbance in one or more of the following areas:

Perception

The senses of people with schizophrenia may be either enhanced or blunted. The filtering and selection processes that allow most people to concentrate on whatever they choose are impaired, and sensory stimulation is jumbled and distorted. People with schizophrenia may also experience **hallucinations**, meaning false, imaginary sensory perceptions that occur without external stimuli. Auditory hallucinations (hearing voices and sounds) is one of the most commonly noted and reported symptoms of schizophrenia.

On rare occasions, people with schizophrenia hurt others in response to their distorted perceptions. But a person with schizophrenia is more likely to be self-destructive and suicidal than violent toward others.

Language and Thought

For people with schizophrenia, words lose their usual meanings and associations, logic is impaired, and thoughts are disorganized and bizarre. When language and thought disturbances are mild, the individual jumps from topic to topic. With more severe disturbances, the person jumbles phrases and words together (into a "word salad") or creates artificial words. The most common—and frightening—thought disturbance experienced by people with schizophrenia is lack of contact with reality (psychosis).

Delusions, false or irrational beliefs that are maintained despite clear evidence to the contrary, are also common in people with schizophrenia. We all experience exaggerated thoughts from time to time, such as thinking a friend is trying to avoid us, but the delusions of schizophrenia are much more extreme. For example, if someone falsely believed that the postman who routinely delivered mail to his house every afternoon was a co-conspirator in a plot to kill him, it would likely qualify as a *delusion of persecution* or *paranoia*. In *delusions of grandeur*, people believe that they are someone very important, perhaps Jesus Christ or the Queen of England. In *delusions of control*, the person believes his or her thoughts or actions are being controlled by outside and/or alien forces, such as "the CIA is controlling my thoughts."

Emotion

Changes in emotion also occur in people with schizophrenia. In some cases, emotions are exaggerated and fluctuate rapidly. At other times, they become blunted. Some people with schizophrenia have *flattened affect*—meaning almost no emotional response of any kind.

Behavior

Disturbances in behavior may take the form of unusual actions that have special meaning to the sufferer. For example, one person massaged his head repeatedly to "clear it" of unwanted thoughts. People with schizophrenia also may become *cataleptic* and assume a nearly immobile stance for an extended period.

Symptoms of Schizophrenia

For many years, researchers divided schizophrenia into five subtypes: *paranoid*, *catatonic*, *disorganized*, *undifferentiated*, and *residual*. Critics suggested that this system does not differentiate in terms of prognosis, cause, or response to treatment, and that the undifferentiated type was merely a catchall for cases that are difficult to diagnose (Black & Grant, 2014; Castle & Buckley, 2015; Grossman & Walfish, 2014). For these reasons, researchers have proposed an alternative classification system:

1. Positive schizophrenia symptoms are additions to or exaggerations of normal functions. Delusions and hallucinations are examples of positive symptoms. (Note that "positive" is NOT pleasant or good! Similar to our discussion in Chapter 6, it means that "something is added," and in this case it's beyond normal levels.)

2. Negative schizophrenia symptoms include the loss or absence of normal functions. Impaired attention, limited or toneless speech, flat or blunted affect, and social withdrawal are all classic negative symptoms of schizophrenia. (Again note that "negative" is NOT the same as unpleasant or bad! As in Chapter 6, it means that "something is taken away," and in this case daily functioning is "taken away" because it's so far below normal levels.)

Positive symptoms are more common when schizophrenia develops rapidly, whereas negative symptoms are more often found in slow-developing schizophrenia. In addition, positive symptoms are associated with better adjustment before the onset, and a better prognosis for recovery.

Explaining Schizophrenia

Because schizophrenia comes in many different forms, it probably has multiple biological and psychosocial bases. Let's look at biological contributions first.

Biological Theories

Prenatal stress and viral infections, birth complications, immune responses, maternal malnutrition, and advanced paternal age all play suspected roles in the development of schizophrenia (Aberg et al., 2014; Debnath et al., 2015; Reddy & Keshavan, 2015). However, most biological theories of schizophrenia focus on genetics, neurotransmitters, and brain abnormalities:

- **Genetics** Current research indicates that the risk for schizophrenia increases with genetic similarity (Arnedo et al., 2015; Castellani et al., 2014; Gottesman, 1991). This means that people who share more genes with a person who has schizophrenia are more likely to develop the disorder (**Figure 12.7**).

- **Neurotransmitters** According to the *dopamine hypothesis*, overactivity of certain dopamine neurons in the brain causes some forms of schizophrenia (Gilani et al., 2014; Howes et al., 2015; Stopper & Floresco, 2015). This hypothesis is based on two observations. First, administering amphetamines increases the amount of dopamine, and can produce (or worsen) some symptoms of schizophrenia, especially in people with a genetic predisposition to the disorder. Second, drugs that reduce dopamine activity in the brain reduce or eliminate some symptoms of schizophrenia.

- **Brain abnormalities** A third area of research in schizophrenia explores links to abnormalities in brain function and structure. Researchers have found larger cerebral ventricles (fluid-filled spaces in the brain) and right hemisphere dysfunction in some people with schizophrenia (Chakrabarty et al., 2014; Guo et al., 2015; Woodward & Heckers, 2015). Also, some people with chronic schizophrenia have a lower level of activity in their frontal and temporal lobes—areas we use in language, attention, and memory (**Figure 12.8**). Can you see how damage in these regions might explain the thought and language disturbances that characterize schizophrenia?

See **FIGURE 12.7, Genetics and schizophrenia**

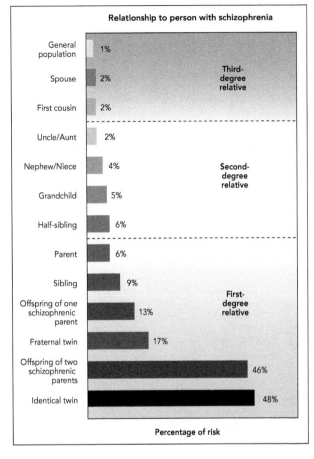

Relationship to person with schizophrenia

Relationship	Percentage of risk	
General population	1%	*Third-degree relative*
Spouse	2%	
First cousin	2%	
Uncle/Aunt	2%	*Second-degree relative*
Nephew/Niece	4%	
Grandchild	5%	
Half-sibling	6%	
Parent	6%	*First-degree relative*
Sibling	9%	
Offspring of one schizophrenic parent	13%	
Fraternal twin	17%	
Offspring of two schizophrenic parents	46%	
Identical twin	48%	

Percentage of risk

See **FIGURE 12.8, Brain activity in schizophrenia**

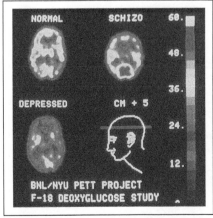

Science Source/Photo Researchers, Inc.

Psychosocial Factors

Clearly, biological factors play a key role in schizophrenia. But the fact that the heritability of schizophrenia is only 48%, even in identical twins—who share identical genes—tells us that nongenetic factors must contribute the remaining percentage. Most psychologists believe there are at least two possible psychosocial contributors to the other 52%.

According to the **diathesis-stress model** of schizophrenia, stress plays an essential role in triggering schizophrenic episodes in people with an inherited predisposition (or *diathesis*) toward the disease (Burns et al., 2014; Frau et al., 2015; Howes & Murray, 2014). In line with this model, children who experience severe trauma before age 16 are three times more likely than other people to develop schizophrenia (Bentall et al., 2012; DeRosse et al., 2014). Similarly, people who experience stressful living environments, including poverty, unemployment, and crowding, are also at increased risk (Burns et al., 2014; Kirkbride et al., 2014).

Some investigators suggest that communication disorders in family members may also be a predisposing factor for schizophrenia. Such disorders include unintelligible speech, fragmented communication, and parents frequently sending severely contradictory messages to children.

See **FIGURE 12.9, The biopsychosocial model and schizophrenia**

Several studies have shown greater rates of relapse, and worsening of symptoms, among hospitalized patients who went home to families that were critical and hostile toward them, or overly involved in their lives emotionally (McFarlane, 2011; Tsai et al., 2014; Wasserman et al., 2013).

How should we evaluate the differing possible explanations for the cause(s) of schizophrenia? Critics of the biological approach argue that the research only fits some cases of schizophrenia. Moreover, with both biological and psychosocial explanations, it's difficult to determine cause and effect. As we've seen throughout this text, the biopsychosocial model suggests that schizophrenia is probably the result of a combination of known and unknown interacting factors (▣ **Figure 12.9**).

12.5 OTHER DISORDERS

Having discussed anxiety disorders, mood disorders, and schizophrenia, we now explore three additional disorders: *obsessive-compulsive, dissociative,* and *personality disorders.*

Obsessive-Compulsive Disorder (OCD)

Q Answer the **Concept Check** questions.

Almost everyone occasionally worries about whether or not they locked their doors, and feel compelled to run back and check. However, people with **obsessive-compulsive disorder (OCD)** experience persistent, unwanted, fearful thoughts (obsessions), and/or irresistible urges to perform repetitive and/or ritualized behaviors (compulsions) to help relieve the anxiety created by the obsession. In adults, women are affected at a slightly higher rate than men, whereas men are more commonly affected in childhood (American Psychiatric Association, 2013).

Common examples of obsessions are fear of germs, fear of being hurt or of hurting others, and troubling religious or sexual thoughts. Examples of compulsions include repeatedly checking, counting, cleaning, washing all or specific body parts, or putting things in a certain order. As mentioned before, everyone worries and sometimes double-checks, but people with OCD have these thoughts, and perform these rituals, for at least an hour or more each day, often longer (Essau & Ozer, 2015; Foa & Yadin, 2014; McKay & Storch, 2014).

See **VIRTUAL FIELD TRIP: Obsessive-Compulsive Disorder**

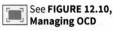

See **FIGURE 12.10, Managing OCD**

Kevin Winter/Getty Images

Imagine what it would be like to worry so obsessively about germs that you compulsively wash your hands hundreds of times a day, until they are raw and bleeding. Most OCD sufferers realize that their actions are senseless. But when they try to stop the behavior, they experience mounting anxiety, which is relieved only by giving in to the compulsions. Given that numerous biological and psychological factors contribute to OCD, it is most often treated with a combination of drugs and cognitive-behavior therapy (CBT) (Abramowitz & Mahaffey, 2014; Grant et al., 2014). (See Chapter 13 and ▣ **Figure 12.10**).

Dissociative Disorders

If you've ever been daydreaming, while driving home from your college campus, and then could not remember making one single turn, you may have experienced a normal form of *dissociation*, meaning a mild disconnection from your immediate surroundings.

The most dramatic extremes of this type of detachment are the **dissociative disorders**, characterized by a sudden break (*dissociation*) in conscious awareness, self-identity, and/or memory. Note that this is a disconnection or detachment from immediate surroundings, or from physical or emotional experience. It is very different from the loss of contact with reality seen in *psychosis* (▣ **Figure 12.11**). There are several forms of dissociative disorders, including dissociative amnesia and dissociative identity disorder (DID). However, all are characterized by a splitting apart (a *dis-association*) of significant aspects of experience from memory or consciousness. Unlike most other

psychological disorders, dissociative disorders appear to be primarily caused by environmental variables, such as extreme stress or abuse (Beidel et al.,
2014; Brand et al., 2014; Spiegel & Simeon, 2015).

See **FIGURE 12.11**,
Dissociation in everyday life

Image Source/Getty Images

The most controversial, and least common, dissociative
disorder is **dissociative identity disorder (DID)**—previously
known as multiple personality disorder (MPD). An individual
with this disorder has at least two separate and distinct personalities (or *identities*) (▣ **Figure 12.12**). Each personality has
unique memories, behaviors, and social relationships. Tran-
sition from one personality to another occurs suddenly, and
is often triggered by psychological stress (Chalavi et al, 2015;
Spiegel & Simeon, 2015). Typically, there is a "core" personality,
who has no knowledge or awareness of the alternate personalities, but is often aware of lost memories and lost periods of
time. The disorder is diagnosed about equally among men and
women (American Psychiatric Association, 2013).

DID is a controversial diagnosis. Some experts suggest that
many cases are faked, or result from false memories, and/or an unconscious need to please
the therapist (Boysen & VanBergen, 2013; Dalenberg et al., 2014; Lilienfeld &
Lynn, 2015).

See **FIGURE 12.12**,
A personal account of DID

Personality Disorders

What would happen if the characteristics of a personality were so chronic, in-
flexible, and maladaptive that they significantly impaired someone's ability
to function? This is what occurs with **personality disorders**. Several types
of personality disorders are included in this category in the fifth edition of
the *DSM*, but here we will focus on antisocial personality disorder (ASPD) and
borderline personality disorder (BPD).

Antisocial Personality Disorder

People with **antisocial personality disorder (ASPD)**—also sometimes mistakenly confused with the nonclinical terms of *psychopathy* or *sociopathy*—
show a pervasive pattern of disregard for, and violation of, the rights of others.
These behaviors typically begin in childhood or early adolescence and continue until adulthood. They also lie so far outside the ethical and legal standards of society that many consider it the most serious of all psychological
disorders. Unlike people with anxiety, mood disorders, and schizophrenia,
those with this diagnosis feel little personal distress (and may not be motivated
to change). Yet their maladaptive traits generally bring considerable harm
and suffering to others (Black & Blum, 2014; Cummings, 2015; Paris, 2015).
Although serial killers are often seen as classic examples of people with ASPD,
most people who have this disorder generally harm others in less dramatic
ways—for example, as ruthless businesspeople and crooked politicians.

Stephen Dunn/Allsport/Getty Images

Unlike most other adults, individuals with ASPD act impulsively, without giving thought to
the consequences. They are usually poised when confronted with their destructive behavior,
and feel contempt for anyone they are able to manipulate. They also change jobs and relationships suddenly, and often have a history of truancy from school or of being expelled for
destructive behavior. People with antisocial personalities can be charming and persuasive,
and they often have remarkably good insight into the needs and weaknesses of other people.

Twin and adoption studies suggest a possible genetic predisposition to ASPD (Kendler
et al., 2015; Werner et al., 2015). Researchers also have found abnormally low autonomic activity during stress, right hemisphere abnormalities, reduced gray matter in the frontal lobes, and
biochemical disturbances (Jiang et al., 2015; Kumari et al., 2014; Schiffer et al., 2014). MRI brain
scans of criminals currently in prison for violent crimes, such as rape, murder, or attempted
murder, and showing little empathy and remorse for their crimes, revealed reduced gray matter volume in the prefrontal cortex (Gregory et al., 2012). (Recall from Chapter 2 that this is the
area of the brain responsible for emotions, such as fear, empathy, and/or guilt.)

Evidence also exists for environmental or psychological causes. People with ASPD often come from homes characterized by abusive parenting styles, emotional deprivation, harsh and inconsistent disciplinary practices, and antisocial parental behavior (Gobin et al., 2015; Krastins et al., 2014; Waxman et al., 2014). Still other studies show a strong interaction between both heredity and environment (Haberstick et al., 2014; Paris, 2015; Trull et al., 2013).

Borderline Personality Disorder

"Mary's troubles first began in adolescence. She began to miss curfew, was frequently truant, and her grades declined sharply. Mary later became promiscuous and prostituted herself several times to get drug money. . . . She also quickly fell in love, and overly idealized new friends. But when they quickly (and inevitably) disappointed her, she would angrily cast them aside. . . . Mary's problems, coupled with a preoccupation with inflicting pain on herself (by cutting and burning), and persistent thoughts of suicide, eventually led to her admittance to a psychiatric hospital at age 26."

—KRING ET AL., 2010, PP. 354–355

Mary's experiences are all classic symptoms of **borderline personality disorder (BPD)**. The core features of this disorder include a pervasive pattern of instability in emotions, relationships, and self-image, along with impulsive and self-destructive behaviors, such as truancy, promiscuity, drinking, gambling, and eating sprees. In addition, people with BPD may attempt suicide and sometimes engage in self-mutilating ("cutting") behaviors (Greenfield et al., 2015; Matusiewicz et al., 2014; Miller et al., 2014).

Those with BPD also tend to see themselves and everyone else in absolute terms—as either perfect or worthless. Constantly seeking reassurance from others, they may quickly erupt in anger at the slightest sign of disapproval. The disorder is also typically marked by a long history of broken friendships, divorces, and lost jobs.

In short, people with this disorder appear to have a deep well of intense loneliness and a chronic fear of abandonment. Unfortunately, given their troublesome personality traits, friends, lovers, and even family members and therapists often do "abandon" them—thus creating a tragic self-fulfilling prophecy. The good news is that BPD can be reliably diagnosed, and it does respond to professional intervention—particularly in young people (Cameron et al., 2014; Gunderson & Links, 2014; Rizvi & Salters-Pedneault, 2013).

Sadly, this disorder is among the most commonly diagnosed, and most functionally disabling, of all psychological disorders (Arntz, 2015; Gunderson & Links, 2014; Rizvi & Salters-Pedneault, 2013). Originally, the term implied that the person was on the *borderline* between neurosis and schizophrenia, but the modern conceptualization no longer has this connotation (Napal, 2015; Schroeder et al., 2013).

What causes BPD? Some research points to environmental factors, such as a childhood history of neglect, emotional deprivation, and physical, sexual, or emotional abuse (Chesin et al., 2015; Hunt et al., 2015; Vermetten & Spiegel, 2014). From a biological perspective, BPD also tends to run in families, and some data suggest that it is a result of impaired functioning of the brain's frontal lobes and limbic system, areas that control impulsive behaviors (Amad et al., 2014; Few et al., 2014; Stone, 2014). Regarding their emotional volatility, research using neuroimaging reveals that people with BPD show more activity in parts of the brain associated with the experience of negative emotions, coupled with less activity in areas that help suppress negative emotion (Ruocco et al., 2013). As in almost all other psychological disorders, most researchers agree that BPD results from an interaction of biopsychosocial factors (Carpenter et al., 2013; Ripoll et al., 2013; Stone, 2014).

Q Answer the **Concept Check** questions.

12.6 GENDER AND CULTURAL EFFECTS

Among the Chippewa, Cree, and Montagnais-Naskapi Indians in Canada, there is a disorder called *windigo*—or *wiitiko*—*psychosis*, characterized by delusions and cannibalistic impulses. Believing they have been possessed by the spirit of a windigo, a cannibal giant with a heart and entrails of ice, victims become severely depressed (Faddiman, 1997). As the malady begins, the individual typically experiences loss of appetite, diarrhea, vomiting, and insomnia. He or she also may see people turning into beavers and other edible animals. In later stages, the victim becomes obsessed with cannibalistic thoughts, and may even attack and kill loved ones in order to devour their flesh (Berreman, 1971; Thomason, 2014).

If you were a therapist, how would you treat this disorder? Does it fit neatly into any of the categories of psychological disorders that we've just discussed? We began this chapter by discussing the complexities and problems with defining, identifying, and classifying abnormal behavior. Before we close, we need to add two additional confounding factors: gender and culture. In this section, we explore a few of the many ways in which men and women differ in their experience of abnormal behavior. We also look at cultural variations in abnormal behavior.

Gender Differences

When you picture someone suffering from depression, anxiety, alcoholism, or antisocial personality disorder, what is the gender of each person? Most people tend to visualize a woman for the first two, and a man for the last two. There is some truth to these stereotypes. Around the world, the rate of severe depression for women is about double that for men (Navarro & Hurtado, 2015; Pérez & Gaviña, 2015; World Health Organization, 2011).

Why is there such a striking gender difference? Certain risk factors for depression (such as genetic predisposition, marital problems, pain, and medical illness) are common to both men and women. However, poverty is a well-known contributor to many psychological disorders, and women are far more likely than men to fall into the lowest socioeconomic groups. Women also experience more sexual trauma, partner abuse, and chronic stress in their daily lives, and these are all well-known contributing factors in depression (Doornbos et al., 2013; Jausoro Alzola & Marino, 2015; Oram et al., 2013).

Research also suggests that some gender differences in depression may reflect differences in the way women and men internalize or externalize their emotions. For instance, women have been found to be more inhibited, and more likely to focus on their *internal* negative emotions and problems. In contrast, men tend to be more disinhibited, and more likely to *externalize* their emotions and problems (Kendler & Gardner, 2014; Rice et al., 2014).

Can you see how these gender differences in depression may result from misapplied gender role stereotypes? For example, the most common symptoms of depression, such as crying, low energy, dejected facial expressions, and withdrawal from social activities, are more socially acceptable for women than for men. In contrast, men in Western societies are typically socialized to suppress their emotions, and to show their distress by acting out (being aggressive), acting impulsively (driving recklessly and committing petty crimes), and/or engaging in substance abuse. Given these differences in socialization and behaviors, combined with the fact that gender differences in depression are more pronounced in cultures with traditional gender roles, male depression may "simply" be expressed in less stereotypical ways, and therefore be *underdiagnosed* (Fields & Cochran, 2011; Pérez & Gaviña, 2015; Seedat et al., 2009).

To further examine different expectations about depression as a function of gender, researchers in one study asked participants to read a story about a fictitious person (Kate in one version, Jack in the other). The story was exactly the same in both conditions, and included the following information: "For the past two weeks, Kate/Jack has been feeling really down. S/he wakes up in the morning with a flat, heavy feeling that sticks with her/him all day. S/he isn't enjoying things the way s/he normally would. S/he finds it hard to concentrate on anything." Although all participants read the same story, those who read about Kate rated her symptoms as more distressing, deserving of sympathy, and difficult to treat (Swami, 2012).

Can you see how understanding the importance of genetic predispositions, external environmental factors (like poverty), and cognitive factors (like internalizing versus externalizing emotions) may help mental health professionals better understand individual and gender-related differences in depression?

Culture and Schizophrenia

People of different cultures experience psychological disorders, particularly schizophrenia, in a variety of ways. It is unclear whether these differences result from actual differences in

See multipart **FIGURE 12.13,**
What is stressful?

David Alan Harvey/Magnum Photos, Inc.

the prevalence of the disorder, or from differences in definition, diagnosis, or reporting (Hsu, 2016; Luhrmann et al., 2015; McLean et al., 2014). The symptoms of schizophrenia also vary across cultures (Barnow & Balkir, 2013; Burns, 2013), as do the particular stressors that may trigger its onset (■ **Figure 12.13**).

Interestingly, despite the advanced treatment facilities and methods in industrialized nations, the prognosis for people with schizophrenia is actually better in many nonindustrialized societies. Why? It may be that the core symptoms of schizophrenia (poor rapport with others, incoherent speech, and so on) make it more difficult to survive in highly industrialized countries. In addition, in most industrialized nations, families and other support groups are less likely to feel responsible for relatives and friends who have schizophrenia (Akyeampong et al., 2015; Burns et al., 2014; Eaton et al., 2012).

Avoiding Ethnocentrism

Given that most research on psychological disorders originates, and is conducted primarily in Western cultures, can you see how such a restricted sampling can limit our understanding of psychological disorders? Furthermore, how might this limited view lead to increased *ethnocentrism*—viewing one's own culture or group as central and "correct"?

See **TABLE 12.2,** Culture-General Symptoms of Mental Health Difficulties

Fortunately, cross-cultural researchers have devised ways to overcome these difficulties (Bernal et al., 2014; Hsu, 2016; Sue et al., 2016). For example, Robert Nishimoto (1988) has found several *culture-general symptoms* that are useful in diagnosing disorders across cultures (■ **Table 12.2**).

Nishimoto also found several *culture-bound symptoms* for mental health difficulties, which are unique to different groups. Vietnamese and Chinese respondents reported "fullness in head," Mexican respondents noted "problems with [their] memory," and Anglo-American respondents reported "shortness of breath" and "headaches." Apparently, people learn to express their problems in ways that are acceptable to others in the same culture (Brislin, 2000; Hsu, 2016; Shannon et al., 2015).

This division between culture-general and culture-bound symptoms also helps us better understand depression. Certain symptoms of depression (such as intense sadness, poor concentration, and low energy) seem to exist across all cultures (Walsh & Cross, 2013; World Health Organization, 2011). But there is also evidence of some culture-bound symptoms. For

See **FIGURE 12.14,** Culture-bound disorders

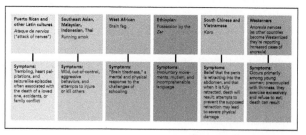

example, feelings of guilt are found more often in North America and Europe than in other parts of the world. And in China, *somatization* (the conversion of depression into bodily complaints) occurs more frequently than it does in other parts of the world (Grover & Ghosh, 2014; Lim et al., 2011).

Just as there are culture-bound *symptoms*, researchers also have found culture-bound *disorders* (■ **Figure 12.14**). The earlier example of windigo psychosis, a disorder limited to a few groups of Canadian Indians, illustrates just such a case. As you can see, culture has a strong effect on psychological disorders. Studying the similarities and differences across cultures can lead to better diagnosis and understanding. It also helps all of us avoid, or at least minimize, our ethnocentrism.

Q   Answer the **Concept Check** questions.

WPLS Go to your WileyPLUS Learning Space course for video episodes, examples, art, tables, Concept Checks, practice, and other pedagogical resources that will help you succeed in this course.

THERAPY

WP LS Go to your WileyPLUS Learning Space course for video episodes, examples, art, tables, Concept Checks, practice, and other pedagogical resources that will help you succeed in this course.

13.1 TALK THERAPIES

Throughout this text, we have emphasized the *science* of psychology, and this chapter is no exception. Now we'll explore how therapists apply this science during **psychotherapy** to help us improve our overall psychological functioning and adjustment to life, and to assist people suffering from one or more psychological disorders. Due to the common stereotype and stigma that therapy is only for deeply disturbed individuals, it's important to note that therapy provides an opportunity for everyone to have our specific problems addressed, as well as to learn better thinking, feeling, and behavioral skills useful in our everyday lives.

In this chapter, we'll discuss the three major approaches to psychotherapy (**Study Organizer 13.1**), its general goals and effectiveness, and other important issues and topics in the field. Along the way, we'll work to demystify and destigmatize its practice, and dispel some common myths.

See **STUDY ORGANIZER 13.1: An Overview of the Three Major Approaches to Therapy**

See **TUTORIAL VIDEO: Myths About Therapy**

We begin our discussion of professional *psychotherapy* with traditional psychoanalysis, and its modern counterpart, psychodynamic therapies. Then we explore humanistic and cognitive therapies. Although these therapies differ significantly, they're often grouped together as "talk therapies" because they emphasize communication between the therapist and client, as opposed to the behavioral and biomedical therapies that we'll discuss later in this chapter.

See **TUTORIAL VIDEO: A Guide to Psychotherapy**

Psychoanalysis/Psychodynamic Therapies

In **psychoanalysis**, a person's *psyche* (or mind) is *analyzed*. Traditional psychoanalysis is based on Sigmund Freud's central belief that abnormal behavior is caused by unconscious conflicts among the three parts of the psyche—the id, the ego, and the superego (Chapter 11).

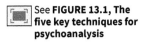

See **FIGURE 13.1, The five key techniques for psychoanalysis**

During psychoanalysis, these conflicts are brought to consciousness. The individual comes to understand the reasons for his or her dysfunction, and realizes that the childhood conditions under which the conflicts developed no longer exist. Once this realization or insight occurs, the conflicts can be resolved, and the person can develop more adaptive behavior patterns (Altman, 2013; Newirth, 2015; Safran, 2014).

Unfortunately, according to Freud, the ego has strong *defense mechanisms* that block unconscious thoughts from coming to light. Thus, to gain insight into the unconscious, the ego must be "tricked" into relaxing its guard. To meet that goal, psychoanalysts employ five major methods: *free association, dream analysis, analyzing resistance, analyzing transference*, and *interpretation* (**Figure 13.1**).

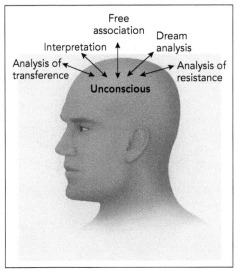

Free Association

During psychoanalysis, when you let your mind wander and remove conscious censorship over thoughts—a process called **free association**—interesting and even bizarre connections seem to spring into awareness. Freud believed that the first thing to come to a person's mind is often an important clue to what the person's unconscious wants to conceal. Having

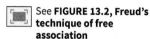 See **FIGURE 13.2, Freud's technique of free association**

someone recline on a couch, with only the ceiling to look at, is believed to encourage free association (▣ **Figure 13.2**).

"The way this works is that you say the first thing that comes to your mind..."

Dream Analysis

Recall from Chapter 5 that, according to Freud, our psychological defenses are lowered during sleep. Therefore, our forbidden desires and unconscious conflicts are supposedly more freely expressed during dreams. Even while dreaming, however, we recognize these feelings and conflicts as unacceptable, and must disguise them as images that have deeper symbolic meaning. Thus, using Freudian **dream analysis**, a dream of riding a horse or driving a car might be analyzed as just the surface description, or *manifest content*. In contrast, the hidden, underlying meaning, or *latent content*, might be analyzed as a desire for, or concern about, sexual intercourse.

Analysis of Resistance

During free association and dream analysis, Freud found that people often show an inability or unwillingness to change their behaviors or to discuss certain memories, thoughts, motives, or experiences, a process called **resistance**. If someone suddenly "forgets" what he or she was saying, or completely changes the subject, it is the therapist's job to identify these possible cases of *resistance*. Understanding that resistance is the barrier that we put up when we fear that awareness will overwhelm us, the therapist's analysis of this resistance helps the individual face his or her problems and learn to deal with them more realistically.

Analysis of Transference

Freud believed that during psychoanalysis, people disclose intimate feelings and memories, and the relationship between the therapist and client may become complex and emotionally charged. As a result, individuals often apply, or *transfer*, emotional reactions that are related to someone in their life onto the therapist. A client might interact with the therapist as if the therapist were a lover or parent. The therapist uses this process of **transference** to help the client "relive" painful past relationships in a safe, therapeutic setting, so that he or she can gain insight and move on to healthier relationships.

Interpretation

The core of all psychoanalytic therapy is **interpretation**. During free association, dream analysis, resistance, and transference, the analyst listens closely, and tries to find patterns and hidden conflicts. At the right time, the therapist explains or *interprets* the underlying meanings to the client.

Evaluating Psychoanalysis

As you can see, psychoanalysis is largely rooted in the assumption that repressed memories and unconscious conflicts actually exist. But, as we noted in Chapters 7 and 11, this assumption is the subject of heated, ongoing debate. Critics also point to two other problems with psychoanalysis (Kernberg, 2014; Miltenberger, 2011; Ng et al., 2015).

- **Limited applicability** Psychoanalysis is time-consuming (often lasting several years with four to five sessions a week) and expensive. In addition, critics suggest that it applies only to a select group of highly motivated, articulate individuals with less severe disorders, and not to those with more complex disorders, such as schizophrenia.

- **Lack of scientific credibility** According to critics, it is difficult, if not impossible, to scientifically document the major tenets of psychoanalysis. How do we prove or disprove the existence of an unconscious mind, or the meaning of unconscious conflicts and symbolic dream images?

Despite these criticisms, research shows that traditional psychoanalysis can be effective for those who have the time and money (Busch, 2014; Lambert, 2013; Weitkamp et al., 2014).

Psychodynamic Therapies

A modern derivative of Freudian psychoanalysis, **psychodynamic therapies**, include both Freud's theories, and those of his major followers—Carl Jung, Alfred Adler, Karen Horney, and Erik Erikson. In contrast to psychoanalysis, psychodynamic therapy is shorter and less intensive (once or twice a week versus several times a week, and only a few weeks or months versus years). In addition, the client is treated face-to-face rather than reclining on a couch, and the therapist takes a more directive approach, rather than waiting for unconscious memories and desires to slowly be uncovered.

Contemporary psychodynamic therapists also focus less on unconscious, early-childhood experiences, and sexual issues. Instead, they emphasize conscious processes and current problems (Gelso et al., 2014; Göttken et al., 2014; Short & Thomas, 2015). Such refinements have helped make treatments shorter, more available, and more effective for an increasing number of people. One of the most popular modern forms of psychodynamic therapy, *interpersonal therapy (IPT)*, focuses on current relationships, with the goal of relieving immediate symptoms, and developing better ways to solve interpersonal problems. Research shows that it's effective for a variety of disorders, including depression, marital conflict, eating disorders, and drug addiction (Driessen et al., 2015; Normandin et al., 2015; Weitkamp et al., 2014).

Humanistic Therapies

In contrast to the psychoanalytic and psychodynamic focus on the unconscious, the humanistic approach emphasizes conscious processes and present versus past experiences. The name, **humanistic therapies**, reflects their focus on the *human* characteristics of a person's potential for self-actualization, free will, and self-awareness. Humanistic therapists assume that people with problems are suffering from a disruption of their normal growth potential, and, hence, their self-concept. When obstacles are removed, the individual is free to become the self-accepting, genuine person everyone is capable of being (Gelso et al., 2014; Schneider et al., 2015).

One of the best-known humanistic therapists is Carl Rogers, who developed an approach that provides a warm, supportive atmosphere that encourages self-actualization, and improves the client's self-concept (Rogers, 1961, 1980). His approach is referred to as **client-centered therapy** (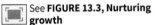 **Figure 13.3**). (Rogers used the term *client* because he believed the label *patient* implied that someone was sick or mentally ill, rather than responsible and competent.)

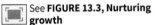 See **FIGURE 13.3, Nurturing growth**

Client-centered therapy, like psychoanalysis and psychodynamic therapies, explores thoughts and feelings as a way to obtain insight into the causes for behaviors. For Rogerian therapists, however, the focus is on providing an accepting atmosphere, and encouraging healthy emotional experiences. Clients are responsible for discovering their own maladaptive patterns.

Rogerian therapists create a therapeutic relationship by focusing on four important qualities of communication: *empathy*, *unconditional positive regard*, *genuineness*, and *active listening*.

Comstock/SUPERSTOCK

Empathy

Using the technique of **empathy**, a sensitive understanding and sharing of another person's inner experience, therapists pay attention to body language, and listen for subtle cues to help them understand the emotional experiences of clients. To further help clients explore their feelings, the therapist uses open-ended statements such as, "You found that upsetting" or "You haven't been able to decide what to do about this," rather than asking questions or offering explanations.

Unconditional Positive Regard

Regardless of the clients' problems or behaviors, humanistic therapists offer **unconditional positive regard**, a genuine caring and nonjudgmental attitude toward people based on

See **FIGURE 13.4,**
Unconditional versus
conditional positive regard

their innate value as individuals, with no "strings" (*conditions*) attached. Client-centered therapists avoid evaluative statements such as, "That's good" and "You did the right thing," because such comments imply that the therapist is judging the client. Rogers believed that most of us receive *conditional* acceptance from our parents, teachers, and others, which leads to poor self-concepts and psychological disorders (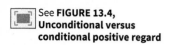 **Figure 13.4**).

"*Just remember, son, it doesn't matter whether you win or lose—unless you want Daddy's love.*"

P. BYRNES.

Pat Byrnes/The Cartoon Bank, Inc

Genuineness

Humanists believe that when therapists use **genuineness**, meaning being personally authentic and sincere, with their clients, the clients will, in turn, develop self-trust and honest self-expression.

Active Listening

Using **active listening**, which includes reflecting, paraphrasing, and clarifying what the client is saying, the clinician communicates that he or she is genuinely interested and paying close attention.

Evaluating Humanistic Theories

See **TUTORIAL VIDEO:**
Therapeutic Techniques to
Improve Your Relationships

Supporters say that there is empirical evidence for the efficacy of client-centered therapy. However, critics argue that outcomes such as self-actualization and self-awareness are difficult to test scientifically, and research on specific humanistic techniques has had mixed results (Decker et al., 2013; Erekson & Lambert, 2015; Schneider et al., 2015).

Cognitive Therapies

Cognitive therapies assume that faulty thought processes—beliefs that are irrational, overly demanding, and/or fail to match reality—create problem behaviors and emotions (Dobson, 2014; Evans, 2015; Pomerantz, 2013).

See **FIGURE 13.5, Using**
cognitive restructuring to
improve sales

Like psychoanalysts and humanists, cognitive therapists believe that exploring unexamined beliefs can produce insight into the reasons for disturbed thoughts, feelings, and behaviors.

Internal Self-Talk and Beliefs

Lost important sales account

- "I hate sales."
- "I'm a shy person, and I'll never be any good at this."
- "I have to find another job before they fire me."

- "Selling can be difficult, but hard work pays off."
- "I'm shy but people respect my honesty and low-key approach."
- "I had the account before, and I'll get it back."

Possible Outcomes

Decreased efforts
Low energy
Depression

Increased efforts
Increased energy
No depression

However, cognitive therapists suggest that negative *self-talk*, the unrealistic things a person tells himself or herself, is most important (Arora et al., 2015; Kross et al., 2014; Zourbanos et al., 2014). Research with women suffering from eating disorders found that changing irrational thoughts and self-talk, such as "If I eat that cake, I will become fat instantly" or "I'll never have a dating relationship if I don't lose 20 pounds," resulted in their having fewer negative thoughts about their bodies (Bhatnagar et al., 2013).

Through a process called **cognitive restructuring**, clients are first taught the three Cs—to *Catch (identify), Challenge, and Change* their irrational or maladaptive thought patterns. Can you see how if we first identify and catch our irrational thoughts, then we can logically challenge them, which then enables us to change our maladaptive behaviors ( **Figure 13.5**)?

Ellis's Rational-Emotive Behavior Therapy

One of the best-known cognitive therapists, Albert Ellis, suggested that irrational beliefs are the primary culprit in problem emotions and behaviors. He proposed that most people mistakenly believe they are unhappy or upset because of external, outside events, such as receiving a bad grade on an exam. Ellis suggested that, in reality, these negative emotions result from faulty interpretations, and irrational beliefs (such as interpreting the bad grade as a sign of your incompetence, and an indication that you'll never qualify for graduate school or a good job). To deal with these irrational beliefs, Ellis developed

rational-emotive behavior therapy (REBT) (Carlson & Knaus, 2014; Ellis & Ellis, 2011, 2014; Evans, 2015). (See **Process Diagram 13.1**.)

This Process Diagram contains essential information NOT found elsewhere in the text, which is likely to appear on quizzes and exams. Be sure to study it CAREFULLY!

Process Diagram 13.1

Ellis's Rational-Emotive Behavior Therapy (REBT)

If you receive a poor performance evaluation at work, you might immediately attribute your bad mood to the negative feedback. Psychologist Albert Ellis would argue that your self-talk ("I always mess up"), between the event and the feeling, is what actually upsets you. Furthermore, ruminating on all the other times you've "messed up" maintains your negative emotional state, and may even lead to serious anxiety, depression, and other psychological disorders.

To treat these problems, Ellis developed an A–B–C–D approach: A stands for *activating event* (Step 1), B the person's *belief system* (Step 2), C the emotional *consequences* (Step 3), and D the act of *disputing* erroneous beliefs (Step 4). During therapy, Ellis helped his clients identify the A, B, Cs underlying their irrational beliefs by actively arguing with, cajoling, and teasing them—sometimes in very blunt, confrontational language. Once clients recognized their self-defeating thoughts, he worked with them on D (Step 4)—how to *dispute* those beliefs, and create and test out new, rational ones. These new beliefs then changed the original maladaptive emotions—thus breaking the vicious cycle. (Note the arrow under Step 4 that goes backwards to Step 2.)

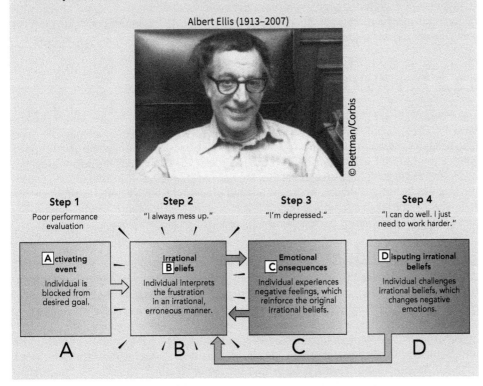

Albert Ellis (1913–2007)

© Bettman/Corbis

Step 1
Poor performance evaluation

Step 2
"I always mess up."

Step 3
"I'm depressed."

Step 4
"I can do well. I just need to work harder."

Activating event

Individual is blocked from desired goal.

Beliefs (Irrational)

Individual interprets the frustration in an irrational, erroneous manner.

Consequences (Emotional)

Individual experiences negative feelings, which reinforce the original irrational beliefs.

Disputing irrational beliefs

Individual challenges irrational beliefs, which changes negative emotions.

A   B   C   D

Beck's Cognitive Therapy (CT) and Cognitive-Behavior Therapy (CBT)

Another well-known cognitive therapist, Aaron Beck, also believes psychological problems result from illogical thinking and destructive self-talk (Beck, 1976, 2000; Craske, 2014; Hofmann et al., 2014). But Beck believes psychological problems, particularly depression, are caused by distorted thinking. Like other cognitive therapists, Beck encourages his clients to directly confront and change the behaviors associated with destructive cognitions, and his **cognitive-behavior therapy (CBT)** is designed to reduce *both* self-destructive thoughts *and* self-destructive behaviors.

Using cognitive-behavior therapy, clients are first taught to recognize and keep track of their thoughts. Next, the therapist trains the client to develop ways to test these automatic thoughts against reality. This approach helps depressed people discover that negative attitudes are largely a product of faulty thought processes.

At this point, Beck introduces the second phase of therapy—persuading the client to actively pursue pleasurable activities. Depressed individuals often lose motivation, even for experiences they once found enjoyable. Using CBT techniques, such as taking an active rather than a passive role and reconnecting with enjoyable experiences, can help in recovering from depression.

Evaluating Cognitive Therapies

Cognitive therapies are highly effective treatments for depression, as well as anxiety disorders, bulimia nervosa, anger management, addiction, PTSD, and even some symptoms of schizophrenia and insomnia (Hyland et al., 2015; Poulsen et al., 2014; Sankar et al., 2015). However, both Beck and Ellis have been criticized for ignoring or denying the client's unconscious dynamics, overemphasizing rationality, and minimizing the importance of the client's past.

Perhaps the most important criticism is that cognitive therapies are successful because they employ behavior techniques, not because they change the underlying cognitive structure (Bandura, 1969, 2008; Granillo et al., 2013; Walker et al., 2014). Imagine that you sought treatment for depression, and learned to construe events more positively, and to curb your all-or-nothing thinking. Further imagine that your therapist also helped you identify activities and behaviors that lessened your depression. Obviously, it's difficult to identify whether changing your cognitions, or changing your behavior, was the most significant therapeutic factor. But from the client's perspective, it doesn't matter. CBT combines both, and it has a proven track record for lifting depression.

Q Answer the **Concept Check** questions.

13.2 BEHAVIOR THERAPIES

The previously discussed talk therapies are often called "insight therapies" because they focus on self-awareness, but sometimes having insight into a problem does not automatically solve it. In **behavior therapies**, the focus is on the problem behavior itself, rather than on any underlying causes (Duncan, 2014; Spiegler, 2016; Stoll & Brooks, 2015). Although the person's feelings and interpretations are not disregarded, they are also not emphasized. The therapist diagnoses the problem by listing maladaptive behaviors that occur and adaptive behaviors that are absent. The therapist then attempts to shift the balance of the two, drawing on the learning principles of *classical conditioning, operant conditioning,* and *observational learning* (Chapter 6).

Classical Conditioning

Behavior therapists use the principles of classical conditioning to decrease maladaptive behaviors by creating new, more adpative associations to replace the faulty ones. We will explore two techniques based on these principles: *systematic desensitization* and *aversion therapy*.

Recall from Chapter 6 that classical conditioning occurs when a neutral stimulus (NS) becomes associated with an unconditioned stimulus (US) to elicit a conditioned response (CR). Sometimes a classically conditioned fear response becomes so extreme that we call it a "phobia." As you can see in **Process Diagram 13.2**, to treat phobias, behavior therapists often use **systematic desensitization**, which begins with relaxation training, followed by imagining or directly experiencing various versions of a feared object or situation, while remaining deeply relaxed (Garber, 2015; Schare et al., 2015; Wolpe & Plaud, 1997). If you or a friend suffers from a spider phobia, you may be amazed to know that after just two or three hours of therapy, starting with simply looking at photos of spiders, and then moving next to a tarantula in a glass aquarium, clients are able to eventually pet and hold the spider with their bare hands (Cowdrey & Walz, 2015; Hauner et al., 2012)!

Process Diagram 13.2

Systematic Desensitization

In systematic desensitization, the therapist and client together construct a *fear hierarchy*, a ranked listing of 10 or so fearful images—from the least anxiety producing to the most anxiety producing. Then, while in a state of relaxation, the client mentally visualizes, or physically experiences, mildly anxiety-producing items at the lowest level of the hierarchy. After becoming comfortable with the mild stimulus, the client then works his or her way up to the most anxiety-producing items at the top. In sum, each progressive step on the fear hierarchy is repeatedly paired with relaxation, until the fear response or phobia is extinguished.

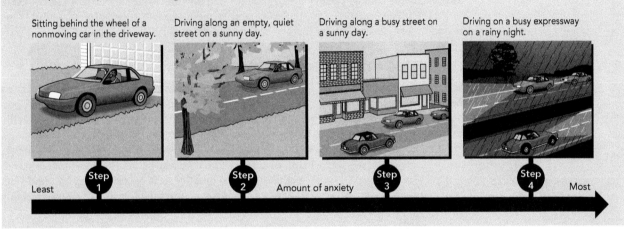

Sitting behind the wheel of a nonmoving car in the driveway.

Driving along an empty, quiet street on a sunny day.

Driving along a busy street on a sunny day.

Driving on a busy expressway on a rainy night.

Least | Step 1 | Step 2 | Amount of anxiety | Step 3 | Step 4 | Most

How does relaxation training desensitize someone? Recall from Chapter 2 that the parasympathetic nerves control autonomic functions when we're relaxed. Because the opposing sympathetic nerves are dominant when we're anxious, it is physiologically impossible to be both relaxed and anxious at the same time. The key to success is teaching the client how to replace his or her fear response with relaxation when *exposed* to the fearful stimulus. Can you see why these and related approaches are often referred to as *exposure therapies*? Modern virtual reality technology also uses systematic desensitization to expose clients to feared situations right in a therapist's office (■ **Figure 13.6**).

In contrast to systematic desensitization, **aversion therapy** uses classical conditioning techniques to create unpleasant (*aversive*) associations and responses, rather than to extinguish them. People who engage in excessive drinking, for example, build up a number of pleasurable associations with alcohol. These pleasurable associations cannot always be prevented. However, aversion therapy provides *negative associations* to compete with the pleasurable ones (■ **Figure 13.7**).

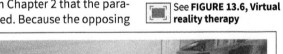

See **FIGURE 13.6, Virtual reality therapy**

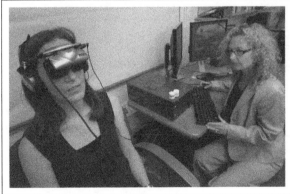

©Syracuse Newspapers/D. Lassman/The Image Works

See **VIRTUAL FIELD TRIP: Kicking the Habit**

Operant Conditioning

Shaping is a common operant conditioning technique used for bringing about a desired or target behavior. Recall from Chapter 6 that shaping provides immediate rewards for successive approximations of the target behavior. Therapists have found this technique particularly successful in developing language skills in children with autism. First, the child is rewarded for

See multipart **FIGURE 13.7,**
Aversion therapy

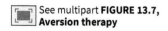

1 **During conditioning**
Someone who wants to stop drinking, for example, could take a drug called Antabuse that causes vomiting whenever alcohol enters the system.

US (drug)
+
Neutral stimulus
(alcoholic drink) → UR (nausea)

connecting pictures, or other devices with words; later, rewards are given only for using the pictures to communicate with others.

Shaping also can be used to help people acquire greater social skills and assertiveness. If you are painfully shy, a clinician might first ask you to role-play simply saying hello to someone you find attractive. Then you might practice behaviors that gradually lead you to suggest a get-together or date. During such role-playing, or behavior rehearsal, the clinician would give you feedback and reinforcement for each successive step you take toward the end goal.

For clients in an inpatient treatment facility, adaptive behaviors can be taught or increased with techniques that provide immediate reinforcement in the form of *tokens*, which are objects or symbols that can be later exchanged for primary rewards, such as food, treats, TV time, a private room, or outings. Clients might at first be given tokens for merely attending group therapy sessions. Later they will be rewarded only for actually participating in the sessions. Eventually, the tokens can be discontinued when the client receives the reinforcement of being helped by participation in the therapy sessions (Fiske et al., 2015; Mullen et al., 2015).

Observational Learning

We all learn many things by observing others. Therapists use this principle in **modeling therapy**, in which clients are asked to observe and imitate appropriate models as they perform desired behaviors. In one study, researchers successfully treated 4- and 5-year-old children with severe dog phobias by asking them first to watch other children play with dogs, and then to gradually approach and get physically closer to the dogs themselves (May et al.,

See **FIGURE 13.8,**
Observational learning

Digital Vision/Getty Images

2013). When this type of therapy combines live modeling with direct and gradual practice, it is called *participant modeling*. This type of modeling is also effective in social skills training and assertiveness training (■ **Figure 13.8**).

Evaluating Behavior Therapies

Criticisms of behavior therapy fall into two major categories:

- **Generalizability** Critics argue that in the real world, people are seldom consistently reinforced, and their newly acquired behaviors may disappear. To deal with this possibility, behavior therapists work to encourage clients to better recognize existing, external, real world rewards, and to generate their own internal reinforcements, which they can then apply at their own discretion.

- **Ethics** Although some critics contend that it is unethical for one person to control another, behaviorists would argue that we're already controlled by rewards and punishments. Furthermore, behavior therapy actually increases our freedom by making these controls overt, and by teaching people how to change their own behavior.

Despite these criticisms, behavior therapy is generally recognized as one of the most effective treatments for numerous problems, including phobias, obsessive-compulsive disorder, eating disorders, sexual dysfunctions, autism, intellectual disabilities, and delinquency (Grossman & Walfish, 2014; Spiegler, 2016; Stoll & Brooks, 2015).

Q Answer the
Concept Check questions.

13.3 BIOMEDICAL THERAPIES

Some problem behaviors seem to be caused, at least in part, by chemical imbalances or disturbed nervous system functioning, and, as such, they can be treated with **biomedical therapies**. Psychiatrists or other medical personnel are generally the only ones who use biomedical therapies. However, in some states, licensed psychologists can prescribe certain medications, and they often work with clients receiving biomedical therapies. In this section, we will discuss three aspects of biomedical therapies: *psychopharmacology*, *electroconvulsive therapy* (*ECT*), and *psychosurgery*.

Psychopharmacology

Since the 1950s, the field of **psychopharmacology** has effectively used drugs to relieve or control the major symptoms of psychological disorders. In some instances, using a psychotherapeutic drug is similar to administering insulin to people with diabetes, whose own bodies fail to manufacture enough. In other cases, drugs have been used even when the underlying cause was not thought to be biological. As shown in **Table 13.1**, psychotherapeutic drugs are classified into four major categories: *antianxiety, antipsychotic, mood stabilizer,* and *antidepressant*.

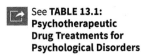
See **TABLE 13.1:** Psychotherapeutic Drug Treatments for Psychological Disorders

See multipart **FIGURE 13.9,** How antidepressants work

How do drug treatments actually work? For most psychotherapeutic medications, the best-understood action of the drugs is to correct an imbalance in the levels of neurotransmitters in the brain (**Figure 13.9**). For example, researchers know that the drug *ketamine* works for major depression, suicidal behaviors, and bipolar disorders because it changes the levels of brain neurotransmitters. Widely known in the medical field for its anesthetic properties, ketamine is also a dangerous date rape/party drug, called "Special K." Ironically, it's also been found to be helpful for diminishing the cravings for alcohol. Therapists are currently very excited about the promise of ketamine treatment, particularly due to its rapid onset, high efficacy, and good tolerability (Coyle & Laws, 2015; Lee et al., 2015; Reardon, 2015). On the other hand, it can have tragic—or even—fatal effects when abused.

Although we don't fully understand all the mechanisms at play, some studies also suggest that antidepressants, may relieve depression in three additional ways. They increase *neurogenesis*, the production of new neurons, or *synaptogenesis*, the production of new synapses, and/or they stimulate activity in various areas of the brain (Kriesal et al., 2014; Miller & Hen, 2015; Walker et al., 2015).

MediaImages/Photodisc/Getty Images, Inc.

Electroconvulsive Therapy and Psychosurgery

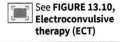
See **FIGURE 13.10,** Electroconvulsive therapy (ECT)

There is a long history of using electrical stimulation to treat psychological disorders. In **electroconvulsive therapy (ECT)**, also known as electroshock therapy (EST), a moderate electrical current is passed through the brain. This can be done by placing electrodes on the outside of both sides of the head (bilateral ECT), or on only one side of the head (unilateral ECT). The current triggers a widespread firing of neurons, or brief seizures. Fortunately, ECT can quickly reverse symptoms of certain mental illnesses, and often works when other treatments have been unsuccessful. The electric current produces many changes in the central and peripheral nervous systems, including activation of the autonomic nervous system, increased secretion of various hormones and neurotransmitters, and changes in the blood–brain barrier (**Figure 13.10**).

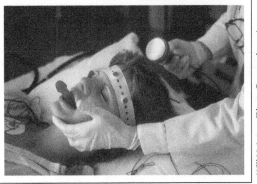
Will McIntyre/Photo Researchers, Inc.

Despite not knowing exactly how ECT works, and that it may cause some short-term memory loss immediately after treatment, and possible long-term memory problems, the risks of untreated, severe depression are generally considered greater than the risks of ECT. Therefore, it is used primarily on those who have not responded to drugs and psychotherapy, and on occasionally suicidal clients, because it works faster than antidepressant drugs (Bumb et al., 2015; Kellner et al., 2015; Vallejo-Torres et al., 2015).

See **VIRTUAL FIELD TRIP:** ECT Treatment Center

The most extreme, and least used, biomedical therapy is **psychosurgery**, brain surgery performed to reduce serious, debilitating, psychological problems. Attempts to change disturbed thoughts, feelings, and behavior by altering the brain have a long history. In Roman times, for example, it was believed that a sword wound to the head could relieve insanity. In 1936, Portuguese neurologist Egaz Moniz first treated uncontrollable psychoses with a form of *psychosurgery* called a *lobotomy*, in which he cut the nerve fibers between the frontal lobes (where association areas for monitoring and planning behavior are found) and the thalamus and hypothalamus. Although a small percentage of people supposedly got better, many were left with profound and permanent brain damage.

Two of the most notable examples of the damage from early lobotomies are Rosemary Kennedy, the sister of President John F. Kennedy, and Rose Williams, sister of American playwright Tennessee Williams. Both women were permanently incapacitated from lobotomies performed in the early 1940s. Thankfully, in the mid-1950s, when antipsychotic drugs came into use, psychosurgery virtually stopped.

Evaluating Biomedical Therapies

Like all other forms of therapy, biomedical therapies have both proponents and critics.

Psychopharmacology

Clinicians have long known that the use of drugs is more effective when it's combined with talk therapy. For example, researchers have examined whether children and teenagers experiencing clinical depression would benefit from receiving cognitive behavioral therapy along with medication to treat this disorder (Kennard et al., 2014). In this study, 75 youths (ages 8 to 17) received either an antidepressant alone, or an antidepressant along with cognitive behavioral therapy (CBT) for six months. Of those who received only the drug, 26.5% experienced depression, compared to only 9% of those who received the drug as well as CBT.

However, drug therapy poses several potential problems. First, although drugs may relieve symptoms for some people, they seldom provide cures. In addition, some individuals become physically dependent on the drugs. Furthermore, psychiatric medications can cause a variety of side effects, ranging from mild fatigue to severe impairments in memory and movement.

Despite the problems associated with them, psychotherapeutic drugs have led to revolutionary changes in mental health. Before the use of drugs, some patients were destined to spend a lifetime in psychiatric institutions. Today, most improve enough to return to their homes and lead successful lives—if they continue to take their medications to prevent relapse.

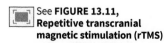
See **FIGURE 13.11, Repetitive transcranial magnetic stimulation (rTMS)**

Richard T. Nowitz/Science Source

ECT

ECT currently serves as a valuable last-resort treatment for severe depression. However, similar benefits may be available through the latest advances in *repetitive transcranial magnetic stimulation (rTMS)*, which uses an electromagnetic coil placed on the scalp. Unlike ECT, which uses electricity to stimulate parts of the brain, rTMS uses magnetic pulses (**Figure 13.11**). To treat depression, the coil is usually placed over the prefrontal cortex, a region linked to deeper parts of the brain that regulate mood. Currently, rTMS's advantages over ECT are still unclear, but studies have shown marked improvement in depression, and clients experience fewer side effects (Bakker et al., 2015; Caruso & Perez, 2015; Zhang et al., 2015).

Psychosurgery

Because all forms of psychosurgery are potentially dangerous and have serious or even fatal side effects, some critics say that it should be banned altogether. Furthermore, the consequences are generally irreversible. For these reasons, psychosurgery is considered experimental and remains a highly controversial treatment.

Recently, however, psychiatrists have been experimenting with a much more limited and precise neurosurgical procedure called *deep brain stimulation* (*DBS*). The surgeon drills two tiny holes into the skull, and implants electrodes in the area of the brain believed to be associated with a specific disorder (■ **Figure 13.12**). These electrodes are then connected to a "pacemaker" implanted in the chest or stomach that sends low-voltage electricity to the problem areas in the brain. Over time, this repeated stimulation can bring about significant improvement in Parkinson's disease, epilepsy, major depression, and other disorders (Fields, 2015; Lipsman et al., 2015; Vedam-Mai et al., 2014). Research has also shown that clients who receive DBS along with antidepressants show lower rates of depression than those who receive either treatment alone (Brunoni et al., 2013).

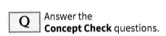
See **FIGURE 13.12, Deep brain stimulation (DBS)**

Photo Researchers/Getty Images

13.4 PSYCHOTHERAPY IN PERSPECTIVE

It's currently estimated that there are more than 1,000 approaches to psychotherapy, and the number is continuing to rise (Gaudiano et al., 2015; Magnavita & Anchin, 2014). Given this high number, and wide variety of approaches, how would you choose one for yourself or someone you know? In the first part of this section, we discuss five goals common to all psychotherapies. Then we explore specific formats for therapy as well as considerations about culture and gender. Our aim is to help you synthesize the material in this chapter, and to put what you have learned about each of the major forms of therapy into a broader context.

Q Answer the **Concept Check** questions.

Therapy Goals and Effectiveness

All major forms of therapy are designed to help the client in five specific areas (**Study Organizer 13.2**).

Although most therapists work with clients in several of these areas, the emphasis varies according to the therapist's training, and whether it is psychodynamic, cognitive, humanistic, behaviorist, or biomedical. Clinicians who regularly borrow freely from various theories are said to take an **eclectic approach**.

See **STUDY ORGANIZER 13.2: The Five Most Common Goals of Therapy**

Does therapy work? After years of controlled studies and *meta-analysis*—a method of statistically combining and analyzing data from many studies—researchers have fairly clear evidence that it does. As you can see in ■ **Figure 13.13**, an early meta-analytic review combined studies of almost 25,000 people and found that the average person who received treatment was better off than 75% of the untreated control clients.

See **FIGURE 13.13, Is therapy generally effective?**

Studies also show that short-term treatments can be as effective as long-term treatments, and that most therapies are equally effective for various disorders (Battino, 2015; Braakmann, 2015; Chan et al., 2015). However, research does show that certain approaches work better for specific disorders. For example, phobias and marital problems seem to respond best to behavioral therapies, whereas depression and obsessive-compulsive disorders can be significantly relieved with cognitive-behavior therapy (CBT) accompanied by medication (Craighead et al., 2013; McAleavey et al., 2014; Short & Thomas, 2015).

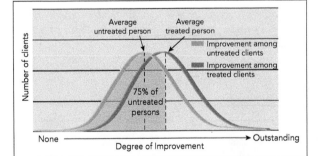

Finally, in recent years, the *empirically supported*, or *evidence-based practice* (*EBP*), *movement* has been gaining momentum because it seeks to identify which therapies have received the clearest research support for particular disorders (Gaudiano et al., 2015; Gray et al., 2015; Midgley et al., 2014). Like many movements, it has been criticized, but this scientific, empirical approach promises to be helpful for therapists and clients alike in their treatment decisions.

Therapy Formats

The therapies described earlier in this chapter are conducted primarily in a face-to-face, therapist-to-client format. In this section, we focus on several major alternatives: *group*, *family*, and *marital therapies*, which treat multiple individuals simultaneously, and *telehealth/electronic therapy*, which treats individuals via the Internet, e-mail, and/or smart phones.

Group Therapy

In **group therapy**, multiple people meet together to work toward therapeutic goals. Typically, a group of 8 to 10 people meet with a therapist on a regular basis to talk about problems in their lives.

A variation on group therapy is the **self-help group**. Unlike other group approaches, self-help groups are not guided by a professional. They are simply circles of people who share a common problem, such as alcoholism, obesity, or breast cancer, and who meet to give and receive support. Faith-based, twelve-step programs, such as Alcoholics Anonymous, Narcotics Anonymous, and Spenders Anonymous, are examples of self-help groups. Although group members don't get the same level of individual attention found in one-on-one therapies, group and self-help therapies provide their own unique advantages (Bateganya et al., 2015; DeLucia et al., 2015; Dyregrov et al., 2014). They are far less expensive than one-on-one therapies and provide a broader base of social support. Group members also can learn from each other's experiences, share insights and coping strategies, and role-play social interactions together.

For example, researchers have studied group sessions in twelve-step programs, like Alcoholics Anonymous, and they've found that youths suffering from a combination of social anxiety disorders and substance abuse disorders, as well as recovering alcoholics, all show lower rates of relapse—particularly if they provide service to others (Pagano et al., 2013, 2015). In sum, research on self-help groups for alcoholism, obesity, and other disorders suggests that they can be very effective, either alone or in addition to individual psychotherapy (Bernecker, 2014; Kendra et al., 2015; McGillicuddy et al., 2015). Keep in mind that therapists often recommend these alternative formats, such as group therapy, to their clients as an additional resource, while continuing their individual therapy.

Marital and Family Therapies

Given that a family or marriage is a system of interdependent parts, the problem of any one individual inevitably affects everyone. Therefore, all members are potential beneficiaries of therapy (Gunn et al., 2015; McGeorge et al., 2015; Wetchler & Hecker, 2015). The line between family and marital or couples therapy is often blurred. Here, our discussion will focus on *family therapy*, in which the primary aim is to change maladaptive family interaction patterns (▣ **Figure 13.14**). All members of the family attend therapy sessions, though at times the therapist may see family members individually or in twos or threes.

See **FIGURE 13.14, Family therapy**

© vgajic/iStockphoto

Family therapy not only helps resolve interpersonal relationship problems, it also is useful in treating a number of disorders and clinical problems. As discussed in Chapter 12, individuals with schizophrenia are at increased risk of relapsing if their family members express emotions, attitudes, and behaviors that suggest criticism, hostility, or emotional overinvolvement (Hooker et al., 2014; Lobban et al., 2013; O'Brien et al., 2014). Family therapy can help family members modify their behavior toward the troubled family member, and it's particularly helpful in the treatment of adolescent drug abuse (Carr, 2014; Horigian & Szapocnik, 2015; Priest et al., 2015).

Telehealth/Electronic Therapy

Today millions of people are receiving advice and professional therapy in newer, electronic formats, such as the Internet, e-mail, virtual reality (VR), and interactive web-based conference systems such as Skype. This latest form of electronic therapy, often referred to as *telehealth*, allows clinicians to reach more clients and provide them with greater access to information regarding their specific problems. Studies have long shown that therapy outcomes improve with increased client contact, and the electronic/telehealth format may be the easiest and most cost-effective way to increase this contact (Bush et al., 2015; Gray et al., 2015; Lokkerbol et al., 2014).

Using electronic options such as the Internet and smart phones does provide alternatives to traditional one-on-one therapies, but, as you might expect, these unique approaches also raise concerns (Barrett & Gershkovich, 2014; Kasckow et al., 2014; Luxton et al., 2014). Professional therapists fear, among other things, that without interstate and international licensing, or a governing body to regulate this type of therapy, there are no means to protect clients from unethical practices or incompetent therapists. What do you think? Would you be more likely to participate in therapy if it were offered via your smart phone, e-mail, or a website? Or is this too impersonal for you?

Cultural Issues in Therapy

The therapies described in this chapter are based on Western European and North American culture. Does this mean they are unique? Or do these psychotherapists accomplish some of the same things that, say, a native healer or shaman does?

When we look at therapies cross-culturally, we do find that they have certain key features in common (Barnow & Balkir, 2013; Braakmann, 2015; Paniagua, 2014).

- **Naming the problem** Troubled people around the world often feel better just knowing that others share a similarly named problem, and that the therapist has had experience treating it.

- **Demonstrating the right qualities** In all cultures, clients must feel that the therapist is caring, competent, approachable, and concerned with finding solutions to their problems.

- **Establishing credibility** Word-of-mouth testimonials and status symbols, such as diplomas on the wall, establish a therapist's credibility, which is key to all forms of successful therapy. A native healer may earn credibility by serving as an apprentice to a revered healer.

- **Placing the problem in a familiar framework** Some cultures believe evil spirits cause psychological disorders, so therapy is directed toward eliminating these spirits. Similarly, in cultures that emphasize the importance of early childhood experiences and the unconscious mind as the cause of mental disorders, therapy will be framed around these familiar issues.

- **Applying techniques to bring relief** In all cultures, therapy includes action—either the client or the therapist must do something. Moreover, what therapists do must fit the client's expectations—whether it is performing a ceremony to expel demons, or talking with the client about his or her thoughts and feelings.

- **Meeting at a special time and place** The fact that therapy occurs in a unique way, outside the client's everyday experiences, seems to be an important feature of all therapies.

Although there are basic similarities in therapies across cultures, there are also important differences. In the traditional Western European and North American model, the emphasis is on the client having independence and control over his or her life—qualities that are highly valued in individualistic cultures. In collectivist cultures, however, the focus of therapy is on interdependence and the acceptance of life realities (Epstein et al., 2014; Lee et al., 2015; Seay & Sun, 2016). For example, in Japanese Naikan therapy, people sit quietly from 5:30 a.m. to 9:00 p.m. for seven days, and are visited by an interviewer every 90 minutes. During this time, they reflect on their relationships with others in order to discover personal guilt for having been ungrateful and troublesome, and to develop gratitude toward those who have helped them (Itoh & Hikasa, 2014; Shinfuku & Kitanishi, 2010).

Not only does culture affect the types of therapy that are developed, it also influences the perceptions of the therapist. What one culture considers abnormal behavior may be quite common—and even healthy—in others. For this reason, recognizing cultural differences is very important for building trust between therapists and clients and for effecting behavioral change (La Roche et al., 2015; Strauss et al., 2015; Weiler et al., 2015).

Gender and Therapy

In our individualistic Western culture, men and women often present different needs and problems to therapists. Research has identified four unique concerns related to gender and psychotherapy (Mendrek et al., 2014; Sáenz Herrero, 2015; Zerbe Enns et al., 2015):

1. Rates of diagnosis and treatment of mental disorders Women are diagnosed and treated for mental illness at a much higher rate than men. Are women "sicker" than men as a group, or are they just more willing to admit their problems? Or are the categories of illness biased against women? More research is needed to answer these questions.

2. Stresses of poverty Women are disproportionately more likely to be poor. And poverty contributes to stress, which is directly related to many psychological disorders.

3. Violence against women Rape, incest, and sexual harassment—which are much more likely to happen to women than to men—may lead to depression, insomnia, posttraumatic stress disorders, eating disorders, and other problems.

4. Stresses of multiple roles Modern men and women today serve in many roles, as family members, students, wage earners, and so forth. The conflicting demands of their multiple roles often create special stresses unique to each gender.

Therapists must be sensitive to possible connections between clients' problems and their gender. Rather than just emphasizing drugs to relieve depression, it may be more appropriate for therapists to also explore ways to relieve the stresses of multiple roles or poverty. Can you see how helping a single parent identify resources, such as play groups, parent support groups, and high-quality child care, might be just as effective at relieving depression as prescribing drugs?

If you've enjoyed this section on cultural and gender issues in therapy, as well as the earlier description of the various forms and formats of psychotherapy, you may be considering a possible career as a therapist. If so, see ▣ **Table 13.2**.

See **TABLE 13.2: Careers in Mental Health**

Q Answer the **Concept Check** questions.

> **WP LS** Go to your WileyPLUS Learning Space course for video episodes, examples, art, tables, Concept Checks, practice, and other pedagogical resources that will help you succeed in this course.

Reading for
SOCIAL PSYCHOLOGY

14

WP LS Go to your WileyPLUS Learning Space course for video episodes, examples, art, tables, Concept Checks, practice, and other pedagogical resources that will help you succeed in this course.

14.1 SOCIAL COGNITION

For many students and psychologists, your authors included, *social psychology* is the most exciting of all fields because it's about you and me, and because almost everything we do is *social*! Unlike earlier chapters that focused on individual processes, like sensation and perception, memory, or personality, this chapter studies how large social forces, such as groups, social roles, and norms, bring out the best and worst in all of us.

Social psychology, one of the largest branches in the field of psychology, focuses on how others influence our thoughts, feelings, and actions. In turn, one of its largest and most important subfields, **social cognition**, examines the way we think about and interpret ourselves and others (Mikulincer et al., 2015; Smith et al., 2015). In this section, we'll look at two of the most important topics in social cognition—*attributions* and *attitudes*.

Attributions

Have you ever been in a serious verbal fight with a loved one—perhaps a parent, close friend, or romantic partner—and he or she said something unfairly harsh to you? If so, how did you react? Were you overwhelmed with feelings of anger? Did you attribute the fight to the other person's ugly, mean temper, and even consider possibly ending the relationship? Or did you calm yourself with thoughts of how he or she is normally a rational person, and therefore must be unusually upset by something that happened at work or college?

Can you see how these two alternative **attributions** (or explanations) for the causes of behavior or events can either destroy or maintain relationships? The study of attributions is a major topic in social cognition and social psychology. Everyone wants to understand and explain why people behave as they do, and why events occur as they do. In fact, humans are known to be the only reason-seeking animals! But social psychologists have discovered another explanation: Developing logical attributions for people's behavior makes us feel safer and more in control (Heider, 1958; Lindsay et al., 2015; Stalder & Cook, 2014). Unfortunately, our attributions are frequently marred by several attributional biases and errors.

Attributional Errors and Biases

Think back to the example above. Can you see how attributing the fight to the bad character of the other person, without considering possible situational factors (like pressures at work) may be misguided? Similarly, imagine a new student who joins your class and seems distant, cold, and uninterested in interaction. It's easy to conclude that she's unfriendly, and maybe even "stuck-up"—a *personality attribution*. However, if you later saw her in a one-to-one interaction with close friends, you might be surprised to find that she's very warm and friendly. In other words, her behavior depends on the situation. This bias toward personality factors rather than situational factors in our explanations for others' behavior is very common, and it's called the **fundamental attribution error (FAE)** (Hooper et al., 2015; Moran et al., 2014; Ross, 1977). (See  **Figure 14.1**.)

 See **FIGURE 14.1, Attribution in action**

AFP/Getty Images

Unlike the FAE, which commonly occurs when we're explaining others' behaviors, the **self-serving bias** more often happens when we explain our own behavior. Ironically, in this case, we tend to favor internal (personality) attributions for our successes and blame external (situational) attributions for our failures. This bias is motivated by our desire to maintain positive self-esteem, and a good public image (Sharma & Shakeel, 2015; Wiggin & Yalch, 2015). Have you noticed how students often take personal credit for doing well on exam, but when they fail a test they tend

See **FIGURE 14.2, The actor-observer effect (blaming others and excusing ourselves)**

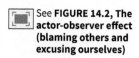

Actor		Observer
Situational attribution		**Personality attribution**
Focuses on external factors in the environment		Focuses on internal, personality of the actor
"I don't even like drinking beer, but it's the best way to meet women."		"He seems to always have a beer in his hand; he must have a drinking problem."

© Christine Schneider/Corbis

to blame the instructor, the textbook, or the "tricky" questions? Similarly, elite Olympic athletes more often attribute their wins to internal (personal) causes, such as their skill and effort, while attributing their losses to external (situational) causes, such as bad equipment or poor officiating (Aldridge & Islam, 2012; Mackinnon et al., 2015). Of course, not everyone shows the self-serving bias all the time. Surprisingly, people with some forms of mental illness (such as anxiety or bipolar disorder), or who feel ashamed of their bodies, may be more likely to blame themselves, or to or feel guilty over perceived failures (Kestemont et al., 2015; Stange et al., 2014).

How do we explain the discrepancy between the attributions we make for ourselves, and those we make for others? According to the **actor–observer effect** (Jones & Nisbett, 1971), when examining our own behaviors, we are the *actors* in the situation (Figure 14.2). And because we can't see our own behavior, our attention is directed toward the environment, which leads us to attribute our own actions to external factors. "I didn't tip the waiter because I got really bad service." In contrast, when explaining the behavior of others, we are *observing* the actors. Therefore, we tend to focus blame on them, and to attribute their actions to internal, personality factors. "She didn't tip the waiter because she's cheap."

Culture and Attributional Biases

Interestingly, both the fundamental attribution error and the self-serving bias may depend in part on cultural factors (Cullen et al., 2015; Kastanakis & Voyer, 2014; Khandelwal et al., 2014). In highly individualistic cultures, like the United States, people are more often defined and understood as individual selves, largely responsible for their own successes and failures. In contrast, people in collectivistic cultures, like China and Japan, are primarily defined as members of their social network, responsible for doing as others expect. Accordingly, they tend to be more aware of situational constraints on behavior, making the FAE less likely (Bond, 2015; Matsumoto & Juang, 2013; Tang et al., 2014).

The self-serving bias is also much less common in collectivistic cultures because self-esteem is related not to doing better than others, but to fitting in with the group (Anedo, 2014; Berry et al., 2011; Morris, 2015). In Japan, for instance, the ideal person is more often aware of his or her shortcomings, and continually works to overcome them—rather than thinking highly of himself or herself (Heine & Renshaw, 2002; Shand, 2013).

Reducing Attributional Errors

The key to making more accurate attributions begins with determining whether a given action stems mainly from personality factors, or from the external situation. Unfortunately, we too often focus on internal, dispositional (personality) factors. Why? We all naturally take cognitive shortcuts (Chapter 8), and we each tend to have unique and enduring personality traits (Chapter 11). To offset this often misguided preference for internal attributions, we can ask ourselves these four questions:

1. *Is the behavior unique or shared by others*? If a large, or increasing, number of people are engaging in the same behavior, such as rioting or homelessness, it's most likely the result of external, situational factors.

2. *Is the behavior stable or unstable?* If someone's behavior is relatively enduring and permanent, it may be correct to make a stable, personality attribution. However, before giving up on a friend who is often quick-tempered and volatile, we may want to consider his or her entire body of personality traits. If he or she is also generous, kind, and incredibly devoted, we could overlook these imperfections.

3. *Was the cause of the behavior controllable or uncontrollable?* Innocent victims of crime, like rape or robbery, are too often blamed for their misfortune because they shouldn't have "been in that part of town," "walking alone," and/or "dressed in expensive clothes." Can you see how these are all undoubtedly inaccurate and unfair personality attributions, as well as examples of "blaming the victim"?

4. *What would I do in the same situation?* Given our natural tendency toward the *self-serving bias* and the *actor-observer effect,* if we conclude that we would behave in the same way, the behavior is most likely the result of external, situational factors.

In sum, given our natural tendency to make internal, personality attributions, we can improve our judgments of others by erring in the opposite direction—looking first for external causes. Furthermore, "giving others the benefit of the doubt" will not only help us avoid attributional errors, it may even save, or at least improve, our relationships.

See **FIGURE 14.3, Attitude formation**

Attitudes

When we observe and respond to the world around us, we are seldom completely neutral. Rather, our responses toward subjects as diverse as pizza, gun control, and abortion reflect our **attitudes**, which are *learned* predispositions to respond positively or negatively to a particular object, person, or event. Social psychologists generally agree that most attitudes have three ABC components: *affect*, or feelings; *behavior*, or actions; and *cognitions*, or thoughts and beliefs (**Figure 14.3**).

Attitude Formation

As mentioned, we tend to *learn* our attitudes, and this generally occurs from direct instruction, personal experiences, and by watching others. In some cases, these sources may differ depending on our gender. For example, researchers have found that teenage boys are more likely to learn sexual attitudes from media representations of sexual behavior, whereas teenage girls tend to learn their sexual attitudes from their mothers, as long as they feel close to their mothers (Vandenbosch & Eggermont, 2011).

Attitude Change

Although attitudes begin to form in early childhood, they're obviously not permanent, a fact that advertisers and politicians know and exploit. As you can see in **Figure 14.4**, we can sometimes change attitudes through experiments. However, a much more common method is to make direct, persuasive appeals, such as in ads that say, "Friends Don't Let Friends Drive Drunk!" Surprisingly, psychologists have identified an even more efficient strategy than persuasion. The strongest personal change comes when we experience contradictions between our thoughts, feelings, or actions—the three components of all attitudes. These contradictions then lead to feelings of discomfort and arousal, known as **cognitive dissonance** (Greenberg et al., 2015; Mikulincer et al., 2015), which makes us highly motivated to change one or more parts of our attitudes. A good example comes from a clever experiment focusing on the pain caused by cognitive dissonance, which found that taking a simple pain killer, like acetaminophen, versus a placebo, can significantly reduce the amount of attitude change (DeWall et al., 2015)!

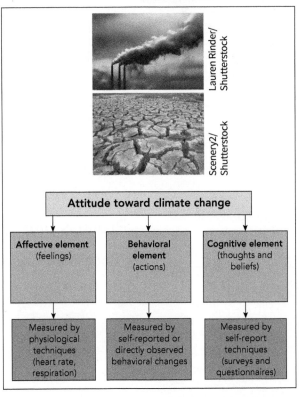

Attitude toward climate change

Affective element (feelings)	Behavioral element (actions)	Cognitive element (thoughts and beliefs)
Measured by physiological techniques (heart rate, respiration)	Measured by self-reported or directly observed behavioral changes	Measured by self-report techniques (surveys and questionnaires)

See **FIGURE 14.4, Using photos to change attitudes**

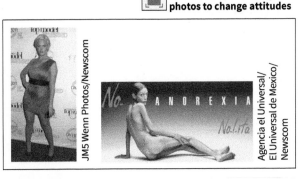

Given that cognitive dissonance is such a key component of attitude change in all our lives, it's important to understand it. To do so, let's closely examine the classic study by Leon Festinger and J. Merrill Carlsmith (1959). These experimenters asked college students to perform several very boring tasks, such as turning wooden pegs or sorting spools into trays. They were then paid either $1 or $20 to lie to NEW research participants by telling them that the boring tasks were actually very enjoyable and fun. Surprisingly, those who were paid just $1 to lie subsequently changed their minds about the task, and actually reported more positive attitudes toward it, than those who were paid $20.

Why was there more attitude change among those who were paid only $1? All participants who lied to other participants presumably recognized the discrepancy between their initial beliefs and feelings (the task was boring) and their behavior (telling others it was enjoyable and fun). However, as you can see in **Figure 14.5**, the participants who were given insufficient monetary justification for lying (the $1 liars), apparently experienced greater *cognitive dissonance*. Therefore, to reduce their discomfort, they expressed more liking for

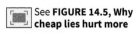See **FIGURE 14.5, Why cheap lies hurt more**

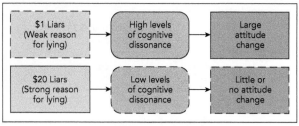

the dull task, compared to those who received sufficient monetary justification (the $20 liars). This second group had little or no motivation to change their attitude—they lied for the money! (Note that in 1959, when the experiment was conducted, $20 would have been the equivalent of about $200 today.)

Can you see the potential danger in how easily some participants in this classic study changed their thoughts and feelings about the boring task in order to match their behavior? More importantly, consider how cognitive dissonance might help explain why military leaders keep sending troops to a seemingly endless war zone. They obviously can't change the behavior that led to the initial loss of lives, so they may reduce their cognitive dissonance by becoming even more committed to a belief that the war is justified. Given the importance of this theory to your everyday life, be sure to carefully study **Process Diagram 14.1**.

This Process Diagram contains essential information NOT found elsewhere in the text, which is likely to appear on quizzes and exams. Be sure to study it CAREFULLY!

Process Diagram 14.1

Understanding Cognitive Dissonance

We've all noticed that people often say one thing, but do another. For example, why do some health professionals, who obviously know the dangers of smoking, continue to smoke?

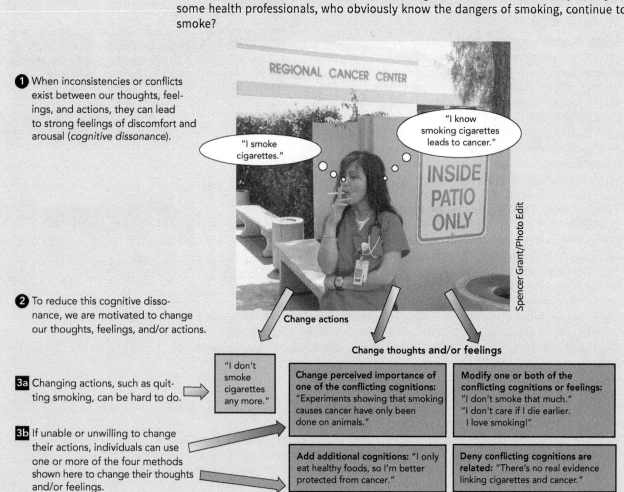

1 When inconsistencies or conflicts exist between our thoughts, feelings, and actions, they can lead to strong feelings of discomfort and arousal (*cognitive dissonance*).

2 To reduce this cognitive dissonance, we are motivated to change our thoughts, feelings, and/or actions.

3a Changing actions, such as quitting smoking, can be hard to do.

3b If unable or unwilling to change their actions, individuals can use one or more of the four methods shown here to change their thoughts and/or feelings.

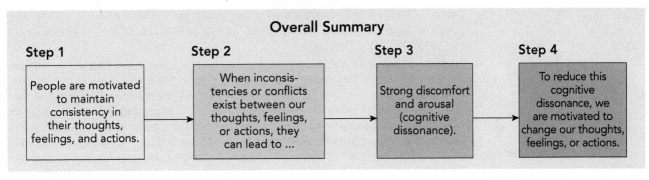

Overall Summary

Step 1
People are motivated to maintain consistency in their thoughts, feelings, and actions.

Step 2
When inconsistencies or conflicts exist between our thoughts, feelings, or actions, they can lead to ...

Step 3
Strong discomfort and arousal (cognitive dissonance).

Step 4
To reduce this cognitive dissonance, we are motivated to change our thoughts, feelings, or actions.

Q | Answer the **Concept Check** questions.

14.2 SOCIAL INFLUENCE

In the previous section, we explored the way we think about and interpret ourselves and others through *social cognition*. We now focus on **social influence**: how situational factors and other people affect us. In this section, we explore three key topics—*conformity*, *obedience*, and *group processes*.

Conformity

Imagine that you have volunteered for a psychology experiment on visual perception. You and all the other participants are shown two cards (■ **Figure 14.6**). The first card has only a single vertical line on it, while the second card has three vertical lines of varying lengths. Your task is to determine which of the three lines on the second card (marked A, B, or C) is the same length as the single line on the first card (marked X).

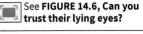 See **FIGURE 14.6, Can you trust their lying eyes?**

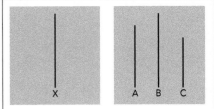

You are seated around a table with six other people, and everyone is called on in order. Because you are seated to the left of the seventh participant, you are always next to last to provide your answers. On the first two trials, everyone agrees on the correct line. However, on the third trial, the first participant chooses line A as the closest in length to line X, an obvious wrong answer! When the second, third, fourth, and fifth participants also say line A, you really start to wonder: "What's going on here? Are they wrong, or am I?"

What do you think you would do at this point in the experiment? Would you stick with your convictions and say line B, regardless of what the others have answered? Or would you go along with the group? In the original version of this experiment, conducted by Solomon Asch (1951), six of the seven participants were actually *confederates* of the experimenter (that is, they were working with the experimenter and purposely gave wrong answers). Their incorrect responses were designed to test the participant's degree of **conformity**, which is defined as changes in thoughts, feelings, or actions because of real or imagined group pressure.

More than one-third of Asch's participants conformed—they agreed with the group's obviously incorrect choice. (Participants in a control group experienced no group pressure, and almost always chose correctly.) Asch's study has been conducted dozens of times, in at least 17 countries, and generally with similar results (Mori et al., 2014; Tennen et al., 2013). Surprisingly, this conformity effect even occurs in online experiments in which participants conform even when they can only see the names of other participants, and also during video games when gamers conform to the vote of computers (Beran et al., 2015; Weger et al., 2015)!

Why are we so likely to conform? To the onlooker, conformity is often difficult to understand. Even the conformer sometimes has a hard time explaining his or her behavior. Let's look at three factors that drive conformity:

- **Normative social influence** Have you ever asked what others are wearing to a party, or copied your neighbor at a dinner party to make sure you picked up the right fork? One of the first reasons we conform is that we want to go along with group *norms*, which are expected behaviors generally adhered to by members of a group. We generally submit to **normative social influence** out of our need for approval and acceptance by the group,

209

because conforming to group norms makes us feel good, and/or because it's adaptive to do so (Baldry & Pagliaro, 2014; Higgs, 2015; Oarga et al., 2015). Most often norms are quite subtle and only implied.

- **Informational social influence** Have you ever bought a specific product simply because of a friend's recommendation? In this case, you probably conformed not to gain your friend's approval, a case of normative social influence, but because you assumed that he or she had more information than you did, a case of **informational social influence**. This may hold true even if we don't personally know the recommender, as in the case of national polling results and online reviewers of movies, products or restaurants (Binning et al., 2015; Lim & Van Der Heide, 2015; Zhu & Huberman, 2014). Given that participants in Asch's experiment observed all the other participants giving unanimous decisions on the length of the lines, can you see how they may have conformed because they simply believed the others had more information than they did?

- **Reference groups** The third major factor in conformity is the power of **reference groups**—the people we most admire, like, and want to resemble. Attractive actors and popular sports stars are paid millions of dollars to endorse products because advertisers know that we want to be as cool as LeBron James, or as beautiful as Natalie Portman (Arsena et al., 2014; Schulz, 2015). Of course, we also have more important reference groups in our lives—parents, friends, family members, teachers, religious leaders, and classmates—all of whom also affect our willingness to conform. For example, popular high school students' attitudes about alcohol use have been shown to have a substantial influence on alcohol consumption by other students in their high school (Teunissen et al., 2012). Surprisingly, popular peers who had *negative* attitudes toward alcohol use were even more influential in determining rates of teenage drinking than those with positive attitudes! This and other research findings suggest that capitalizing on popular peers who have negative drinking attitudes (even though they may be in the minority) might be a very effective strategy for reducing underage drinking (Rees & Wallace, 2014).

Obedience

As we've seen, conformity means going along with the group. A second form of social influence, **obedience**, involves going along with direct commands, usually from someone in a position of authority. From very early childhood, we're socialized to respect and obey our parents, teachers, and other authority figures.

See **FIGURE 14.7, When it pays to conform and obey**

David McNew/Getty Images

Conformity and obedience aren't always bad (▣**Figure 14.7**). In fact, we generally conform and obey most of the time because it's in our own best interests (and everyone else's) to do so. Like most other Westerners, we stand in line at a movie theater instead of pushing ahead of others. This allows an orderly purchasing of tickets. Conformity and obedience allow social life to proceed with safety, order, and predictability.

However, on some occasions, it is important not to conform or obey. We don't want teenagers (or adults) engaging in risky sex or drug use just to be part of the crowd. And we don't want soldiers (or anyone else) mindlessly following orders, just because they were told to do so by an authority figure. Recognizing and resisting destructive forms of obedience are particularly important to our society—and to social psychology. Let's start with an examination of a classic series of studies on obedience by Stanley Milgram (1963, 1974).

Imagine that you have responded to a newspaper ad seeking volunteers for a study on memory at the Yale University laboratory. Upon your arrival, an experimenter explains to you and another participant that he is studying the effects of punishment on learning and memory. You are selected to play the role of the "teacher." The experimenter leads you into a room, in which he straps the other participant—the "learner"— into a chair. He applies electrode paste to the learner's wrist "to avoid blisters and burns," and attaches an electrode that is connected to a "shock generator."

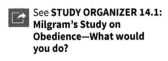
See **STUDY ORGANIZER 14.1:** Milgram's Study on Obedience—What would you do?

Next, you're led into an adjacent room, and told to sit in front of the shock generator's control panel, which is wired through the wall to the chair of the learner. As you can see in **Study Organizer 14.1**, *the control panel for the shock generator consists of 30 switches (each representing successively higher levels of shock, from 15 volts to 450 volts). Written labels appear below each group of switches, ranging from "Slight Shock" to "Danger: Severe Shock," all the way to "XXX." The experimenter explains that it is your job to teach the learner a list of word pairs, and to punish any errors by administering a shock. With each wrong answer, you are to increase the shock by one level.*

You begin teaching the word pairs, but the learner's responses are often wrong. Before long, you are inflicting shocks that you can only assume must be extremely painful. After you administer 150 volts, the learner begins to protest: "Get me out of here . . . I refuse to go on."

You hesitate, and the experimenter tells you to continue. He insists that even if the learner refuses to answer, you must keep increasing the shock levels. But the other person is obviously in pain. What will you do?

It's important to note that the "learners" were confederates, who helped with the experiment, and never received actual shocks. They only pretended to be in pain. Milgram provided specific scripts that they followed at every stage of the experiment. However, the real participants, the "teachers," believed they were delivering shocks to a person in obvious pain. Although the "teachers" sweated, trembled, stuttered, laughed nervously, and repeatedly protested that they did not want to hurt the learner, most still obeyed!

The psychologist who designed this study, Stanley Milgram, was actually investigating not punishment and learning, but obedience to authority: Would participants obey the experimenter's prompts and commands to shock another human being? In Milgram's public survey, fewer than 25% thought they would go beyond 150 volts. And no respondents predicted that they would go past the 300-volt level. Yet 65% of the teacher-participants in this series of studies obeyed completely—going all the way to the end of the scale (450 volts), even beyond the point when the "learner" (Milgram's confederate) stopped responding altogether.

Even Milgram was surprised by his results. Before the study began, he polled a group of psychiatrists, and they predicted that most people would refuse to go beyond 150 volts, and that fewer than 1% of those tested would "go all the way." But, as Milgram discovered, 65% of his participants—men and women of all ages and from all walks of life—administered the highest voltage. The study has been fully or partially replicated many times, and in many other countries, with similarly high levels of obedience (Corti & Gillespie, 2015; Graupmann & Frey, 2014; Haslam et al., 2015).

Despite these replications, Milgram's full, original setup could never be undertaken today due to ethical and moral considerations. Deception is a necessary part of some research, but the degree of it in Milgram's research, and the discomfort of the participants, would never be allowed under today's research standards (see Chapter 1). Keep in mind, however, that immediately following the study, Milgram carefully debriefed each volunteer, and then

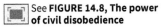
See **FIGURE 14.8, The power of civil disobedience**

followed up with the participants for several months. Most of his "teachers" reported finding the experience personally informative and valuable. One final, important reminder: The *"learner" was an accomplice of the experimenter and only pretended to be shocked.*

Why did the teachers in Milgram's study obey the orders to shock a fellow participant, despite their moral objections? Are there specific circumstances that increase or decrease obedience? Look again at Study Organizer 14.1, and note that in a series of follow-up studies, Milgram found several important factors that influenced obedience: *legitimacy and closeness of the authority figure, remoteness of the victim, assignment of responsibility,* and *modeling or imitation of others* (Auzoult, 2015; Greenberg et al., 2015; Hollander, 2015; Reicher, 2014). Note also in ▣ **Figure 14.8** how civil disobedience is a powerful form of modeling that can be used to offset destructive forms of obedience.

© Bettman/Corbis

In addition to the four factors that Milgram identified, other social psychologists have emphasized several additional influences:

- **Socialization** Can you see how socialization might help explain many instances of mindless and sometimes destructive obedience? From an early age, we're all taught to listen to, and respect, people in positions of authority. In this case, participants in Milgram's study came into the research lab with a lifetime of socialization toward the value of scientific research—and respect for the experimenter's authority. They couldn't suddenly step outside themselves, and question the morality of this particular experimenter and his orders.

- **The foot-in-the-door technique** The gradual, increasing steps in many obedience situations may also help explain why so many people were willing to give the maximum shocks in Milgram's study. The initial mild level of shocks may have worked as a **foot-in-the-door technique**, in which a first, small request is used as a setup for later, larger requests. Once Milgram's participants complied with the initial request, they might have felt obligated to continue.

- **Adherence to ideologies** Some film critics and political commentators have suggested that popular movies, like *American Sniper*, with their heavy emphasis on unwavering obedience to authority, might be encouraging a military ideology that justifies the war time killing of others (e.g., Frangicetto, 2015). In support of this position, archival research on Milgram's original study (Haslam et al., 2015) found that the "teachers" were actually happy to participate—in spite of the emotional stress. Why? The participants believed they were contributing to a valuable enterprise with virtuous goals. Do you agree with these archival researchers who suggest that the major ethical problem with Milgram's study lies not with the stress generated for the "teachers," but with the ideology used to justify harming others?

- **Relaxed moral guard** One common intellectual illusion that hinders critical thinking about obedience is the belief that only evil people do evil things, or that evil announces itself. The experimenter in Milgram's study looked and acted like a reasonable person, who was simply carrying out a research project. Because he was not seen as personally corrupt and evil, the participants' normal moral guard was down, which can maximize obedience. As philosopher Hannah Arendt has suggested, the horrifying thing about the Nazis was not that they were so deviant, but that they were so "terrifyingly normal."

"I was just following orders." Adolph Eichmann (Nazi lieutenant colonel responsible for the deportation of millions of Jews to concentration camps, and a major organizer of the Holocaust).

Group Processes

Although we seldom recognize the power of group membership, social psychologists have identified several important ways that groups affect us (Mikulincer et al., 2015).

Group Membership

How do the roles that we play within groups affect our behavior? This question fascinated social psychologist Philip Zimbardo. In his famous study at Stanford University, 24 carefully screened, well-adjusted young college men were paid $15 a day for participating in a two-week simulation of prison life (Haney et al., 1978; Zimbardo, 1993).

The students were randomly assigned to the role of either prisoner or guard. Prisoners were "arrested," frisked, photographed, fingerprinted, and booked at the local police station. They were then blindfolded and driven to the "Stanford Prison." There, they were given ID numbers, deloused, issued prison clothing (tight nylon caps, shapeless gowns, and no underwear), and locked in cells. In contrast, participants assigned to be guards were outfitted with official-looking uniforms, official police nightsticks (billy clubs), and whistles. Most importantly, they were given complete control over their "prisoners."

Not even Zimbardo foresaw how the study would turn out. Although some guards were nicer to the prisoners than others, they all engaged in some abuse of power. In some cases, the slightest disobedience was punished with degrading tasks, or the loss of "privileges" (such as eating, sleeping, and washing). As demands increased and abuses began, the prisoners became passive and depressed. One prisoner did fight back with a hunger strike, but it ended with a forced feeding by the guards.

Eventually, four prisoners had to be released because of severe psychological reactions. The study was stopped after only six days because of the alarming psychological changes in the participants.

Note that this was not a true experiment in that it lacked a control group, an operational definition, and clear measurements of the dependent variable (Chapter 1). However, it did provide valuable insights into the potential effects of roles on individual behavior (■ **Figure 14.9**). According to interviews conducted after the study, the students became so absorbed in their roles that they forgot they were participants in a psychological study (Zimbardo et al., 1977).

In addition to the potentially devastating effects of roles and situations on our behavior, Zimbardo's study also demonstrates **deindividuation**. To be deindividuated means that we feel less self-conscious, less inhibited, and less personally responsible as a member of a group than when we're alone. This is particularly true when we feel anonymous (■ **Figure 14.10**). Groups sometimes actively promote deindividuation by requiring members to wear uniforms, which are known to increase allegiance and conformity.

Group Decision Making

We've just seen how group membership affects the way we think about ourselves, but how do groups affect our decisions? Are two heads truly better than one?

Most people assume that group decisions are more conservative, cautious, and middle-of-the-road than individual decisions. But is this true? Initial investigations indicated that after discussing an issue, groups actually supported riskier decisions than decisions they made as individuals before the discussion (Stoner, 1961). Subsequent research on this *risky-shift phenomenon*, however, shows that some groups support riskier decisions, while others support more conservative decisions (Atanasov & Kunreuther, 2015; Liu & Latané, 1998; Markovitch et al., 2014).

How can we tell whether a given group's decision will be risky or conservative? A group's final decision depends primarily on its dominant *preexisting* tendencies. If the dominant initial position is risky, the final decision will be even riskier, and the reverse is true if the initial position is conservative. This tendency for the decisions and opinions of members of a group to become more extreme (either riskier or more conservative), depending on the members' initial dominant tendency, is called **group polarization** (Mikulincer et al., 2015; Sobel, 2014; Spears & Postmes, 2015).

What causes group polarization? It appears that as individuals interact and share their opinions, they pick up new and more persuasive information that supports their original opinions, which may help explain why American politics has become so polarized in recent years (Gruzd & Roy, 2014; Suhay, 2015; Westfall et al., 2015). Can you see how if we only interact and work with like-minded people, or only watch news programs that support our views, we're likely to become even more polarized? An interesting study in Washington, DC found that interns who worked in a partisan workplace became more polarized in their opinions than those who worked in less partisan environments (Jones, 2013).

Group polarization is also important within the legal system. Imagine yourself as a member of a jury (■ **Figure 14.11**). In an ideal world, attorneys from both sides would present the essential facts of the case. Then, after careful deliberation, each

See **FIGURE 14.9, Power corrupts**

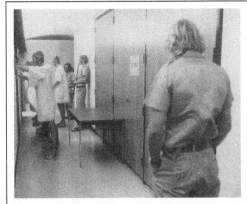

See **FIGURE 14.10, Lost in the crowd**

See **FIGURE 14.11, Juries and group polarization**

individual juror would move from his or her initially neutral position toward the defendant, to a more extreme position—either conviction or acquittal. In a not-so-ideal world, the quality of legal arguments from opposing sides may not be equal, you and the other members of the jury may not be neutral at the start, and group polarization may cause jurors to make riskier or more conservative judgments than they would have on their own.

A related phenomenon is **groupthink**, which occurs when maintaining harmony among group members becomes more important than making a good decision (Apfelbaum et al., 2014; Brodbeck & Guillaume, 2015; Janis, 1972; Vaughn, 1996). As you can see in **Study Organizer 14.2**, there are many factors that explain groupthink, but the two most important might be the pressure for uniformity, and the unwillingness to hear dissenting information.

See **STUDY ORGANIZER 14.2:** How Groupthink Occurs

Many highly-publicized tragedies—from our failure to anticipate the attack on Pearl Harbor in 1941, to the terrorist attacks of September 11, and the war in Iraq—have been blamed on groupthink (Janis, 1972, 1989; Karrasch & Gunther, 2014; Tsintsadze-Maass & Maass, 2014). Groupthink might also help explain why so few coaches or other staff members responded to allegations of child abuse by Jerry Sandusky, former assistant football coach at Penn State University.

How can we prevent, or at least minimize, groupthink? As a critical thinker, first study the list of the antecedent conditions and symptoms of groupthink provided in Study Organizer 14.2, then try generating your own ideas for possible solutions. For example, you might suggest that group leaders either absent themselves from discussions, or remain impartial and silent. Second, can you see how group members should avoid isolation, and be encouraged to voice their dissenting opinions, in addition to seeking advice and input from outside experts? A third option is to suggest that members should generate as many alternatives as possible, and that voting be done by secret ballot versus a show of hands. Finally, group members should be reminded that they will be held responsible for their decisions, which will help offset the illusion of invulnerability, collective rationalizations, stereotypes, and so on. Some of these recommendations for avoiding groupthink were found in the decisions that led to the 2011 assassination raid on Osama bin Laden's compound. Before the final call, each member of President Obama's decision-making team was polled, and even vice-president Joe Biden felt free to disagree (Landler, 2012). For an in-depth, fascinating look at groupthink, watch the classic 1957 film, *Twelve Angry Men*.

Answer the **Concept Check** questions.

14.3 SOCIAL RELATIONS

Kurt Lewin (1890–1947), often considered the "father of social psychology," was among the first to suggest that all behavior results from interactions between the individual and the environment. In this final section, on **social relations**, we explore how we develop and are affected by interpersonal relations, including prejudice, aggression, altruism, and interpersonal attraction.

See **FIGURE 14.12,** Clarifying prejudice versus discrimination

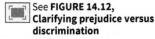

		Prejudice	
		Yes	No
Discrimination	Yes	A person of color is *denied* a job because the owner of a business is prejudiced.	A person of color is *denied* a job because the owner of a business fears white customers won't buy from a person of color.
	No	A person of color is *given* a job because the owner of a business hopes to attract a wider variety of customers.	A person of color is *given* a job because he or she is the best suited for it.

Prejudice

Prejudice is a learned, unjustified negative attitude toward members of a particular group. It literally means "pre-judgment." Like all other attitudes, it's composed of three ABC elements: *affective* (emotions about the group), *behavioral tendencies* (**discrimination** against group members), and *cognitive* (**stereotypes** about group members). Although prejudice is composed of three separate elements, in everyday usage *prejudice* generally refers to thoughts and feelings, whereas *discrimination* is used to describe actions. To avoid confusion, we'll use the term "prejudice" to refer to all three components. But it's still important to point out that prejudice and discrimination often coincide, but not always (■ **Figure 14.12**).

Common Sources of Prejudice

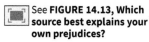 See **FIGURE 14.13, Which source best explains your own prejudices?**

How does prejudice originate? Why does it persist? Five commonly cited sources are *learning*, *personal experience*, *limited resources*, *displaced aggression*, and *mental shortcuts* (■ **Figure 14.13**).

Learning

People learn prejudice the same way they learn other attitudes—primarily through *classical* and *operant conditioning* and *social learning* (Chapter 6). For example, repeated exposure to stereotypical portrayals of persons of color and women in movies, online, magazines, and TV teach children that such images are correct (Gattino & Tartaglia, 2015; Killen et al., 2015; Sigalow & Fox, 2014). Similarly, hearing parents, friends, and teachers express their prejudices can create or reinforce prejudice (Abolmaali et al., 2014; Brown, 2014; Miklikowska, 2015). *Ethnocentrism*, believing our own culture represents the norm or is superior to others, is another form of a learned prejudice.

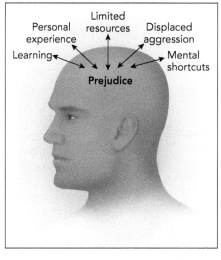

Personal Experience

We also develop prejudice through direct experience. When people make prejudicial remarks or "jokes," they often gain attention and even approval from others. Sadly, denigrating others seems to boost some people's self-esteem (Fein & Spencer, 1997; Ho & O'Donohoe, 2014; Nelson, 2013). In addition, once someone has one or more negative interactions or experiences with members of a specific group, he or she may generalize the resulting prejudice to all members of that group.

Limited Resources

Most of us understand that prejudice exacts a high price on its victims, but few appreciate the significant economic and political advantages it offers to the dominant group (Achiume, 2014; Dreu et al., 2015; Wilkins et al., 2015). Can you see how stereotypes about Blacks and Latinos being inferior to Whites helps justify and perpetuate a social order in the United States, in which Whites hold disproportionate power and resources?

Displaced Aggression

As a child, did you ever feel like hitting a sibling who was tormenting you? Frustration sometimes leads people to attack the perceived cause of that frustration. But, as history has shown, when the source is ambiguous, or too powerful and capable of retaliation, people often redirect their aggression toward an alternative, innocent target, known as a *scapegoat*. Blacks, Jews, Native Americans, and other less empowered groups have a long and tragic history of being tortured, enslaved, killed, and scapegoated. Recent examples include blaming gay men in the 1980's for the AIDS epidemic, and attributing the housing and banking collapse of 2008 to minority groups or the working class for buying houses they could not afford. Similarly, during campaigns for the 2016 presidential election, Latino immigrant groups were used by politicians as popular scapegoats for numerous problems (■ **Figure 14.14**).

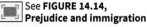 See **FIGURE 14.14, Prejudice and immigration**

Mental Shortcuts

Prejudice is also believed to develop from everyday *mental shortcuts* that we create to simplify our complex social world (McFarlane, 2014; Prati et al., 2015). Stereotypes allow quick judgments about others, thereby freeing up mental resources for other activities. Note, however, that these mental shortcuts too often lead to negative outcomes (Reuben et al., 2014). One study found that when science faculty were sent identical résumés from a male and a female student applying for a job as a laboratory manager, both male and female faculty members rated the male applicant as more competent and worthy of employment than the female (Moss-Racusin et al., 2012).

Sean Thew/EPA/Corbis

Can you see how such stereotypes might have serious real-world consequences in the work force—especially for those trying to enter careers in stereotypically male or female fields?

People also use stereotypes as mental shortcuts when they create ingroups and outgroups. An *ingroup* is any category to which people see themselves as belonging; an *outgroup* is any other category. Research finds that ingroup members judge themselves more positively (as being more attractive, having better personalities, and as more deserving of resources) compared to outgroup members—a phenomenon known as **ingroup favoritism** (Effron & Knowles, 2015; Hoogland et al., 2015; Sierksma et al., 2015). Members of the ingroup also tend to judge members of the outgroup as more alike and less diverse than members of their own group, a phenomenon aptly known as the **outgroup homogeneity effect** (Kang & Lau, 2013; Ratner & Amodio, 2013). A danger of this erroneous belief is that when members of specific groups are not recognized as varied and complex individuals, it's easier to treat them in discriminatory ways.

A sad example occurs during wars and international conflicts. Viewing people on the other side as simply faceless enemies makes it easier to kill large numbers of soldiers and civilians. This type of dehumanization and facelessness also help perpetuate our current high levels of fear and anxiety associated with terrorism (Greenwald & Pettigrew, 2014; Haslam, 2015; Lee et al., 2014).

Like all attitudes, prejudice can operate even without a person's conscious awareness or control. This process is known as automatic bias, or **implicit bias** (Blair et al., 2015; March & Graham, 2015; Reuben et al., 2014). For example, when symphony orchestras decided to use "blind" auditions, using a screen to hide the candidates identities, the hiring of female musicians increased (Goldin & Rouse, 2000). Why do we develop these hidden, implicit biases? Recall from Chapter 8 that we naturally put things into groups or categories to help us make sense of the world around us. Unfortunately, the prototypes and hierarchies we develop are sometimes based on incorrect stereotypes of various groups.

How do we identify these hidden, implicit biases? The most prominent method is the *Implicit Association Test (IAT)*. You can "try this yourself" by going to: https://implicit.harvard.edu/implicit.

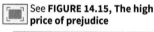

See **TUTORIAL VIDEO: Implicit Attitudes**

The Price of Prejudice

Having discussed the definitions and sources of prejudice, let's look at the toll it takes on individuals—both the victims and perpetrators—and the world around us. As you can see in

See **FIGURE 14.15, The high price of prejudice**

Figure 14.15, prejudice has a long, sad global history. The atrocities committed against the Jews and other groups during the Holocaust, as well as the current crises in the Middle East and Africa, offer stark reminders of the cost of human hatred. Within the United States, our history of slavery, current racial and gender disparities in employment, wealth, education, and healthcare, the current immigration controversy, and the stigma against mental illness (Chapters 12 and 13) all provide further troubling evidence of the ongoing costs of prejudice (Glaser, 2015; Jones & Corrigan, 2014; Marks et al., 2015).

As mentioned earlier, prejudice also severely limits the perpetrator's ability to accurately judge others, and to correctly process information. A startling, example is the U.S. public's reaction to the naming of hurricanes. Archival research on more than six decades of death rates from U.S. hurricanes reveals that more people have died from feminine-named hurricanes (like Betsy) than from masculine-named hurricanes (such as Charlie) (Jung et al., 2014)! Why? Given that this is an obvious illusory correlation, and third-variable problem (Chapter 1), how would you explain this difference? Follow up laboratory experiments suggest that a hurricane's name taps into our hidden, gender-based stereotypes. Men are generally perceived as being strong and aggressive, whereas women are seen as being weak and passive. These gender stereotypes then apparently lead respondents to perceive a lower risk, and to take fewer precautions, to "Hurricane Betsy" than to "Hurricane Charlie"—with deadly consequences! Hurricanes with a relatively masculine name caused an average of 15.15 deaths, whereas those with a feminine name caused an estimated 41.85 deaths! [Note that two particularly destructive storms with feminine names, Audrey and Katrina, were removed from the data set as outliers (Jacobs, 2014).]

AP Photo

If prejudice is so costly, why does it persist? As you may remember from Chapter 6, learning theorists believe that human behavior only continues if it's reinforced. Where are the reinforcers? As noted earlier, perhaps the most important, but seldom acknowledged, payoffs for prejudice are the economic and political benefits to the perpetrators. If certain groups of people are considered less important or valuable, they can be paid less for equal work. Another possibility is *ethnocentrism*, the tendency to favor our own group over others (Bausch, 2015; Castillo, 2015).

Overcoming Prejudice

What can we do to reduce and combat prejudice? Four major approaches have been suggested: *cooperation and common goals, intergroup contact, cognitive retraining,* and *cognitive dissonance* (▣ **Figure 14.16**).

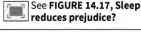

See **FIGURE 14.16, How can we reduce prejudice?**

Cooperation and Common Goals

Muzafer Sherif and his colleagues (1966, 1998) conducted an ingenious study to show the role of cooperation in reducing prejudice. The researchers artificially created strong feelings of ingroup and outgroup identification in a group of 11- and 12-year-old boys at a summer camp. They did this by physically separating the boys into different cabins, and assigning different projects to each group, such as building a diving board, or cooking out in the woods.

© KaliNine LLC/iStockphoto

Once each group developed strong feelings of group identity and allegiance, the researchers set up a series of competitive games, including tug-of-war and touch football. They awarded desirable prizes to the winning teams. Because of this treatment, the groups began to pick fights, call each other names, and raid each other's camps. Researchers pointed to these behaviors as evidence of the experimentally produced prejudice.

The good news is that after creating competition between the two groups, the researchers created "mini-crises," and tasks that required expertise, labor, and cooperation from both groups. Prizes were awarded to all, and prejudice between the groups slowly began to dissipate. By the end of the camp, the earlier hostilities and *ingroup favoritism* had vanished. Sherif's study showed not only the importance of cooperation as opposed to competition, but also the importance of *superordinate goals* (the "mini-crises") in reducing prejudice. Modern research agrees with Sherif's findings regarding the value of cooperation and common goals (Rutland & Killen, 2015; Sierksma et al., 2015; Zhang, 2015).

Intergroup Contact

A second approach to reducing prejudice is to increase contact and positive experiences between groups (Dickter et al., 2015; Masciadrelli, 2014; Tropp & Page-Gould, 2015). However, as you just discovered with Sherif's study of the boys at the summer camp, contact can sometimes increase prejudice. Increasing contact works only under certain conditions that provide for *close interaction, interdependence* (superordinate goals that require cooperation), and *equal status*.

See **FIGURE 14.17, Sleep reduces prejudice?**

Cognitive Retraining

One of the most promising strategies for reducing prejudice requires *empathy*—taking another's perspective (Buchtel, 2014; Liaw et al., 2014; Prati et al., 2015). Can you see how televised specials and Hollywood movies, like *42* and *Selma,* help us understand and empathize with the pressures and heroic struggles of Blacks to gain equal rights? We can also learn to be less prejudiced when we selectively pay attention to *similarities* rather than *differences* (Gaertner & Dovidio, 2014; Phillips & Ziller, 1997; West et al., 2014). Unfortunately, emphasizing gender differences (as in the book *Men Are From Mars*, *Women Are From Venus*) may increase and perpetuate gender stereotypes. Interestingly, a study (Hu et al., 2015) found that cognitive training can even reduce prejudice while we sleep (▣ **Figure 14.17**).

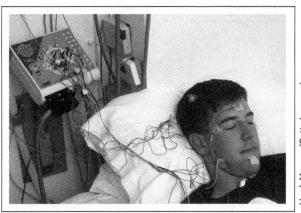

Hank Morgan/Getty Images, Inc.

See **FIGURE 14.18, Breaking the "Gay Barrier"**

Cognitive Dissonance

As you recall, one of the most efficient methods to change an attitude is with *cognitive dissonance*. Each time we meet someone who does not conform to our prejudiced views, we experience dissonance—"I thought all gay men were effeminate. This guy is a deep-voiced, professional athlete. I'm confused." To resolve the dissonance, we can maintain our stereotypes by saying, "This gay man is the exception to the rule." However, if we continue our contact with a large variety of gay men, or when the media shows more instances of non-stereotypical gay individuals, this "exception to the rule" defense eventually breaks down, and attitude change occurs ( **Figure 14.18**).

Aggression

Like prejudice, violence toward others is an unfortunate part of our social relations. When we intentionally try to inflict psychological or physical harm on another, psychologists define it as **aggression**. In this section, we explore its multiple causes, and possible ways to reduce it.

Biological Explanations

Because aggression has such a long history and is found in all cultures, some scientists believe that humans are instinctively aggressive (Buckholtz & Meyer-Lindenberg, 2015; Buss & Duntley, 2014; Peper et al., 2015). Most social psychologists, however, reject this "instinct" argument, but do accept the fact that biology plays a role. Twin studies suggest that some individuals are genetically predisposed to have hostile, irritable temperaments, and to engage in aggressive acts (Lacourse et al., 2014; Rhee & Waldman, 2011). Research with brain injuries and other disorders also has identified possible aggression circuits in the brain (Haller, 2014; Sabaz et al., 2014; Wood & Thomas, 2013). In addition, studies have linked the hormone testosterone, and lowered levels of some neurotransmitters, with aggressive behavior (Malik et al., 2014; Mazur, 2013; Pfattheicher & Keller, 2014).

Psychosocial Explanations

Substance abuse (particularly alcohol abuse) is a major factor in many forms of aggression (Abbey et al., 2014; Heinz et al., 2015; Kose et al., 2015). Similarly, aversive stimuli, such as loud noise, heat, pain, bullying, insults, and foul odors, also may increase aggression (Anderson, 2001; DeWall et al., 2013). (See  **Figure 14.19**.)

See **FIGURE 14.19, Aggression in sports**

In addition, social-learning theory suggests that people raised in a culture with aggressive models will develop more aggressive responses (Farrell et al., 2014; Kirk & Hardy, 2014; Suzuki & Lucas, 2015). For example, the United States has a high rate of violent crime, and there are widespread portrayals of violence on TV, the Internet, movies, and video games, which may contribute to aggression in both children and adults (Breuer et al., 2015; Busching et al., 2015; Strasburger et al., 2014). However, the research is controversial (Ferguson, 2010, 2015; Hoffman, 2014), and the link between violent media and aggression appears to be at least a two-way street. Laboratory studies, correlational research, and cross-cultural studies have found both that exposure to violence increases aggressiveness, and that aggressive children tend to seek out violent programs (Kalnin et al., 2011; Krahé, 2013; Qian et al., 2013).

Reducing Aggression

How can we control or eliminate aggression? Some people suggest that we should release our aggressive impulses by engaging in harmless activities, such as exercising vigorously, punching a pillow, or watching competitive sports. But studies suggest that this type of supposed *catharsis* doesn't really help. In fact, it may even intensify the negative feelings (Bushman, 2002; Kuperstok, 2008; Seebauer et al., 2014).

A more effective approach is to introduce *incompatible responses*. Because certain emotional responses, such as empathy and humor, are incompatible with aggression, purposely making a joke, or showing some sympathy for an opposing person's point of view, can reduce anger and frustration (Baumeister & Bushman, 2014; Gottman, 2015; Maldonado et al., 2014).

Before going on, it's important to talk about the numerous sociocultural factors that contribute to aggression, including gender differences, developmental issues, and socioeconomic factors. In addition, the presence and use of guns greatly increases aggression—and the odds that aggression will be fatal (Cornell & Guerra, 2013; Williamson et al., 2014). Given that the rate of gun homicides in the U.S. remains substantially higher than almost every other nation in the world, the American Psychological Association (APA) commissioned a panel of experts to investigate the best methods for preventing gun violence. Consider their three key recommendations:

- *Primary (or universal) prevention* involves healthy development in the general population, such as teaching better social and communication skills to all ages.
- *Secondary (or selective) prevention* consists of providing assistance for at-risk individuals, including mentoring programs and conflict-mediation services.
- *Tertiary (or indicated) prevention* involves intensive services for individuals with a history of aggressive behavior, to prevent a recurrence or escalation of aggression, such as programs that rehabilitate juvenile offenders (American Psychological Association, 2013).

Altruism

After reading about prejudice and aggression, you will no doubt be relieved to discover that human beings also behave in positive ways. People help and support one another by donating blood, giving time and money to charities, aiding stranded motorists, and so on. There are also times when people do not help. Let's consider both responses.

When and Why Do We Help?

Altruism, or *prosocial behavior*, consists of actions designed to help others, with no obvious benefit to the helper. There are three key approaches predicting when and n why we help (▥ **Figure 14.20**).

See **FIGURE 14.20, Three models for helping**

The **evolutionary theory of helping** suggests that altruism is an instinctual behavior that has evolved because it favors survival of the helper's genes (Kurzban et al., 2015; Richardson, 2015; Wilson, 2015). By helping our own child, or other relative, we increase the odds of our own genes' survival.

Other research suggests that altruism may actually be self-interest in disguise. According to this **egoistic model of helping**, we help others only because we hope for later reciprocation, because it increases our self-esteem, and/or because it helps us avoid feeling distressed and guilty (Dickert et al., 2015; Schroeder & Graziano, 2015).

Opposing the evolutionary and egoistic models is the **empathy–altruism hypothesis** (Batson, 1991, 2014; Lebowitz & Dovidio, 2015; Lemmon & Wayne, 2014). This perspective holds that simply seeing or hearing another person's suffering can create *empathy*—a subjective grasp of that person's feelings or experiences. And when we feel empathic toward another, we are motivated to help that person for his or her own sake. For example, middle school students who had been bullied are more likely to say that they would help another student who was being bullied (Batanova et al., 2014). The ability to empathize may even be innate. Research with infants in the first few hours of life are more likely to cry, and become distressed, at the sound of another infant's cries, but not to tape recordings of their own cries, or after hearing the cries of an infant chimpanzee (Geangu et al., 2010; Hay, 1994; Laible & Karahuta, 2014).

Why Don't We Help?

In 1964, a young woman, Kitty Genovese, was brutally stabbed to death near her apartment building in New York City. The

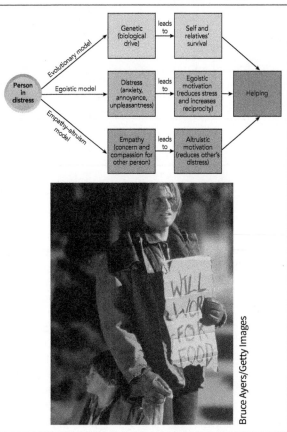
Bruce Ayers/Getty Images

attack occurred about 3:00 A.M. and lasted for over half an hour. According to news reports at the time, 38 of her neighbors supposedly watched the repeated attacks, and heard her screams for help–yet no one came to help. Finally, one neighbor called the police, but it was too late. Kitty Genovese had died.

The story of Kitty Genovese's murder gained national attention, with many people attributing her neighbor's alleged lack of responsiveness to the callousness of big city dwellers—New York City residents, in particular. It's important to note, however, that later investigations found the early news reports to be filled with errors (Griggs, 2015; Seedman & Hellman, 2014).

Despite the inaccuracies, this case inspired psychologists John Darley and Bibb Latané (1968) to conduct a large number of studies investigating exactly when, where, and why we do or don't help our fellow human beings. They found that whether or not someone helps depends on a series of interconnected events and decisions: The potential helper must notice what is happening, interpret the event as an emergency, accept personal responsibility for helping, decide how to help, and then actually initiate the helping behavior (**Process Diagram 14.2**).

This Process Diagram contains essential information NOT found elsewhere in the text, which is likely to appear on quizzes and exams. Be sure to study it CAREFULLY!

Process Diagram 14.2

When and Why Don't We Help?

According to Latané and Darley's five step decision process (1968), if our answer at each step is "yes," we will help others. If our answer is "no" at any point, the helping process ends.

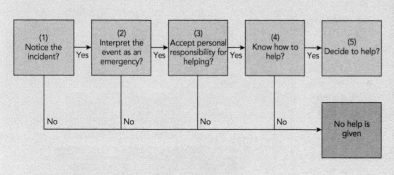

Sam Sarkis/Photodisc/Getty Images

See **TUTORIAL VIDEO: The Bystander Effect**

How does this sequence explain television news reports and "caught on tape" situations in which people are robbed or attacked, and no one comes to their aid? The breakdown generally comes at the third stage—*taking personal responsibility for helping*. In follow-up interviews, most onlookers report that they failed to intervene because they were certain that someone must already have called the police. This so-called **bystander effect**, in which the presence of others discourages an individual from feeling responsible to help, is a well-known problem that affects our helping behaviors.

In one of the earliest studies of the power of the bystander effect, participants were asked to complete a questionnaire, either alone or with others. While they were working, a large amount of smoke was pumped into the room through a wall vent to simulate an emergency. As you might expect, most of the participants working alone, about 75%, quickly reported the smoke. In contrast, fewer than 40% reported smelling smoke when three other participants were in the room. And only 10% reported the smoke when they were with passive participants who ignored the smoke (Latané & Darley, 1968).

Why are we less likely to help when others are around? According to the principle of **diffusion of responsibility**, we assume the responsibility for acting is shared, or diffused, among all onlookers. In contrast, when we're the lone observer, we recognize that we have the sole responsibility for acting.

As a critical thinker, can you also see how *informational social influence*, which we discussed earlier, also plays a role? Given that people in a group monitor the behavior of others to determine how to behave, we often fail to act because we assume others have more information than we do. (Please keep these studies in mind when you're in a true emergency situation. Do not simply rely on others for information. Make your own quick decisions to act. It may save your life!)

How can we promote helping? Considering what we've just learned about the *bystander effect*, *diffusion of responsibility*, and *informational social influence*, the answers are to first clarify when help is needed, and then assign responsibility. For example, most parents teach their children to scream as loudly as they can if they're being abducted by a stranger. Given that children often scream, for a variety of reasons, can you see why this might not work? Instead, they should be taught to make eye contact with anyone who may be watching, and then to shout something like: "This isn't my parent. Help me!" On the other hand, if you notice a situation in which it seems unclear whether someone needs help or not, simply ask: "Do you need help?" Note, however, that there are some occasions when someone in desperate need of help can't verbally respond to questions, and we may need to take immediate action. For example, during the final stages of drowning (versus just distressed swimming), victims are trying so hard to inhale and stay afloat that they're unable to call or signal for help. (For more information on the *instinctive drowning response*, see http://mariovittone.com/2010/05/154/)

Highly publicized television programs, like ABC's "What Would You Do?" and "CNN Heroes," which honor and reward altruism, also increase helping. Finally, enacting laws that protect helpers from legal liability, so-called "good Samaritan" laws, also encourages helping behavior.

Interpersonal Attraction

What causes us to feel admiration, liking, friendship, intimacy, lust, or love? All these social experiences are reflections of **interpersonal attraction**, meaning our positive feelings toward another. Psychologists have found three compelling factors that predict interpersonal attraction: *physical attractiveness*, *proximity*, and *similarity*. Each influences our attraction in different ways.

Physical Attractiveness

The way we look—including facial characteristics, body size and shape, and manner of dress—is one of the most important factors in our initial attraction, liking, or loving of others (Buss, 2003, 2011; Fletcher et al., 2014; Sprecher et al., 2015). In addition, attractive individuals are seen as being more poised, interesting, cooperative, achieving, sociable, independent, intelligent, trustworthy, healthy, and sexually warm than unattractive people (Khan & Sutcliffe, 2014; Mattes & Milazzo, 2014; Sofer et al., 2015).

Evolutionary psychologists have long argued that men prefer attractive women because youth and good looks generally indicate better health, sound genes, and high fertility (Cloud & Perilux, 2014; Nedelec & Beaver, 2014). In contrast, women supposedly prefer men with maturity and resources because they would be better providers, and the responsibility of rearing and nurturing children has historically fallen primarily on women's shoulders (Buss, 1989, 2011; Jones, 2014; Meltzer et al., 2014; Valentine et al., 2014).

However, recent studies, particularly in the industrialized Western countries, have shown physical attractiveness is equally important to men and women. How do those of us who are less than maximally attractive ever find a mate? Research (and experience) shows that both sexes generally don't hold out for partners who are ideally attractive. Instead, according to the *matching hypothesis*, we tend to select partners whose physical attractiveness approximately matches our own (McClintock, 2014; Prichard et al., 2014; Regan, 1998, 2011).

See **FIGURE 14.21, Culture and attraction**

Furthermore, evidence also suggests that beauty is in "the eye of the beholder." What we judge as beautiful varies somewhat from era to era and culture to culture (■ **Figure 14.21**). For example, the Chinese once practiced foot binding because small feet were considered beautiful in women. All the toes except the big one were bent under a young girl's foot and into the sole. Tragically, the incredible pain and physical distortion made it almost impossible for her to walk. She also suffered chronic bleeding and frequent infections throughout her life (Dworkin, 1974). Even in modern times, cultural demands for attractiveness encourage an increasing number of women (and men) to undergo expensive, and often painful, surgery to *increase* the size of their eyes, breasts, lips, chest, and penis. They also undergo surgery to *decrease* the size of their nose, ears, chin, stomach, hips, and thighs.

So how do those of us who are not "superstar beautiful" manage to find mates—without these extreme measures? In addition to the "matching hypothesis" mentioned earlier, researchers have found human mating involves a host of mental skills and attributes, including charisma, humor, personality, intelligence, compassion, and even clever pick-up lines (Dillon et al., 2015; Watson et al., 2014). Perhaps the least recognized, but one of the most effective, ways to increase attractiveness and signal interest to a potential romantic partner is with verbal and nonverbal flirting. Given that almost everyone fears rejection, flirting provides positive cues of your interest, and thereby increases attractiveness (Hall & Xing, 2015; Kurzban, 2014; McBain et al., 2013).

Proximity

Obviously, attraction also depends on two people generally being in the same place at the same time. Thus, *proximity*, or geographic nearness, is another major factor in attraction (Finkel et al., 2015; Greenberg et al., 2015; Sprecher et al., 2015). Surprisingly, this may also be true in a social network virtual world! The physically closer an avatar is to another avatar, the more likely she or he is to receive a friend invitation (Chesney et al., 2014). In addition, an examination of over 300,000 Facebook users found that even though people can have relationships with people throughout the world, the likelihood of a friendship decreases as distance between people increases (Nguyen & Szymanski, 2012).

If you're wondering if these are examples of correlation being confused with causation, that's great! You're becoming an educated consumer of research—and a good critical thinker. However, in this case, there is experimental evidence supporting a causative link between proximity and attraction. For example, oxytocin, a naturally occurring bodily chemical, is known to be a key facilitator of interpersonal attraction and parental attachment (Goodson, 2013; Preckel et al., 2014; Weisman et al., 2012). In one very interesting experiment, the intranasal administration of oxytocin stimulated men in monogamous relationships, but not single ones, to keep a much greater distance between themselves and an attractive woman during a first encounter (Scheele et al., 2012). The researchers concluded that oxytocin may help maintain monogamous relationships, by making men avoid close physical proximity to other women.

See **FIGURE 14.22, Repeated exposure— why we hate photos of ourselves**

Why is proximity so important? It's due in part to *repeated exposure*. Just as familiar people become more physically attractive over time, repeated exposure also increases overall liking (■ **Figure 14.22**). This makes sense from an evolutionary point of view. Things we have seen before are less likely to pose a threat than novel stimuli (Kongthong et al., 2014; Monin, 2003; Yoshimoto et al., 2014). It also explains why modern advertisers tend to run highly redundant ad campaigns with familiar faces and jingles.

Similarity

The major cementing factor for long-term relationships, whether liking or loving, is *similarity*. We tend to prefer, and stay with, people who are most like us—those who share our ethnic background, social class, interests, and attitudes (Brown & Brown, 2015; Sanbonmatsu et al., 2011; Watson et al., 2014). In other words, "birds of a feather flock together."

What about the old saying "opposites attract"? The term *opposites* here probably refers to personality traits rather than to social background or values. An attraction to a seemingly opposite person is more often based on the recognition that in one or two important personality traits, that person offers something we lack. In sum, lovers can enjoy some differences, but the more alike people are, the more both their loving and their liking endure.

Loving Others

It's easy to see why interpersonal attraction is a fundamental building block of our feelings about others. But how do we make sense of love? Why do we love some people and not others? Many people find the subject to be alternately mysterious, exhilarating, comforting—and even maddening. In this section, we explore *Sternberg's triangular theory* and *consummate love*, *romantic love*, and *companionate love*.

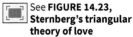
See **FIGURE 14.23,** Sternberg's triangular theory of love

Robert Sternberg, a well-known researcher on creativity and intelligence (Chapter 8), produced a **triangular theory of love** (Sternberg, 1986, 2006) (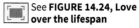 **Figure 14.23**), which suggests that different types and stages of love result from three basic components:

- **Intimacy**—emotional closeness and connectedness, mutual trust, friendship, warmth, self-disclosure, and forming of "love maps."

- **Passion**—sexual attraction and desirability, physical excitement, a state of intense longing to be with the other.

- **Commitment**—permanence and stability, the decision to stay in the relationship for the long haul, and the feelings of security that go with this intention.

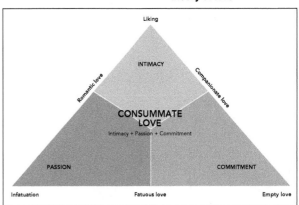

For Sternberg, a healthy degree of all three components in both partners characterizes the strongest and most enduring form of love, **consummate love**. Trouble occurs when one of the partners has a higher or lower need for one or more of the components. Imagine a situation in which one partner has a much higher need for intimacy, whereas the other partner has a stronger interest in passion. Can you see how this lack of compatibility can be fatal to the relationship—unless the partners are willing to compromise and strike a mutually satisfying balance (Sternberg, 2014)?

When you think of romantic love, do you imagine falling in love, a magical experience that puts you on cloud nine? **Romantic love**, which is an intense feeling of attraction to another in an erotic context, has intrigued people throughout history (Acevedo & Aron, 2014; Fehr, 2015; Gottman, 2015). Its intense joys and sorrows also have inspired countless poems, novels, movies, and songs around the world. A cross-cultural study by anthropologists William Jankowiak and Edward Fischer found romantic love in 147 of the 166 societies they studied. They concluded that "romantic love constitutes a human universal or, at the least, a near universal" (1992, p. 154).

Romantic love may be almost universal, but even in the most devoted couples, the intense attraction and excitement of romantic love generally begin to fade 6 to 30 months after the relationship begins. In contrast, **companionate love**, which is characterized by deep trust, caring, tolerance, and friendship, grows and evolves ( **Figure 14.24**) (Fehr et al., 2014; Hatfield & Rapson, 1996; Livingston, 1999). Why? Romantic love is largely based on mystery and fantasy. People often fall in love with what they want another person to be—and these illusions usually fade with the realities of everyday living (Fletcher & Simpson, 2000; Levine, 2001). In contrast, companionate love, like Sternberg's consummate love, slowly develops as couples grow and spend more time together.

How can we keep romantic love alive? One of the most constructive ways is to recognize its fragile nature, and nurture it with carefully planned surprises, flirting, flattery, and special dinners and celebrations. In the long run, however, romantic love's most important function might be to keep us attached long enough to move on to the deeper and more enduring companionate love.

See **FIGURE 14.24, Love over the lifespan**

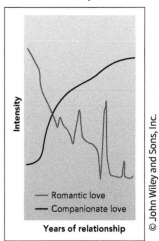

See **VIRTUAL FIELD TRIP: Internet Dating**

Final Note

As the authors of this text, and your tour guides through the fascinating world of psychology, we hope you've enjoyed the journey. For us, the key take-home message, which we hope you'll always remember, is that every human on this planet is an exclusive combination of a physical body, a complex system of mental processes, and large, sociocultural factors. Our deepest wish is that you'll make the most out of your own unique combination, and will apply what you've learned about yourself and others to improve your own life, and the world around you.

Warmest regards,

CATHERINE SANDERSON AND KAREN HUFFMAN

Q Answer the **Concept Check** questions.

WP LS Go to your WileyPLUS Learning Space course for video episodes, examples, art, tables, Concept Checks, practice, and other pedagogical resources that will help you succeed in this course.